Sources and Studies in the History of Mathematics and Physical Sciences

Robert E. Bradley, C. Edward Sandifer

Cauchy's *Cours d'analyse*

An Annotated Translation

Springer

Robert E. Bradley
Department of Mathematics and
Computer Science
Adelphi University
Garden City
NY 11530
USA
bradley@adelphi.edu

C. Edward Sandifer
Department of Mathematics
Western Connecticut State University
Danbury, CT 06810
USA
sandifere@wcsu.edu

Series Editor:
J.Z. Buchwald
Division of the Humanities and Social Sciences
California Institute of Technology
Pasadena, CA 91125
USA
buchwald@its.caltech.edu

ISBN 978-1-4419-0548-2 e-ISBN 978-1-4419-0549-9
DOI 10.1007/978-1-4419-0549-9
Springer Dordrecht Heidelberg London New York

Library of Congress Control Number: 2009932254

Mathematics Subject Classification (2000): 01A55, 01A75, 00B50, 26-03, 30-03

Printed on acid-free paper

Springer is part of Springer Science+Business Media (www.springer.com)

We dedicate this volume to Ronald Calinger, Victor Katz and Frederick Rickey, who taught us the importance and satisfaction of reading original sources, and to our friends in ARITHMOS, with whom we enjoy putting those lessons into practice.

Translators' Preface

Modern mathematics strives to be rigorous. Ancient Greek geometers had similar goals, to prove absolute truths by using perfect deductive logic starting from incontrovertible premises.

Often in the history of mathematics, we see a pattern where the ideas and applications come first and the rigor comes later. This happened in ancient times, when the practical geometry of the Mesopotamians and Egyptians evolved into the rigorous efforts of the Greeks. It happened again with calculus. Calculus was discovered, some say invented, almost independently by Isaac Newton (1642–1727) about 1666 and by Gottfried Wilhelm von Leibniz (1646–1716) about 10 years later, but its rigorous foundations were not established, despite several attempts, for more than 150 years.

In 1821, Augustin-Louis Cauchy (1789–1857) published a textbook, the *Cours d'analyse*, to accompany his course in analysis at the École Polytechnique. It is one of the most influential mathematics books ever written. Not only did Cauchy provide a workable definition of limits and a means to make them the basis of a rigorous theory of calculus, but also he revitalized the idea that all mathematics could be set on such rigorous foundations. Today, the quality of a work of mathematics is judged in part on the quality of its rigor; this standard is largely due to the transformation brought about by Cauchy and the *Cours d'analyse*.

The 17th century brought the new calculus. Scientists of the age were convinced of the truth of this calculus by its impressive applications in describing and predicting the workings of the natural world, especially in mechanics and the motions of the planets. The foundations of calculus, what Colin Maclaurin (1698–1746) and Jean le Rond d'Alembert (1717–1783) later called its *metaphysics*, were based on the intuitive geometric ideas of Leibniz and Newton. Some of their contemporaries, especially Bishop George Berkeley (1685–1753) in England and Michel Rolle (1652–1719) in France, recognized the problems in the foundations of calculus. Rolle, for example, said that calculus was "a collection of ingenious fallacies," and Berkeley ridiculed infinitely small quantities, one of the basic notions of early calculus, as "the ghosts of departed quantities." Both Berkeley and Rolle freely admitted the practicality of calculus, but they challenged its lack of rigorous foundations. We should note that Rolle's colleagues at the Paris Academy eventually convinced him to change his mind, but Berkeley remained skeptical for his entire life.

Later in the 18th century, only a few mathematicians tried to address the questions of foundations that had been raised by Berkeley and Rolle. Over the years, three main schools of thought developed: infinitesimals, limits, and formal algebra of series. We could consider the British ideas of fluxions and evanescent quantities either to be a fourth school or to be an ancestor of these others. Leonhard Euler (1707–1783) [Euler 1755] was the most prominent exponent of infinitesimals, though he devoted only a tiny part of his immense scientific corpus to issues of foundations. Colin Maclaurin [Maclaurin 1742] and Jean le Rond d'Alembert [D'Alembert 1754] favored limits. Maclaurin's ideas on limits were buried deep in his *Treatise of Fluxions*, and they were overshadowed by the rest of the opus. D'Alembert's works were very widely read, but even though they were published at almost the same time as Euler's contrary views, they did not stimulate much of a dialog.

We suspect that the largest school of thought on the foundations of calculus was in fact a pragmatic school – calculus worked so well that there was no real incentive to worry much about its foundations.

In *An V* of the French Revolutionary calendar, 1797 to the rest of Europe, Joseph-Louis Lagrange (1736–1813) [Lagrange 1797] returned to foundations with his book, the full title of which was *Théorie des fonctions analytiques, contenant les principes du calcul différentiel, dégagés de toute considération d'infiniment petits ou d'évanouissans, de limites ou de fluxions, et réduits à l'analyse algébrique des quantités finies* (Theory of analytic functions containing the principles of differential calculus, without any consideration of infinitesimal or vanishing quantities, of limits or of fluxions, and reduced to the algebraic analysis of finite quantities). The book was based on his analysis lectures at the École Polytechnique. Lagrange used power series expansions to define derivatives, rather than the other way around. Lagrange kept revising the book and publishing new editions. Its fourth edition appeared in 1813, the year Lagrange died. It is interesting to note that, like the *Cours d'analyse*, Lagrange's *Théorie des fonctions analytiques* contains no illustrations whatsoever.

Just two years after Lagrange died, Cauchy joined the faculty of the École Polytechnique as professor of analysis and started to teach the same course that Lagrange had taught. He inherited Lagrange's commitment to establish foundations of calculus, but he followed Maclaurin and d'Alembert rather than Lagrange and sought those foundations in the formality of limits. A few years later, he published his lecture notes as the *Cours d'analyse de l'École Royale Polytechnique; Ire Partie. Analyse algébrique*. The book is usually called the *Cours d'analyse*, but some catalogs and secondary sources call it the *Analyse algébrique*. Evidently, Cauchy had intended to write a second part, but he did not have the opportunity. The year after its publication, the École Polytechnique changed the curriculum to reduce its emphasis on foundations [Lützen 2003, p. 160]. Cauchy wrote new texts, *Résumé des leçons données a l'École Polytechnique sur le calcul infinitesimal, tome premier* in 1823 and *Leçons sur le calcul différentiel* in 1829, in which he reduced the material in the *Cours d'analyse* about foundations to just a few dozen pages.

Because it became obsolete as a textbook just a year after it was published, the *Cours d'analyse* saw only one French edition in the 19th century. That first edition, published in 1821, was 568 pages long. The second edition, published as Volume 15 (also identified as Series 2, Volume III) of Cauchy's *Oeuvres complètes*, appeared in 1897. Its content is almost identical to the 1821 edition, but its pagination is quite different, there are some different typesetting conventions, and it is only 468 pages long. The *Errata* noted in the first edition are corrected in the second, and a number of new typographical errors are introduced. At least two facsimiles of the first edition were published during the second half of the 20th century, and digital versions of both editions are available on line, for example, through the Bibliothèque Nationale de France. There were German editions published in 1828 and 1885, and a Russian edition published in Leipzig in 1864. A Spanish translation appeared in 1994, published in Mexico by UNAM. The present edition is apparently the first edition in any other language.

The *Cours d'analyse* begins with a short Introduction, in which Cauchy acknowledges the inspiration of his teachers, particularly Pierre Simon Laplace (1749–1827) and Siméon Denis Poisson (1781–1840), but most especially his colleague and former tutor André Marie Ampère (1775–1836). It is here that he gives his oft-cited intent in writing the volume, "As for the methods, I have sought to give them all the rigor which one demands from geometry, so that one need never rely on arguments drawn from the generality of algebra."

The Introduction is followed by 16 pages of "Preliminaries," what today might be called "Chapter Zero." Here, Cauchy takes pains to define his terms, carefully distinguishing, for example, between *number* and *quantity*. To Cauchy, numbers had to be positive and real, but a quantity could be positive, negative or zero, real or imaginary, finite, infinite or infinitesimal.

Beyond the Preliminaries, the book naturally divides into three major parts and a couple of short topics. The first six chapters deal with real functions of one and several variables, continuity, and the convergence and divergence of series.

In the second part, Chapters 7 to 10, Cauchy turns to complex variables, what he calls *imaginary quantities*. Much of this parallels what he did with real numbers, but it also includes a very detailed study of roots of imaginary equations. We find here the first use of the words *modulus* and *conjugate* in their modern mathematical senses. Chapter 10 gives Cauchy's proof of the fundamental theorem of algebra, that a polynomial of degree n has n real or complex roots.

Chapters 11 and 12 are each short topics, partial fraction decomposition of rational expressions and recurrent series, respectively. In this, Cauchy's structure reminds us of Leonhard Euler's 1748 text, the *Introductio in analysin infinitorum* [Euler 1748], another classic in the history of analysis. In Euler, we find 11 chapters on real functions, followed by Chapters 12 and 13, "On the expansion of real functions into fractions," i.e., partial fractions, and "On recurrent series," respectively.

The third major part of the *Cours d'analyse* consists of nine "Notes," 140 pages in the 1897 edition. Cauchy describes them in his Introduction as "...several notes placed at the end of the volume [where] I have presented the derivations which may be useful both to professors and students of the Royal Colleges, as well as to those who wish to make a special study of analysis."

Though Cauchy was only 32 years old when he published the *Cours d'analyse*, and had been only 27 when he began teaching the analysis course on which it was based, he was already an accomplished mathematician. This should not be surprising, as it was not easy to earn an appointment as a professor at the École Polytechnique. Indeed, by 1821, Cauchy had published 28 memoirs, but the *Cours d'analyse* was his first full-length book.

Cauchy's first original mathematics concerned the geometry of polyhedra and was done in 1811 and 1812. Louis Poinsot (1777–1859) had just established the existence of three new nonconvex regular polyhedra. Cauchy, encouraged to study the problem by Lagrange, Adrien-Marie Legendre (1752–1833) and Étienne Louis Malus (1775–1812), [Belhoste 1991, pp. 25–26] extended Poinsot's results, discovered a generalization of Euler's polyhedral formula, $V-E+F=2$, and proved that a convex polyhedron with rigid faces must be rigid. These results became his earliest papers, the two-part memoir "*Recherches sur les polyèdres*" and "*Sur les polygones et les polyèdres*." [Cauchy 1813] Despite his early success, Cauchy seldom returned to geometry, and these are his only significant results in the field.

After Cauchy's success with the problems of polyhedra, his father encouraged him to work on one of Fermat's (1601–1665) problems, to show that every integer is the sum of at most three triangular numbers, at most four squares, at most five pentagonal numbers, and, in general, at most n n-gonal numbers. He presented his solution to the Institut de France on November 13, 1815 and published it under the title "*Démonstration générale du théorème de Fermat sur les nombres polygones*" [Cauchy 1815]. Belhoste [Belhoste 1991, p. 46] tells us that this was the article "that made him famous," and suggests that "[t]he announcement of his proof may have supported his appointment to the École Polytechnique a few days later."

Just a month later, on December 26, 1815, the Academy's judgment was confirmed when Cauchy won the *Grand Prix de Mathématiques* of the Institut de France, and its prize of 3000 francs, for an essay on the theory of waves.

With his career established, Cauchy married Aloïse de Bure (1795?–1863) in 1818. They had two daughters. It is a measure of Cauchy's later fame and success that one of his daughters married a count, the other a viscount. Indeed, Freudenthal [DSB Cauchy, p. 135] says that Cauchy "was one of the best known people of his time."

The de Bure family were printers and booksellers. The title page of the *Cours d'analyse*, published by de Bure frères, describes them as "*Libraires du Roi et de la bibliothèque du Roi.*"

It seems that Cauchy was an innovative but unpopular teacher at the École Polytechnique. He, along with Ampère and Jacques Binet (1786–1856), proposed substantial revisions in the analysis, calculus and mechanics curricula. Cauchy wrote the *Cours d'analyse* to support the new curriculum.

In 1820, though, before the *Cours d'analyse* was published, but apparently after it had been written and the publisher had committed to printing it, the *Conseil*

Fig. 1 Cauchy, by Susan Petry, 18 × 28 cm, bas relief in tulip wood, 2008. An interpretation of portraits by Boilly (1821) and Roller (∼ 1840). Photograph by Eliz Alahverdian, 2008. Reprinted with permission of Susan Petry and Aliz Alahverdian.

d'Instruction, more or less a Curriculum Committee, largely influenced by de Prony (1755–1839) and Navier (1785–1836), ordered that Cauchy and Ampère change the curriculum again. As a consequence, the *Cours d'analyse* was never used as a textbook. A more complete account of this episode is found in [Belhoste 1991, pp. 61–66].

Lectures at the École Polytechnique were scheduled to be 50 lectures per term, each consisting of 30 minutes "revision" then 60 minutes of lecture. On April 12, 1821, Cauchy was delivering the 65th lecture of the term. When the lecture neared the end of its second hour, students began to jeer, and some walked out. A formal investigation followed, and eventually both the students and Cauchy were found responsible, but nobody was punished. Fuller accounts are found in [Belhoste 1991, pp. 71–74] and [Grattan-Guinness 1990, pp. 709–712].

From 1824 to 1830, Cauchy also taught part-time at the Collège de France, where he presented, among other techniques, methods of differential equations, and gave lectures on the theory of light. At the same time he worked also as a substitute professor on the Faculté des Sciences de Paris, where he replaced Poisson, and lectured on the mechanics of solids, fluid mechanics and on his general theory of elasticity.

By 1826, Cauchy had grown impatient with the time it took for the Academy to publish his articles and memoirs. That year they published only 11 of his memoirs, up from six in 1825, so he founded a private journal, the *Exercises de mathématiques*, published by his in-laws, Debure frères. By 1830, he had published five volumes of the *Exercises*, containing 51 of his articles. These comprise volumes 18 to 21 of the *Oeuvres complètes*.

The July Revolution of 1830 deposed the Bourbon monarch, Charles X. Cauchy refused to take a loyalty oath to his Orleans successor, Louis-Philippe, and went into 8 years of voluntary exile. He taught at the University of Turin from 1831 to 1833, where he continued his journal under the new name, *Résumés analytiques* (*Oeuvres complètes*, volume 22), and then spent the rest of his exile tutoring in Prague in the exile court of Charles X. While in Prague, his king awarded Cauchy the title "Baron."

In 1838, Cauchy returned to Paris, but because he had not taken the loyalty oath, he was not allowed to teach, either at the École Polytechnique or at his part-time jobs. He was still an active member of the Académie des Sciences, though, and over the next 10 years he submitted over 400 items to the *Comptes rendus*, the published notes and articles presented at the weekly meetings of the Academy. Because the Academy took breaks and vacations, "weekly" meetings did not actually take place every week. Over these 10 years, Cauchy averaged an article for each week the Academy was in session. These articles occupy most of volumes 4 to 10 of Cauchy's 27-volume *Oeuvres complètes*. At the same time, he continued his private journal under yet another title, the *Exercises d'analyse et de physique mathématique*. These 47 articles fill volumes 23 to 26 of the *Oeuvres complètes*. During his decade away from the classroom, 1838 to 1848, Cauchy produced about half of his published works by item count, about a third of them by page count. It was a remarkable decade.

The February Revolution of 1848 ended the reign of Louis-Philippe and established the Second Republic. Loyalty oaths were not required, so Cauchy returned to the Faculté des Sciences as professor of mathematical astronomy. When loyalty oaths were reestablished in 1852, Napoleon III made an exception for Cauchy.

Cauchy's last 9 years were active. In 1853, he published one last volume of the *Exercises d'analyse et de physique mathématique*. He did a good deal of research on the theory of light and bickered with his colleagues. He made another 159 contributions to the *Comptes rendus*. The last of his 589 contributions to that journal came on May 4, 1857. [*Oeuvres* 12, p. 435] It was a short note on mathematical astronomy, and he closed it with the words *C'est ce que j'expliquerais plus au long dans un prochain Mémoire*, "I will explain this at greater length in a future Memoir." Clearly, he was not expecting to die just 18 days later.

Many studies give more detailed accounts of Cauchy's life, works and times than we give here. For a full biography of Cauchy, we refer our readers to [Belhoste 1991]. The entry in the *Dictionary of Scientific Biography* [DSB Cauchy] is much briefer; it contains many inaccurate citations to Cauchy's work and in general seems to suffer from "hero worship." For example, we find no other source that describes Cauchy as "one of the best-known people of his time, and must have been often mentioned in newspapers, letters and memoirs." Still, its basic facts are correct.

For accounts of Cauchy's work and its importance, we recommend [Grabiner 2005] and [Grattan-Guinness 2005] as good places to begin. See also [Grattan-Guinness 1990] for a comprehensive account of the French mathematical community in the time of Cauchy.

Grattan-Guinness first presents his case that Cauchy "plagiarized" Bolzano in [Grattan-Guinness 1970a]. This assertion precipitated a controversy that raged through [Grattan-Guinness 1970b], [Freudenthal 1971b], and still echoed in [Grabiner 2005].

Other modern contributions to Cauchy scholarship are more numerous than we wish to describe, but we will mention in particular [Jahnke 2003], [Lützen 2003], [Ferraro 2008] and [Bottazzini 1990]. Starting with these references, the interested reader can find a great many more.

As we translated the *Cours d'analyse*, we laid out the text and formulas, used italics, bold face and punctuation, and, as much as possible, adopted the styles of the 1897 edition of the text. We have also added an index (neither the 1821 nor the 1897 editions have indices), and we have used our footnotes to note passages that are quoted, cited or translated in certain important secondary sources. We have not made note of errors cited in the *Errata* of the 1821 edition, all of which were corrected in the 1897 edition, but we have noted errors not mentioned in the *Errata*, as well as new errors introduced in the second edition. We distinguish such footnotes with the signature "(tr.)." Expository footnotes are unsigned.

We believe that the primary purpose of a translation such as this one is to make the work available in English, and not to provide a platform for our opinions on how this work should be interpreted. Towards this end, we have generally limited our commentary to expository remarks rather than interpretative ones. For those passages that are controversial and subject to a variety of interpretations, we try to refer the interested reader to appropriate entry-point sources and do not try to be comprehensive.

For a variety of reasons, we decided to follow Grabiner [Grabiner 2005], Freudenthal [DSB Cauchy] and others, rather than Kline [Kline 1990], and to make our translation, as well as to cite page numbers, from the second edition. Although electronic copies of both editions are freely available on the World Wide Web, bound copies of the 1821 edition are rather hard to find, while the second edition is found in many university libraries. The on-line library catalog WorldCat reports 57 copies of the 1821 edition in North America, and only seven copies of the facsimiles. Yet they report at least 117 copies of the 1897 edition in North America. We say "at least" because there are several different kinds of catalog entries, and it is difficult to tell how much duplication there is. We would estimate at least 200 copies. The two editions are identical in content, notation and format, but differ in pagination, page layout and some punctuation. In general, we found the typography and page layout of the 1821 edition somewhat cluttered, even quirky, particularly in the ways that formulas were cut into many lines to be arranged on the page. Weighing all these circumstances, it seemed more reasonable to follow the more accessible version.

In general, we resisted the temptation to modernize Cauchy's notation and terminology. When he uses the word *limites* to mean both what we call "limits" and what we call "bounds," we translate it as "limits" in both cases. In the index, citations of the word "limit" direct the reader to instances in which the limit process is being used, and not to instances meaning "bounds." Moreover, when he fails to distinguish between open intervals and closed intervals, or between "less than" and "less than or equal to," we translate it as Cauchy wrote it, and do not attempt to force upon Cauchy distinctions he himself did not make.

There are two conspicuous exceptions. Cauchy wrote lx, or sometimes Lx to denote the logarithm of x to a given base A. We modernize this to $\log x$ or $\operatorname{Log} x$ to avoid unnecessary confusion. Likewise, we write $\ln x$ to denote the natural logarithm of x, rather than using Cauchy's lx. Also, Cauchy used periods at the end of the abbreivated names of trignonometric functions (such as cos. x) and denoted the tangent and arctangent functions tang. x and arc tang. x. Following modern usage, we omit the periods and use $\tan x$ and $\arctan x$.

Cauchy did not adopt Euler's innovation of the 1770s, to write i for $\sqrt{-1}$, so we write $\sqrt{-1}$ as well.[1]

Within our translation of the text, numbers in square brackets, like [116], mark where new pages begin in the 1897 edition. Thus, for example, when we find the notation [116] in the midst of the statement of the Cauchy Convergence Criterion, we know that Cauchy's statement of that criterion appeared on pages 115 and 116

[1] Many people attribute Euler's first use of the symbol i to denote $\sqrt{-1}$ to his 1748 text, the *Introductio in analysin infinitorum* [Euler 1748] , but readers who check Volume II, Chapter 21, § 515 will see that the quantity Euler denotes there as i is actually $\ln(-n)$, for some positive value of n, and not the imaginary unit, $\sqrt{-1}$.

of the 1897 edition. We give a page concordance of the two French editions in an appendix.

Cauchy seemed to enjoy choosing his words carefully and precisely, and then once the correct words were chosen, using those very words over and over again. For example, in Chapter VII, § III, he studies the n-th roots of unity, or, as he calls them, 1 to the fractional power $\frac{1}{n}$. He states his theorems and gives his proofs about these objects. Later in that same section, when he studies other fractional powers of 1, $-\frac{1}{n}$, $\frac{m}{n}$, and $-\frac{m}{n}$, the words in his theorems and proofs are almost identical, changing only what must be changed. We have taken care to do the same in our translation.

Our ambition is, as much as the very idea of translation allows, to let Cauchy speak for himself.

We are grateful to Emili Bifet, David Burns, Larry D'Antonio, Ross Gingrich, Andy Perry, Kim Plofker, Fred Rickey, Chuck Rocca and Jeff Suzuki who, as participants in the ARITHMOS reading group, read early drafts of portions of this translation. Likewise, we are grateful to our students Shannon Abernathy, Erik Gundel, Amanda Peterson and Joseph Piraneo, who read parts of this manuscript in a history of mathematics seminar at Western Connecticut State University in the Spring of 2008. Careful proofreading and helpful suggestions by both groups have greatly improved this translation. We also acknowledge the assistance of the editorial staff at Springer, particularly Ann Kostant and Charlene Cruz Cerdas. Most importantly, we thank our wives Susan Petry and Terry Sandifer for supporting and encouraging our efforts, and for being understanding about the many long days that this project occupied.

Garden City, New York,
Danbury, Connecticut,
March 2009

Robert E. Bradley
C. Edward Sandifer

Contents

Introduction

[i] Because several people, who were so good as to guide the first steps of my scientific career, and among whom I would cite with recognition Messieurs Laplace[1] and Poisson[2] have expressed the desire to have me publish the *Cours d'analyse* of the École Royale Polytechnique, I have decided to put this Course in writing for the greatest usefulness to students. I offer here the first part of it[3] known by the name *Algebraic analysis*,[4] and in which I successively treat the various kinds of real and imaginary functions, [ii] convergent and divergent series, the resolution of equations and the decomposition of rational fractions.[5] In speaking of the continuity of functions, I could not dispense with a treatment of the principal properties of infinitely small quantities, properties which serve as the foundation of the infinitesimal calculus.[6] Finally, in the preliminaries and in several notes placed at the end of the volume, I have presented the derivations which may be useful both to professors and students of the Royal Colleges, as well as to those who wish to make a special study of analysis.

As for the methods, I have sought to give them all the rigor which one demands from geometry, so that one need never rely on arguments drawn from the generality of algebra.[7] Arguments of this kind, although they are commonly accepted, especially [iii] in the passage from convergent to divergent series, and from real

---

[1] Pierre-Simon Laplace (1749–1827).

[2] Siméon Denis Poisson (1781–1840).

[3] Cauchy planned a second part of the *Cours d'analyse*, but no such volume was ever published. When Navier replaced Ampère as the second teacher of analysis, the faculty of the École Polytechnique considered revisions to the analysis curriculum. The changes that were made in 1822 as a result of these reforms took most of the emphasis on foundations out of the course in analysis, making Cauchy's planned second volume obsolete before it was written. This also explains why Cauchy produced no subsequent editions of the *Cours d'analyse*.

[4] *Cours d'analyse* is sometimes referred to as *Analyse algébrique*, for example, in the on-line catalog of the *Bibliotèque Nationale de France*.

[5] As we will see in Chapter I, these are "rational functions" in the modern sense.

[6] It is interesting that Cauchy does not also mention limits here.

[7] This sentence is quoted, in translation, in [DSB Cauchy, p. 135].

quantities to imaginary expressions,[8] may be considered, it seems to me, only as examples serving to introduce the truth some of the time, but which are not in harmony with the exactness so vaunted in the mathematical sciences. We must also observe that they tend to grant a limitless scope to algebraic formulas, whereas, in reality, most of these formulas are valid only under certain conditions or for certain values of the quantities involved. In determining these conditions and these values and in establishing precisely the meaning of the notation that I will be using, I will make all uncertainty disappear, so that the different formulas present nothing but relations among real[9] quantities, relations which will always be easy to verify [iv] by substituting numbers for the quantities themselves. It is true that, in order to remain consistently faithful to these principles, I will have to accept several propositions which may appear to be a bit rigid at first. For example, I state in Chapter VI that *a divergent series does not have a sum*;[10] in Chapter VII that *an imaginary equation is nothing but the symbolic representation of two equations involving real quantities*;[11] in Chapter IX that *if the constants or the variables involved in a function, having first been taken to be real, become imaginary, the notation used to express the function can be kept in the calculation only by virtue of a new convention keeping the sense of the notation in the latter hypothesis*;[12] &c. But those who read my book will recognize, [v] I hope, that propositions of this nature, entailing the happy necessity of putting more precision into the theories and of applying useful restrictions to assertions that are too broad, work in favor of analysis and furnish several research topics which are not without importance. Therefore, before summing any series, I must examine the cases in which the series can be summed, or, in other words, the conditions for its convergence; and I have, on this subject, established general rules which appear to me to merit some attention.

Moreover, if I have sought, on the one hand, to perfect mathematical analysis, yet on the other hand I am far from pretending that this analysis ought to be applied to all the rational sciences. Without a doubt, in those sciences we call "natural," the only [vi] method which we may successfully employ consists in observing the facts and then subjecting those observations to calculation. But it would be a grave error to think that we can find certainty only in geometric proofs, or in the evidence of the senses; and even though nobody has yet tried to prove by analysis the existence of Augustus[13] or that of Louis XIV,[14] all sensible people would admit that their existence is as certain to them as the square of the hypotenuse or the theorem of

---

[8] Cauchy never speaks of imaginary *numbers*, but only of imaginary *expressions*. His imaginary expressions correspond to the modern notion of complex numbers. Here and throughout the book, we will use Cauchy's terminology and write "imaginary expressions."

[9] Here, Cauchy is careful to exclude imaginary expressions. As we will see later, imaginary expressions are equal, for example, when their corresponding real quantities are equal.

[10] See p. 85 or [Cauchy 1821, p. 123] or [Cauchy 1897, p. 114].

[11] See p. 119 or [Cauchy 1821, p. 176] or [Cauchy 1897, p. 155].

[12] See p. 159 or [Cauchy 1821, p. 240] or [Cauchy 1897, p. 204].

[13] Probably Caesar Augustus (63 BCE – 14 CE).

[14] Louis XIV (1638–1715). We note that Cauchy, whose given name is Augustin-Louis, may be engaging in a rare dsiplay of humor by choosing these two particular examples.

Maclaurin.[15] Furthermore, the proof of this last theorem is within reach of only a few people, and scientists themselves do not all agree on its scope one ought to attribute to it; whereas everyone knows quite well who ruled France in the 17th century, and no reasonable argument can be raised against this. What I say here [vii] about historical facts applies equally well to a whole range of questions in religion, ethics and politics. We should thus believe that there are truths other than algebraic truths, and realities other than tangible objects. Let us cultivate with ardor the mathematical sciences, without wishing to extend them beyond their domain; and let us not imagine that we are able to attack history with formulas, nor to make moral judgments with theorems of algebra or integral calculus.

In closing this Introduction, I cannot but acknowledge the insights and advice of several people who have been very helpful, particularly Messieurs Poisson, Ampère[16] and Coriolis.[17] I am indebted to this last person for the rule on the convergence of infinite products,[18] among other things, and I have profited many times from [viii] the observations of Monsieur Ampère, as well as from the methods which he develops in his lessons on analysis.[19]

[15] Colin Maclaurin (1698–1746); the reference is probably to Maclaurin series.

[16] André-Marie Ampère (1775–1836).

[17] Gaspard Gustave de Coriolis (1792–1843).

[18] See Note IX, Theorem I, p. 386 or [Cauchy 1821, p. 564] or [Cauchy 1897, p. 460].

[19] These appear to have been collected in *Cours d'analyse et de mécanique l'école polytechnique*, a manuscript of notes taken by G. Vincens of Ampère's course, which is available in the Dibner Collection of the Smtihsonian Institute.

Preliminaries

Cours d'analyse

of

the École Royale Polytechnique

Preliminaries

Review of the various kinds of real quantities which one might consider, be they algebraic or trigonometric, and of the notation we use to represent them. On the averages of several quantities.

---

[17] To avoid any kind of confusion in algebraic language and notation, we shall establish here in these Preliminaries the meanings of various terms and notation that we will use in ordinary algebra and trigonometry. The explanations that we will give for these terms are necessary so that we will be certain of being perfectly understood by those who read this work. First of all, we will indicate what idea will be appropriate to attach to the two words *number* and *quantity*.

We always take the meaning of *numbers* in the sense that is used in arithmetic, where numbers arise from the absolute measure of magnitudes, and we will only apply the term *quantities* to real positive or negative quantities, that is to say to numbers preceded by the signs $+$ or $-$. Furthermore, we regard these quantities as intended to express increase and decrease, so that a given magnitude will simply be represented by a number if we only mean to compare it to another magnitude of the same type taken as a unit, and by the same number preceded by the sign $+$ or the sign $-$, if we consider it [18] as being capable of increasing or decreasing a given magnitude of the same kind. Given this, the signs $+$ or $-$ placed in front of a number modify its meaning, more or less as an adjective modifies the meaning of a noun. We call the *numerical value* of a quantity that number which forms its basis.[1] We say that two quantities are *equal* if they have the same sign and the same numerical value, and two quantities are *opposites*[2] if their numerical values are the same but with opposite signs. From these principles, it is easy to give an account of the various operations that one may perform on these quantities. For example, given two quantities, one may always find a third quantity which, taken as increasing a fixed number, if it is positive, and as decreasing, if it is negative, brings us to the same result as the two given quantities, applied one after the other in the same way. This third quantity, which by itself produces the same effect as the other two is what we call their *sum*. For example, the two quantities -10 and $+7$ have as their sum -3, given that a decrease of 10 units followed by an increase of 7 units is equivalent to a decrease of 3 units.[3] To *add* two quantities is to form their sum. The difference between a first quantity and a second is a third quantity which, added to the second, gives the first. Finally, we say that one quantity is *larger* or *smaller* than another depending on whether the difference between the first and the second is positive or negative. It follows from this definition that positive numbers are always larger than negative numbers, and the latter ought to be considered as being as small as their numerical values are large.[4]

In algebra, we use letters to represent quantities as well as numbers. Since it is customary to classify the a numbers as positive quantities, we may denote the positive quantity that has as its numerical value the number A by $+A$ or just by A, whereas the opposite negative quantity is denoted by $-A$. Likewise, when the letter a represents a quantity, it is customary to regard [19] the two expressions a and $+a$ as synonyms, and to denote by $-a$ the quantity that is opposite to $+a$. These remarks suffice to establish what we call the *rule of signs* (see Note I).

We call a quantity *variable* if it can be considered as able to take on successively many different values. We normally denote such a quantity by a letter taken from the end of the alphabet. On the other hand, a quantity is called *constant*, ordinarily denoted by a letter from the beginning of the alphabet, if it takes on a fixed and determined value. When the values successively[5] attributed to a particular variable indefinitely approach a fixed value in such a way as to end up by differing from it by as little as we wish, this fixed value is called the *limit* of all the other values.[6] Thus, for example, an irrational number is the limit of the various fractions that give better and better approximations to it.[7] In Geometry, the area of a circle is the limit

[1] That is, the absolute value of the quantity.

[2] Later in Note III, Cauchy uses "contrary" rather than "opposite" to represent this idea.

[3] Cauchy discusses arithmetic operations and signs in some detail in Note I.

[4] In the 18th century, opposite numbers were considered to be the same size. Here, when Cauchy proposes that negative numbers are smaller than positive ones, it is a relatively new idea.

[5] This adverb "successively" (*successivement*) seems appropriate to a discussion of convergence of sequences, although perhaps less so in the case of continuous variables.

[6] This passage is translated in [Kline, p. 951].

[7] [DSB Cauchy, p. 136] cites this passage, but incorrectly states that it defines "convergence and absolute convergence of series, and limits of sequences and functions." It clearly does less than that.

towards which the areas of the inscribed polygons converge when the number of their sides grows more and more, etc.

When the successive numerical values of such a variable decrease indefinitely, in such a way as to fall below any given number, this variable becomes what we call *infinitesimal*, or an *infinitely small quantity*. A variable of this kind has zero as its limit.[8]

When the successive numerical values of a given variable increase more and more in such a way as to rise above any given number, we say that this variable has *positive infinity* as its limit,[9] denoted by the symbol ∞, if it is a positive variable, and *negative infinity*, denoted by $-\infty$, if it is a negative variable. The infinities, positive and negative, are designated together by the name of *infinite quantities*.

The quantities that arise in calculation as the result of operations made on one or more constant or variable quantities can be divided into various kinds, depending on the [20] nature of the operations that produce them. In algebra, for example, we distinguish sums and differences, products and quotients, powers and roots, and exponentials and logarithms. In trigonometry, we distinguish sines and cosines, secants and cosecants, tangents and cotangents, and the arcs of a circle for which a trigonometric line is given.[10] To better understand what is meant by these last kinds of quantities, it is necessary to review the following principles.

A length measured along a straight or curved line may, like any kind of magnitude, be represented either by a number or by a quantity. It would be represented as a number when we consider it only as a measure of this length, and as a quantity, that is to say as a number preceded by a $+$ or a $-$ sign, when we consider the length in question as drawn from a fixed point along the given line in one direction or the other, serving as the increase or the decrease of another constant length that ends at this fixed point. The fixed point in question from which we must measure the variable lengths denoted by these quantities is what we call the *origin* of these same lengths. Two lengths measured from a common origin but in opposite directions must be represented by quantities of different sign. We may choose at will the direction in which lengths are denoted by positive quantities, but once that choice is made, we must necessarily consider lengths denoted by negative numbers as going in the opposite direction.

In a circle, whose plane is assumed to be vertical, we ordinarily take for the origin of the arcs the endpoint of the radius drawn horizontally from left to right, and we measure positive arcs as rising above this point, that is to say, those arcs which we measure by positive quantities. On the same circle, when the radius is assumed to be 1, the sine of an arc, that is to say the projection of the radius which passes through the endpoint of the arc onto the vertical diameter is measured [21] positively from bottom to top, and negatively in the opposite direction, taking the center of the circle as the origin of the sines. The tangent is measured positively in the same direction as the sine but measured from the origin of the arcs along the vertical line drawn

[8] This passage is also translated in [Kline, p. 951].

[9] This passage is also translated in [Kline, p. 951].

[10] Cauchy means inverse trigonometric functions. Note that they were still called "lines" and that this is implicitly using Euler's unit circle definition of trigonometric functions.

from this origin. Finally, the secant is measured from the center of the circle along the radius drawn through the endpoint of the arc in question[11] and positively in the direction of this radius.

Frequently, the result of an operation performed on a quantity may have several values, different from one another. When we do not wish to distinguish among the various values, we use a notation in which the quantity is enclosed in doubled symbols or double parentheses and we reserve the ordinary notation for the most simple value, or the one that seems to deserve to be distinguished. Therefore, for example, if a is a positive quantity, the square root of a has two values, numerically equal, but with opposite signs, an arbitrary one of which is expressed by the notation

$$((a))^{\frac{1}{2}} \quad \text{or} \quad \sqrt{\!\!\sqrt{a}}$$

while the positive value alone is written as

$$a^{\frac{1}{2}} \quad \text{or} \quad \sqrt{a},$$

so that we have

(1) $$\sqrt{\!\!\sqrt{a}} = \pm\sqrt{a}$$

or, what amounts to the same thing,

(2) $$((a))^{\frac{1}{2}} = \pm a^{\frac{1}{2}}.$$

Similarly, if we represent a positive or negative quantity by a, the notation[12]

$$\arcsin((a)) \quad \text{or} \quad \arctan((a))$$

denotes an arbitrary arc having the quantity a for its sine or for its tangent, respectively, while the notation

$$\arcsin(a) \quad \text{or} \quad \arctan(a)$$

[22] indicates only that particular arc with the smallest numerical value. With the aid of these conventions, we avoid the confusion that could result from the use of symbols, the values of which have not been determined precisely. In order to remove all difficulty in this matter, I give here the table of notations which we will use for expressing the results of algebraic and trigonometric operations.

The sum of two quantities is usually denoted by the juxtaposition of these two quantities, each of which is expressed by a letter preceded by the sign $+$ or $-$, which we may suppress (if the sign is $+$) in front of the first letter only. And so,

---

[11] That is, along the radius from the center through the circumference and to the vertical line along which tangents are measured.

[12] Cauchy writes "arc sin$((a))$" and "arc tang$((a))$." Here and subsequently we will use the more modern notation and write "arcsin" and "arctan." (tr.)

$$+a+b, \quad \text{or simply} \quad a+b,$$

denotes the sum of the two quantities $+a$ and $+b$, and

$$+a-b, \quad \text{or simply} \quad a-b,$$

denotes the sum of the two quantities $+a$ and $-b$, which is equivalent to the difference of the two quantities $+a$ and $+b$.

We indicate the equality of two quantities a and b by the sign $=$, written between them, as follows,

$$a=b,$$

and we indicate that the first is greater than the second, that is to say that the difference $a-b$ is positive, by writing

$$a>b \quad \text{or} \quad b<a.$$

As usual, we represent the product of two quantities $+a$ and $+b$ by[13]

$$+a\times+b, \quad \text{or simply} \quad a.b \quad \text{or} \quad ab$$

and their quotient by

$$\frac{a}{b} \quad \text{or} \quad a:b.$$

[23] Now let m and n be two whole numbers, A an arbitrary number and a and b two arbitrary quantities, positive or negative. Then

$$A^m,\ A^{\frac{1}{n}}=\sqrt[n]{A},\ A^{\pm\frac{m}{n}} \quad \text{and} \quad A^b$$

represent the positive quantities which we obtain by raising the number A to the powers denoted respectively by the exponents

$$m,\ \frac{1}{n},\ \pm\frac{m}{n} \quad \text{and} \quad b,$$

and

$$a^{\pm m}$$

denotes the quantity, positive or negative, that arises from taking the quantity a to the power $\pm m$. We use the notation

$$((a))^{\frac{1}{n}}=\sqrt[n]{\!\sqrt{a}} \qquad \text{and} \quad ((a))^{\pm\frac{m}{n}}$$

to denote not only the positive and negative values, when they exist, of the powers of the quantity a raised to the exponents

[13] In [Cauchy 1821, p. 9, Cauchy 1897, p. 22] Cauchy used a period in $a.b$ rather than a centered dot, as we would today.

$$\frac{1}{n} \quad \text{and} \quad \pm\frac{m}{n},$$

but also the imaginary values[14] of these same powers (see Chap. VII for the meaning of *imaginary expressions*). It is helpful to observe that if we let A be the numerical value of a, and if we assume that the fraction $\frac{m}{n}$ is in lowest terms, then the power

$$((a))^{\frac{m}{n}}$$

has a single positive or negative real value, namely

$$+A^{\frac{m}{n}} \quad \text{or} \quad -A^{\frac{m}{n}},$$

as long as $\frac{m}{n}$ is a fraction with an odd denominator, but if the denominator is even, then it has [24] either the two real values just mentioned, or no real values. We could make a similar remark about the expression

$$((a))^{-\frac{m}{n}}.$$

In the particular case where the quantity a is positive and we let $\frac{m}{n} = \frac{1}{2}$, the expression $((a))^{\frac{m}{n}}$ has two real values, given by formula (2) or, what amounts to the same thing, by formula (1).

The notations[15]

$$l(B),\ L(B),\ L'(B),\ \ldots$$

denote the real logarithms of the number B to different bases, whereas each the following,

$$l((b)),\ L((b)),\ L'((b)),\ \ldots$$

denote, in addition to the real logarithm of the quantity b, when it exists, any of the imaginary logarithms of this same quantity (see Chap. IX for the meaning of *imaginary logarithms*).

In trigonometry,

$$\sin a, \quad \cos a, \quad \tan a, \quad \cot a, \quad \sec a, \quad \csc a, \quad \operatorname{siv} a \quad \text{and} \quad \operatorname{cosiv} a$$

denote, respectively, the sine, cosine, tangent, cotangent, secant, cosecant, versine and vercosine of the arc a.[16] The notations

---

[14] Cauchy does not actually define an "imaginary value," but it is clear that it is what we get when we assign particular real values to the real quantities in an imaginary expression.

[15] Here we have reproduced Cauchy's notation for logarithm. Subsequently, we will always use more modern notation, like $\ln(B)$, $\log(B)$, $\operatorname{Log}(B)$.

[16] We note that Cauchy uses "tang. a" for the trigonometric function as well as the inverse trigonometric function. His notations for secant and cosecant are "séc. *a*" and "coséc. *a*." Note also his use of the obsolete trigonometric functions versed sine and versed cosine. He will later also use the obsolete function chord on p. 45; [Cauchy 1821, p. 63, Cauchy 1897, p. 66].

$$\arcsin((a)),\ \arccos((a)),\ \arctan((a)),$$
$$\operatorname{arccot}((a)),\ \operatorname{arcsec}((a)),\ \operatorname{arccsc}((a))$$

indicate some one of the arcs which have the quantity a as their sine, cosine, tangent, cotangent, secant or cosecant. We use the simple notations

$$\arcsin(a),\ \arccos(a),\ \arctan(a),\ \operatorname{arccot}(a),\ \operatorname{arcsec}(a),\ \operatorname{arccsc}(a),$$

[25] or we may suppress the parentheses and write

$$\arcsin a,\ \arccos a,\ \arctan a,\ \operatorname{arccot} a,\ \operatorname{arcsec} a,\ \operatorname{arccsc} a$$

when, from among the arcs for which a trigonometric function is equal to a,[17] we wish to designate the one with smallest numerical value, or, if there are two such arcs with opposite signs, the one with the positive value. Consequently,

$$\arcsin a,\ \arctan a,\ \operatorname{arccot} a,\ \operatorname{arccsc} a,$$

denote positive or negative arcs between the limits

$$-\frac{\pi}{2} \quad \text{and} \quad +\frac{\pi}{2},$$

where π denotes the semiperimeter of the unit circle, whereas

$$\arccos a \quad \text{and} \quad \operatorname{arcsec} a$$

denote positive arcs between 0 and π.

By virtue of the conventions that we have just established, if we denote by k an arbitrary positive integer, we obviously have, for arbitrary positive or negative values of the quantity a,

$$(3) \qquad \begin{cases} \arcsin((a)) = \frac{\pi}{2} \pm \left(\frac{\pi}{2} - \arcsin a\right) \pm 2k\pi, \\ \arccos((a)) = \pm \arccos a \pm 2k\pi, \\ \arctan((a)) = \arctan a \pm k\pi, \\ \arccos a + \arcsin a = \frac{\pi}{2} \quad \text{and} \\ \operatorname{arccsc} a + \operatorname{arcsec} a = \frac{\pi}{2}. \end{cases}$$

Furthermore, we find that, for positive values of a,

$$(4) \qquad \operatorname{arccot} a + \arctan a = \frac{\pi}{2},$$

[26] and for negative values of a,

---

[17] Here, Cauchy writes "...parmi les arcs dont un ligne trigonométrique est égale à a." This translates literally "...among the arcs for which a trigonometric line is equal to a." Cauchy is treating trigonometric functions as giving lines, which have signed lengths, rather than in the modern view of giving real numbers.

$$\operatorname{arccot} a + \arctan a = -\frac{\pi}{2}. \tag{5}$$

When a variable quantity converges towards a fixed limit, it is often useful to indicate this limit with particular notation. We do this by placing the abbreviation[18]

$$\lim$$

in front of the variable quantity in question. Sometimes, when one or several variables converge towards fixed limits, an expression containing these variables converges towards several different limits at the same time. We therefore denote an arbitrary one of these limits using the doubled parentheses following the abbreviation lim, so as to enclose the expression under consideration. Specifically, suppose that a positive or negative variable denoted by x converges towards the limit 0, and denote by A a constant number. It is easy to see that each of the expressions

$$\lim A^x \quad \text{and} \quad \lim \sin x$$

has a unique value determined by the equation

$$\lim A^x = 1$$

or

$$\lim \sin x = 0,$$

whereas the expression

$$\lim \left(\left(\frac{1}{x}\right)\right)$$

takes two values, $+\infty$ and $-\infty$, and

$$\lim \left(\left(\sin \frac{1}{x}\right)\right)$$

admits an infinity of values between the limits -1 and $+1$.

We will finish these preliminaries by presenting several theorems on average quantities, the knowledge of which will [27] be extremely useful in the remainder of this work. We call an *average* among several given quantities a new quantity between the smallest and the largest of those under consideration. From this definition it is clear that there are an infinity of averages among several unequal quantities, and that the average among several equal quantities is equal to their common value. Given this, we will easily establish, as one can see in Note II, the following propositions:

[18] The notation "Lim." for limit was first used by Simon Antoine Jean L'Huilier (1750–1840) in [L'Huilier 1787, p. 31]. Cauchy wrote this as "lim." in [Cauchy 1821, p. 13]. The period had disappeared by [Cauchy 1897, p. 26].

Theorem I.[19] — *Let* $b, b', b'', \ldots$ *denote* n *quantities of the same sign, and* $a, a', a'', \ldots$ *be the same number of arbitrary quantities. The fraction*

$$\frac{a+a'+a''+\ldots}{b+b'+b''+\ldots}$$

is an average of the following quantities,

$$\frac{a}{b}, \frac{a'}{b'}, \frac{a''}{b''}, \ldots.$$

Corollary. — If we let

$$b = b' = b'' = \ldots = 1,$$

it follows from the preceding theorem that the quantity

$$\frac{a+a'+a''+\ldots}{n}$$

is an average of the quantities

$$a, a', a'', \ldots.$$

This particular kind of average is called the *arithmetic mean.*

Theorem II. — *Let* $A, A', A'', \ldots; B, B', B'', \ldots,$ *be two sequences of numbers taken at will, which we suppose contain* n *terms each. Form from these two sequences the roots*

$$\sqrt[B]{A}, \sqrt[B']{A'}, \sqrt[B'']{A''}, \ldots.$$

[28] *Then* $\sqrt[B+B'+B''+\ldots]{AA'A''\ldots}$ *is a new root which is an average of the other roots.*

Corollary. — If we let

$$B = B' = B'' = \ldots = 1,$$

we find that the positive quantity

$$\sqrt[n]{AA'A''\ldots}$$

is an average of

$$A, A', A'', \ldots.$$

This particular average is called the *geometric mean.*

Theorem III. — *With the same hypotheses as in theorem I, and if* $\alpha, \alpha', \alpha'', \ldots$ *again denote quantities of the same sign, the fraction*

---

19 Cauchy gives a proof of this theorem in Note II, Theorem XII [Cauchy 1821, p. 447, Cauchy 1897, p. 368].

$$\frac{\alpha a+\alpha' a'+\alpha'' a''+\ldots}{\alpha b+\alpha' b'+\alpha'' b''+\ldots}$$

is an average of

$$\frac{a}{b}, \frac{a'}{b'}, \frac{a''}{b''}, \ldots.$$

Corollary. — If we suppose that

$$b=b'=b''=\ldots=1,$$

we conclude from the previous theorem that the sum

$$a\alpha+a'\alpha'+a''\alpha''+\ldots$$

is equivalent to the product of

$$\alpha+\alpha'+\alpha''+\ldots$$

with an average of the quantities a, a', a'', ….

For brevity, when we wish to denote an average of [29] several quantities a, a', a'', …, we use the notation

$$M(a,a',a'',\ldots).$$

Given this, the preceding theorems and their corollaries are included in the following formulas:

$$\frac{a+a'+a''+\ldots}{b+b'+b''+\ldots}=M\left(\frac{a}{b},\frac{a'}{b'},\frac{a''}{b''},\ldots\right), \tag{6}$$

$$\frac{a+a'+a''+\ldots}{n}=M\left(a,a',a'',\ldots\right), \tag{7}$$

$$\sqrt[B+B'+B''+\ldots]{AA'A''\ldots}=M\left(\sqrt[B]{A},\ \sqrt[B']{A'},\ \sqrt[B'']{A''},\ldots\right), \tag{8}$$

$$\sqrt[n]{AA'A''\ldots}=M\left(A,A',A'',\ldots\right), \tag{9}$$

$$\frac{a\alpha+a'\alpha'+a''\alpha''+\ldots}{b\alpha+b'\alpha'+b''\alpha''+\ldots}=M\left(\frac{a}{b},\frac{a'}{b'},\frac{a''}{b''},\ldots\right), \tag{10}$$

$$a\alpha+a'\alpha'+a''\alpha''+\ldots=(\alpha+\alpha'+\alpha''+\ldots)M\left(a,a',a'',\ldots\right). \tag{11}$$

In these formulas,

$$a,\ a',\ a'',\ \ldots;b,\ b',\ b'',\ \ldots;\alpha,\ \alpha',\ \alpha'',\ \ldots$$

denote three sequences of quantities, and

$$A,\ A',\ A'',\ \ldots;B,\ B',\ B'',\ \ldots$$

two sequences of numbers, all five sequences consisting of n terms. The second and third sequences consist of quantities of the same sign. The notation that we have just adopted gives a way to express that a quantity is included between two given limits. In fact, any quantity between the limits a and b is an average of those two limits, and may be denoted by

$$M(a,b).$$

And so, for example, any positive quantity can be represented by $M(0,\infty)$, and any negative quantity by $M(-\infty,0)$, and any real quantity by $M(-\infty,+\infty)$. When we do not want to indicate [30] any particular one of the quantities contained between the limits a and b, we double the parentheses and write

$$M((a,b)).$$

For example, if we suppose that the variable x converges to zero, we have[20]

$$\lim\left(\left(\sin\frac{1}{x}\right)\right) = M((-1,+1)),$$

given that the expression $\lim\left(\left(\sin\frac{1}{x}\right)\right)$ admits an infinity of values between the extreme values -1 and $+1$.

[20] Note that here $M((-1,+1))$ is meant to include both endpoints, but that above, $M(0,\infty)$ was meant to exclude the endpoint 0. Apparently, Cauchy does not see a need to distinguish between open and closed intervals.

Chapter 1
On real functions.

First Part[1]
ALGEBRAIC ANALYSIS

Chapter I. ON REAL FUNCTIONS.

1.1 General considerations on functions.

[31] When variable quantities are related to each other such that the value of one of the variables being given one can find the values of all the other variables, we normally consider these various quantities to be expressed by means of the one among them, which therefore takes the name the *independent variable*. The other quantities expressed by means of the independent variable are called *functions* of that variable.

When variable quantities are related to each other such that the values of some of them being given one can find all of the others, we consider these various quantities to be expressed by means of several among them, which therefore take the name *independent variables*. The other quantities expressed by means of the independent variables are called *functions* of those same variables.

The various expressions that are used in algebra and trigonometry, when they involve variables that are considered to be independent, are also functions of these same variables. And so, for example,

$$\log(x), \ \sin x, \ \ldots$$

[32] are functions of the variable x, while

[1] Cauchy had originally planned for the *Cours d'analyse* to consist of two volumes. He never wrote the second one; see p. viii.

R.E. Bradley, C.E. Sandifer, *Cauchy's Cours d'analyse*, Sources and Studies in the History of Mathematics and Physical Sciences, DOI 10.1007/978-1-4419-0549-9_1,

$$x+y,\ x^y,\ xyz,\ \ldots$$

are functions of the variables x and y, or of x, y and z, ...

When the functions of one or several variables are directly expressed, as in the preceding examples, by means of those same variables, they are called *explicit functions*. But when they are given only as relations among the functions and the variables, that is to say the equations that the quantities must satisfy, as long as the equations are not solved algebraically, the functions are not expressed directly by means of the variables, then they are called *implicit functions*. To make them explicit, it suffices to solve, when it is possible, the equations that determine them. For example, when y is given by the equation

$$\log(y) = x,$$

then it is an implicit function of x. If we let A be the base of the system of logarithms being considered, the same function made explicit by solving the given equation will be

$$y = A^x.$$

When we want to denote an explicit function of a single variable x or of several variables, x, y, z, ..., without specifying the nature of that function, we use one of the notations

$$f(x),\ F(x),\ \phi(x),\ \chi(x),\ \psi(x),\ \varpi(x),\ \ldots,$$
$$f(x,y,z,\ldots),\ F(x,y,z,\ldots),\ \phi(x,y,z,\ldots),\ldots.$$

For a function of one variable to be completely determined, it is necessary and sufficient that from every particular value assumed by the variable, one can deduce the corresponding value of the function. Sometimes, for each value of the variable, the given function [33] takes on several values different from one another. Conforming to the conventions adopted in the preliminaries, we will usually designate these multiple values of a function by the notations in which the variables are written with doubled symbols or with doubled parentheses. Thus, for example

$$\arcsin((x))$$

indicates one of the arcs that have x as their sines, and

$$\sqrt{\sqrt{x}} = \pm\sqrt{x}$$

indicates one of the two square roots of the variable x, assuming that x is positive.

1.2 On simple functions.

Among the functions of a single variable x, the ones we call *simple* are those that result from just one operation on that variable. There are only a few simple functions that we ordinarily consider in analysis, some of which arise from algebra and the others from trigonometry. Addition and subtraction, multiplication and division, raising to powers and extracting roots, and finally the formation of exponentials and logarithms, produce the simple functions that arise in algebra. Thus, if A denotes a constant number and if $a = \pm A$ is a constant quantity, the simple algebraic functions of the variable x are

$$a+x,\ a-x,\ ax,\ \frac{a}{x},\ x^a,\ A^x \text{ and } \log(x).$$

Here we need not account for roots because we can always write them as powers. There are a great number of simple functions that arise in trigonometry. They include the simple functions of all the trigonometric lines as well as the arcs that correspond to these same lines, but [34] they all reduce to the four following functions:

$$\sin x, \qquad \cos x,$$
$$\arcsin x \text{ and } \arccos x,$$

and we will put the other trigonometric lines, $\tan x$, $\sec x$, ..., along with the corresponding arcs, $\arctan x$, $\operatorname{arcsec} x$, ..., among the composite functions, because these lines can always be expressed by means of the sine and the cosine. We could even, for the sake of rigor, reduce the two simple functions, $\sin x$ and $\cos x$, to a single one because they are related to each other by the equation $\sin^2 x + \cos^2 x = 1$, but we use these two functions so frequently that it is useful to keep both of them among the simple functions.

1.3 On composite functions.

The functions that are given by a single variable by means of several operations are called *composite functions*. We will distinguish among these the *functions of functions* that result from several successive operations, the first operation acting upon the variable and each of the others acting on the result of the preceding operation. By virtue of these definitions,

$$x^x,\ \sqrt[x]{x},\ \frac{\log x}{x},\ \ldots$$

are composite functions of the variable x, and

$$\log(\sin x) \quad \text{and} \quad \log(\cos x),\ \ldots$$

are functions of functions, of which each is the result of two successive operations.

Composite functions are distinguished from each other by the nature of the operations that produce them. It seems that we ought [35] to call by the name *algebraic functions* all those functions that are formed by the operations of algebra, but instead we will reserve that name particularly for those functions formed using only the first algebraic operations, namely addition and subtraction, multiplication and division, and finally the raising to fixed powers. Those functions that involve variable exponents or logarithms we will call *exponential* or *logarithmic functions*.

The algebraic functions are divided into *rational functions* and *irrational functions*. The rational functions are those in which the variable is raised only to integer powers.[2] In particular, an *integer function*[3] is any polynomial function[4] that involves only integer powers[5] of the variable, for example

$$a+bx+cx^2+\dots,$$

and a *fractional function* or *rational fraction* is the quotient of two such polynomials. The *degree* of an integer function of x is the exponent of the highest power of x in that function. A function of the first degree, namely

$$a+bx,$$

is also called a *linear function* because in Geometry we use it to represent the ordinates[6] of a straight line. Every integer or fractional function is at the same time rational, and every other kind of algebraic function is irrational.

The functions produced by the operations of trigonometry are called *trigonometric* or *circular functions*.

The various names that we have just given to composite functions of just one variable apply as well to functions of several variables when these functions enjoy, with respect to each of the variables that they involve, the properties corresponding to the various names. Thus, for example, any polynomial [36] that contains nothing but integer powers of the variables $x, y, z, \dots$ is an integer function of these variables. We call the degree of this integer function the sum of the exponents of the variables in the term where that sum is the largest. An integer function of the first degree, like

$$a+bx+cy+dz+\dots$$

is called a linear function.

---

[2] Cauchy uses the term *puissances entières* here. This is the second appearance of the adjective *entier*, the first being on [Cauchy 1821, p. 9, Cauchy 1897, p. 23], where he spoke of *nombres entiers*. The word translates literally as "entire." Because to Cauchy, "numbers" are positive, there was no question in the first case that he meant to exclude negative values. Here, though, when he writes of *puissances entières*, and because he has never actually defined the word *entier*, it is not clear whether these powers are allowed to be negative, or if they are strictly positive.

[3] Cauchy calls this a *fonction entière*, which translates literally as an "entire function." We translate it as an "integer function" to avoid confusion with the modern definition of "entire" from complex variables. Cauchy's "integer" functions form a subset of the modern set of "entire" functions.

[4] Note that Cauchy does not define "polynomial."

[5] Again, Cauchy writes *puissances entières*, but here it is clear from his example that he means positive integers only.

[6] I.e., y-coordinates.

Chapter 2
On infinitely small and infinitely large quantities, and on the continuity of functions. Singular values of functions in various particular cases.

2.1 On infinitely small and infinitely large quantities.

[37] We say that a variable quantity becomes *infinitely small* when its numerical value decreases indefinitely in such a way as to converge towards the limit zero. It is worth remarking on this point that one ought not confuse a constant decrease with an indefinite decrease. The area of a regular polygon circumscribed about a given circle decreases constantly as the number of sides increases, but not indefinitely, because it has as its limit the area of the circle.[1] Similarly, a variable which takes as successive values only the different terms of the sequence[2]

$$\frac{2}{1}, \frac{3}{2}, \frac{4}{3}, \frac{5}{4}, \frac{6}{5}, \ldots,$$

taken to infinity, would decrease constantly, but not indefinitely because its successive values converge towards the limit 1. On the other hand, a variable which takes as successive values only the different terms of the sequence

$$\frac{1}{4}, \frac{1}{3}, \frac{1}{6}, \frac{1}{5}, \frac{1}{8}, \frac{1}{7}, \ldots,$$

taken to infinity, does not decrease constantly, since the difference between two consecutive terms of this sequence is alternately [38] positive and negative. Nevertheless, it decreases indefinitely because its value ultimately becomes smaller than any given number.

[1] The example of the polygon circumscribed about or inscribed in a circle had been the standard example illustrating the informal definition of the limit for many years. See, for example, [Chapelle 1765].

[2] When Cauchy uses *suite*, he almost always means "sequence." When he uses *series*, he may mean either "sequence" or "series." We translate *suite* as "sequence," *série* as "series" and *progression* as "progression," because we do not want to leave the reader with the impression that Cauchy made modern distinctions that he did not actually make.

R.E. Bradley, C.E. Sandifer, *Cauchy's Cours d'analyse*, Sources and Studies in the History of Mathematics and Physical Sciences, DOI 10.1007/978-1-4419-0549-9_2,

We say that a variable quantity becomes *infinitely large* when its numerical value increases indefinitely in such a way as to converge towards the limit ∞. It is again essential to observe here that one ought not confuse a variable that increases indefinitely with a variable that increases constantly. The area of a regular polygon inscribed in a given circle increases constantly as the number of sides increases, but not indefinitely. The terms of the natural sequence of integer numbers

$$1,\ 2,\ 3,\ 4,\ 5,\ \ldots$$

increase constantly and indefinitely.

Infinitely small and infinitely large quantities enjoy several properties that lead to the solution of important questions, which I will explain in a few words.

Let α be an infinitely small quantity, that is a variable whose numerical value decreases indefinitely. When the various integer powers of α, namely

$$\alpha,\ \alpha^2,\ \alpha^3,\ \ldots,$$

enter into the same calculation, these various powers are called, respectively, infinitely small of the *first*, the *second*, the *third order*, etc. In general, we call any variable quantity infinitely small of the first order if its ratio with α converges to a finite limit different from zero as the numerical value of α diminishes.[3] We call a variable quantity involving α infinitely small of the second order if its ratio with α^2 converges towards a finite limit different from zero, and so forth for higher orders. Given this, if k denotes a finite quantity different from zero and ε denotes a variable number that decreases indefinitely with the numerical value of α, the general form of infinitely small quantities of the first order is

$$k\alpha \quad \text{or at least} \quad k\alpha\,(1\pm\varepsilon)\,.$$

[39] The general form of infinitely small quantities of the second order will be

$$k\alpha^2 \quad \text{or at least} \quad k\alpha^2\,(1\pm\varepsilon)\,,$$
$$\ldots\ldots \quad \ldots\ldots\ldots\ldots \quad \ldots\ldots\ldots\ldots$$

Finally, the general form of infinitely small quantities of order n (where n represents an integer number) will be

$$k\alpha^n \quad \text{or at least} \quad k\alpha^n\,(1\pm\varepsilon)\,.$$

We may easily establish the following theorems concerning these various orders of infinitely small quantities.

---

[3] Here, Cauchy is making the implicit assumption that α is never zero. Contemporaries, such as L'Huilier and Lacroix, were unclear on whether a variable quantity could ever exceed its limit. By explicitly using the numerical value, Cauchy evidently avoided such problems. Nevertheless, he resists allowing the variable quantity to attain its limiting value during the limiting process. See [Grabiner 2005, pp. 84–85] for further discussion.

Theorem I. — *If we compare two infinitely small quantities of different orders with each other, while both converge towards the limit zero, then eventually the one of the higher order will constantly have the smaller numerical value.*

Proof. — Indeed, let

$$k\alpha^n(1\pm\varepsilon) \quad \text{and} \quad k'\alpha^{n'}(1\pm\varepsilon')$$

be two infinitely small quantities, one of order n, the other of order n', and suppose that $n' > n$. The ratio between the second of these infinitely small quantities and the first, namely

$$\frac{k'}{k}\alpha^{n'-n}\frac{1\pm\varepsilon'}{1\pm\varepsilon},$$

converges indefinitely with α towards the limit zero, which cannot occur unless the numerical value of the second eventually becomes constantly less than that of the first.

Theorem II. — *An infinitely small quantity of order n, that is to say of the form*

$$k\alpha^n(1\pm\varepsilon),$$

changes sign with α whenever n is an odd number, and for very small numerical values of α takes the same sign as the quantity k whenever n is an even number.

Proof. — Indeed, under the first hypothesis, α^n changes [40] sign with α, and under the second hypothesis, α^n is always positive. Furthermore, the sign of the product $k(1\pm\varepsilon)$ is the same as that of k, when ε is very small.

Theorem III. — *The sum of several infinitely small quantities of orders*

$$n,\ n',\ n'',\ \ldots$$

(where n', n'', ... denote numbers larger than n) is a new infinitely small quantity of order n.

Proof. — Indeed,

$$\begin{aligned}&k\alpha^n(1\pm\varepsilon)+k'\alpha^{n'}(1\pm\varepsilon')+k''\alpha^{n''}(1\pm\varepsilon'')+\ldots\\ &=k\alpha^n\left[1\pm\varepsilon+\tfrac{k'}{k}\alpha^{n'-n}(1\pm\varepsilon')+\tfrac{k''}{k}\alpha^{n''-n}(1\pm\varepsilon'')+\ldots\right]\\ &=k\alpha^n(1\pm\varepsilon_1),\end{aligned}$$

where ε_1 is a number which converges with α towards the limit zero.

From the principles which we have just stated, we easily deduce, as we will see, several remarkable propositions concerning polynomials ordered according to increasing powers of an infinitely small quantity α.

Theorem IV. — *Any polynomial ordered according to increasing powers of α, for example*

$$a+b\alpha+c\alpha^2+\dots,$$

or more generally

$$a\alpha^n+b\alpha^{n'}+c\alpha^{n''}+\dots,$$

(where the numbers n, n', n'', ... form an increasing sequence), will eventually be constantly of the same sign as its first term

$$a \quad \text{or} \quad a\alpha^n$$

for very small numerical values of α.

Proof. — Indeed, the sum formed by the second term and those that follow is, in the first case, an infinitely small quantity of the first order, whose numerical value will eventually be smaller than the finite quantity a,[4] and, in the second case, an infinitely small quantity [41] of order n', which eventually takes a numerical value that is constantly smaller than that of the infinitely small quantity of order n.

Theorem V. — *When, in the polynomial*

$$a\alpha^n+b\alpha^{n'}+c\alpha^{n''}+\dots,$$

ordered according to increasing powers of α, the degree n' of the second term is an odd number, then for very small numerical values of α, this polynomial is either greater than or less than its first term $a\alpha^n$, depending on whether the variable α and the coefficient b have the same or opposite signs.

Proof. — Indeed, under the given hypothesis, the sum of the terms that follow the first, namely

$$b\alpha^{n'}+c\alpha^{n''}+\dots,$$

has the same sign as each of the two products $b\alpha^{n'}$ and $b\alpha$, for very small numerical values of α.

Theorem VI. — *When, in the polynomial*

$$a\alpha^n+b\alpha^{n'}+c\alpha^{n''}+\dots,$$

ordered according to increasing powers of α, the degree n' of the second term is an even number, then for very small numerical values of α, this polynomial will eventually become constantly greater than its first term whenever b is positive, and constantly less whenever b is negative.

---

[4] Here, Cauchy means the sum to be smaller than the *numerical value* of a, as a may be negative.

Proof. — Indeed, under the given hypothesis, the sum of the terms that follow the first has, for very small numerical values of α, the sign of the product $b\alpha^{n'}$, and consequently the sign of b.

Corollary. — Supposing in the preceding theorem that $n = 0$, we get the following proposition:

Theorem VII. — *If, in the polynomial*[5]

$$a + b\alpha^{n'} + c\alpha^{n''} + \ldots,$$

ordered according to increasing powers of α, n' *denotes an even number,* [42] *then among the values of this polynomial corresponding to infinitely small values of* α, *the one that corresponds to* $\alpha = 0$, *that is* a, *will always be the smallest whenever* b *is positive, and the greatest when* b *is negative.*

This particular value of the polynomial, either larger or smaller than all of its neighboring values, is what we call a *maximum* or a *minimum*.

The properties of infinitely small quantities having been established, we deduce from them the analogous properties of infinitely large quantities by observing that any variable quantity of this last kind may be represented as $\frac{1}{\alpha}$, where α denotes an infinitely small quantity. Thus, for example, when, in the polynomial

$$ax^m + bx^{m-1} + cx^{m-2} + \ldots + hx + k,$$

ordered according to decreasing powers of x, this variable becomes infinitely large, then substituting $\frac{1}{\alpha}$ for x we reduce the given polynomial to

$$\frac{a}{\alpha^m}\left(1 + \frac{b}{a}\alpha + \frac{c}{a}\alpha^2 + \ldots + \frac{h}{a}\alpha^{m-1} + \frac{k}{a}\alpha^m\right).$$

Thus we see immediately that for very small numerical values of α, or what amounts to the same thing, for very large numerical values of x, this polynomial has the same sign as its first term,

$$\frac{a}{\alpha^m} = ax^m.$$

As this remark applies even in the case where some of the quantities $b, c, \ldots, h, k$ reduce to zero, we can state the following theorem:

Theorem VIII. — *When, in a polynomial ordered according to decreasing powers of the variable* x, *we let the numerical value of this variable increase indefinitely, then the polynomial will eventually have the same sign as its first term.*

[5] In [Cauchy 1897, p. 41], there is a typographical error here, writing n where we write n''. This error was not in [Cauchy 1821, p. 32]. (tr.)

2.2 On the continuity of functions.

[43] Among the objects related to the study of infinitely small quantities, we ought to include ideas about the continuity and the discontinuity of functions. In view of this, let us first consider functions of a single variable.

Let $f(x)$ be a function of the variable x, and suppose that for each value of x between two given limits, the function always takes a unique finite value. If, beginning with a value of x contained between these limits, we add to the variable x an infinitely small increment α, the function itself is incremented by the difference[6]

$$f(x+\alpha)-f(x),$$

which depends both on the new variable α and on the value of x. Given this, the function $f(x)$ is a *continuous* function of x between the assigned limits if, for each value of x between these limits, the numerical value of the difference

$$f(x+\alpha)-f(x)$$

decreases indefinitely with the numerical value of α. In other words, *the function $f(x)$ is continuous with respect to x between the given limits if, between these limits, an infinitely small increment in the variable always produces an infinitely small increment in the function itself.*[7]

We also say that the function $f(x)$ is a continuous function of the variable x in a neighborhood of a particular value of the variable x whenever it is continuous between two limits of x that enclose that particular value, even if they are very close together.

Finally, whenever the function $f(x)$ ceases to be continuous in the neighborhood of a particular value of x, we say that it becomes discontinuous, and that there is *solution of continuity*[8] for this particular value.

[44] Having said this, it is easy to recognize the limits between which a given function of a variable x is continuous with respect to that variable. So, for example, the function $\sin x$, which takes a unique finite value for each particular value of the variable x, is continuous between any two limits of this variable, given that the numerical value of $\sin\left(\frac{1}{2}\alpha\right)$, and consequently that of the difference[9]

$$\sin(x+\alpha)-\sin x=2\sin\left(\tfrac{1}{2}\alpha\right)\cos\left(x+\tfrac{1}{2}\alpha\right),$$

---

[6] Cauchy defines continuity only on the interior of a bounded interval, and for the whole interval, not just at a single point. See [Grabiner 2005, p. 87] for more on this point. This passage is also cited in [DSB Cauchy, p. 136].

[7] [Grattan-Guinness 1970b] has suggested that Cauchy "stole" this and other ideas from Bolzano's paper of 1817. See also [Freudenthal 1971b, Jahnke 2003, p. 161, Grabiner 2005, pp. 9–12].

[8] This word "solution" takes an old meaning here; it means that continuity dissolves or disappears.

[9] To verify this formula, let $u=x+\frac{1}{2}\alpha$ and $v=\frac{1}{2}\alpha$, then apply the usual formula for $\sin(a+b)$ to the expression $\sin(u+v)-\sin(u-v)$.

decreases indefinitely with the numerical value of α, whatever finite value is given to the variable x.[10] In general, with respect to the 11 simple functions which we have considered above (Chap. I, § II), namely

$$a+x,\ a-x,\ ax,\ \tfrac{a}{x},\ x^a,\ A^x,\ \log(x),$$

$$\sin x,\ \cos x,\ \arcsin x,\ \arccos x,$$

if we consider the question of the continuity, we find that each of these functions remains continuous between two finite limits of the variable x whenever they are always real[11] between these two limits and they are never infinite on the interval.

It follows that each of these functions is continuous in the neighborhood of any finite value given to the variable x if that finite value is contained:[12]

For the functions

$$\left.\begin{array}{r} a+x \\ a-x \\ ax \\ A^x \\ \sin x \\ \cos x \end{array}\right\} \text{between the limits } x=-\infty \text{ and } x=+\infty,$$

For the function

$$\frac{a}{x} \quad \left\{\begin{array}{l} \text{first, between the limits } x=-\infty \text{ and } x=0, \\ \text{second, between the limits } x=0 \text{ and } x=\infty, \end{array}\right.$$

[45] For the functions

$$\left.\begin{array}{r} x^a \\ \log(x) \end{array}\right\} \text{between the limits } x=0 \text{ and } x=\infty,$$

and finally

---

[10] This proof is somewhat unsatisfying because it relies on the unspoken assumptions that $\sin x$ is continuous at zero and that $\cos x$ is bounded.

[11] This is meant to rule out imaginary quantities that might arise from roots of negative numbers or complex values of logarithm functions. In modern terms, the content of this passage is that the 11 simple functions are continuous on their domains of definition.

[12] Recall from the Preliminaries [Cauchy 1821, p. 9, Cauchy 1897, p. 23] that A is a number, hence positive, so there is no ambiguity about whether A^x is well defined. Because a may be negative, Cauchy avoids problems with a^x by restricting his interval of definition. His treatment of $\frac{a}{x}$ makes it clear that here Cauchy considers an interval not to contain its endpoints.

For the functions
$$\left.\begin{array}{r}\arcsin x\\ \arccos x\end{array}\right\}\text{ between the limits } x=-1 \text{ and } x=+1.$$

It is worth observing that in the case where $a = \pm m$ (where m denotes an integer number), the simple function

$$x^a$$

is always continuous in the neighborhood of a finite value of the variable x, as long as this value is contained:

$$\text{if } a=+m, \quad \text{between the limits } x=-\infty \text{ and } x=+\infty,$$

$$\text{if } a=-m, \quad \left\{\begin{array}{l}\text{between the limits } x=-\infty \text{ and } x=0\\ \quad\text{as well as}\\ \text{between the limits } x=0 \text{ and } x=\infty.\end{array}\right.$$

Among the 11 functions that we have just cited, only two become discontinuous for a value of x contained in the interval between whose limits these functions remain real.[13] The two functions in question are

$$\frac{a}{x} \quad \text{and} \quad x^a \text{ (when } a=-m).$$

Both become infinite and, as a consequence, discontinuous when $x = 0$.

Now let

$$f(x,y,z,\ldots)$$

be a function of several variables, $x,y,z,\ldots$, and suppose that in the neighborhood of particular values $X,Y,Z,\ldots$ of these [46] variables, $f(x,y,z,\ldots)$ is simultaneously a continuous function of x, a continuous function of y, a continuous function of z, ... We prove easily that if we let α, β, γ, ... denote infinitely small quantities, and that if we give x, y, z, ... the values X, Y, Z, ... or values very near to these, the difference[14]

$$f(x+\alpha,y+\beta,z+\gamma,\ldots)-f(x,y,z,\ldots)$$

is itself infinitely small. Indeed, it is clear from the previous hypothesis that the numerical values of the differences

$$\begin{array}{l}f(x+\alpha,y,z,\ldots)-f(x,y,z,\ldots),\\ f(x+\alpha,y+\beta,z,\ldots)-f(x+\alpha,y,z,\ldots),\\ f(x+\alpha,y+\beta,z+\gamma,\ldots)-f(x+\alpha,y+\beta,z,\ldots),\\ \ldots\ldots\ldots\ldots\ldots\ldots\ldots\ldots\ldots\ldots\end{array}$$

---

13 Cauchy does not have the notion of the "domain" of a function. For him, even functions like $\frac{1}{x}$ or $\sqrt{x}$ are always defined, but sometimes the values of those functions are infinite or complex.

14 In [Cauchy 1821, p. 38, Cauchy 1897, p. 46], this was written $f(x+\alpha,y+\beta,z+\gamma)-f(x,y,z,\ldots)$, with no ellipses in the first term. (tr.)

decrease indefinitely with those of the quantities α, β, γ, ..., namely the numerical value of the first difference decreases with the numerical value of α, that of the second difference with the numerical value of β, that of the third with the numerical value of γ, and so on. We must conclude that the sum of all these differences, namely

$$f(x+\alpha,y+\beta,z+\gamma,\ldots)-f(x,y,z,\ldots),$$

converges towards the limit zero if α, β, γ, ... converge to the same limit. In other words,

$$f(x+\alpha,y+\beta,z+\gamma,\ldots)$$

has as its limit

$$f(x,y,z,\ldots).$$

The proposition that we have just proven evidently remains true in the case where we have established certain relations among the variables α, β, γ, ... It is sufficient that these relations permit the new variables to converge all at the same time towards the limit zero.

When, in the same proposition, we replace x, y, z, ... by [47] X, Y, Z, ..., and $x+\alpha$, $y+\beta$, $z+\gamma$, ... by x, y, z, ..., we obtain the following statement:

Theorem I.[15] — *If the variables x, y, z, ... have for their respective limits the fixed and determined quantities X, Y, Z, ..., and the function $f(x,y,z,\ldots)$ is continuous with respect to each of the variables x, y, z, ... in the neighborhood of the system of particular values*

$$x=X,\ y=Y,\ z=Z,\ \ldots,$$

then $f(x,y,z,\ldots)$ has $f(X,Y,Z,\ldots)$ as its limit.

Because in the second statement, the variables α, β, γ, ... are replaced by $x-X$, $y-Y$, $z-Z$, ..., the relations that we were able to establish in the first statement among α, β, γ, ... may be established in the second statement among the quantities $x-X$, $y-Y$, $z-Z$, ...[16] As a result, the function $f(x,y,z,\ldots)$ has $f(X,Y,Z,\ldots)$ as its limit in the case where the variables x, y, z, ... are subject to certain relations, as long as these relations permit them to approach indefinitely the limits X, Y, Z, ...

To clarify these ideas, suppose that x, y, z, ... are functions of the same variable t, considered to be independent and continuous with respect to this variable in the neighborhood of the particular value

$$t=T.$$

If for convenience we let

$$f(x,y,z,\ldots)=u,$$

[15] As stated, this theorem is not true. See [Gelbaum 2003, p. 115 ff] for counterexamples.

[16] [Cauchy 1821, p. 39, Cauchy 1897, p. 47] omitted ellipses here, writing $x-X$, $y-Y$, $z-Z$. (tr.)

then u is a composite function of the variable t. If

$$X,\ Y,\ Z,\ \ldots,\ U,$$

respectively, denote the values of

$$x,\ y,\ z,\ \ldots,\ u$$

in the case where $t = T$, it is clear, on the one hand, that a [48] value of t very close to T gives for u a unique and finite value. On the other hand, it is sufficient to let t converge towards the limit T for the variables x, y, z, ... to converge towards the limits X, Y, Z, ..., and consequently, the function $u = f(x,y,z,\ldots)$ towards the limit $U = f(X,Y,Z,\ldots)$. We prove in absolutely the same way that if we give t a value very close to T, the corresponding value of the function u is the limit towards which this function approaches indefinitely as t converges towards the given value. We must conclude that u is a continuous function of t in the neighborhood of $t = T$. We may therefore state the following theorem:

Theorem II. — *Let*

$$x,\ y,\ z,\ \ldots$$

denote several functions of the variable t, which are continuous with respect to this variable in the neighborhood of the particular value $t = T$. Furthermore, let

$$X,\ Y,\ Z,\ \ldots$$

be the particular values of x, y, z, ... corresponding to $t = T$. Suppose that in the neighborhood of these particular values, the function

$$u = f(x,y,z,\ldots)$$

is simultaneously continuous with respect to x, continuous with respect to y, continuous with respect to z, ... Then u, considered as a function of t, is also continuous with respect to t in the neighborhood of the particular value $t = T$.

If in the previous theorem we reduce the variable quantities x, y, z, ... to a single variable x, we get a new theorem, which can be stated as follows:

Theorem III. — *Suppose that in the equation*

$$u = f(x),$$

the variable x is a function of another variable t. Imagine further that [49] *the variable x is a continuous function of t in the neighborhood of the particular value $t = T$, and that u is a continuous function of x in the neighborhood of the particular value $x = X$ corresponding to $t = T$. The quantity u, considered as a function of t, is also continuous with respect to this variable in the neighborhood of the particular value $t = T$.*

Suppose, for example,

$$u = ax \quad \text{and} \quad x = t^n,$$

where a denotes a constant quantity and n an integer number. We conclude from theorem III that between any arbitrary limits of the variable t,

$$u = at^n$$

is a continuous function of this variable.

Similarly, if we let

$$u = \frac{x}{y}, \quad x = \sin t, \quad \text{and} \quad y = \cos t,$$

we conclude from theorem II that the function

$$u = \tan t$$

is continuous with respect to t in the neighborhood of any finite value of this variable any time the value in question does not have the form

$$t = \pm 2k\pi \pm \frac{\pi}{2},$$

where k denotes an integer number, that is to say any time that this value of t corresponds to a finite value of $\tan t$. On the contrary, the function $\tan t$ admits solution of continuity, by becoming infinite, for each of the values of t given by the preceding formula.

Now let us suppose

$$u = a + x + y + z + \dots,$$
$$x = bt, \quad y = ct^2, \quad \dots,$$

[50] where $a, b, c, \dots$ denote constant quantities. Because u is a continuous function of $x, y, z, \dots$ between any limits of these variables, and because $x, y, z, \dots$ are continuous functions of the variable t between arbitrary limits of t, we conclude from theorem III that the function

$$u = a + bt + ct^2 + \dots$$

is itself continuous with respect to t between arbitrary limits. As a consequence, because $t = 0$ gives $u = a$, if we make t converge towards the limit zero, then the function u converges towards the limit a and eventually takes the same sign as this limit, and this agrees with theorem IV of § I.

A remarkable property of continuous functions of a single variable is that they may be used in Geometry to represent the ordinates of straight or curved continuous lines. From this remark we easily deduce the following proposition:

Theorem IV.[17] — *If the function $f(x)$ is continuous with respect to the variable x between the limits $x = x_0$ and $x = X$, and if b denotes a quantity between $f(x_0)$ and $f(X)$, we may always satisfy the equation*

$$f(x) = b$$

by one or more real values of x contained between x_0 and X.

Proof. — To establish the preceding proposition, it suffices to show that the curve that has as its equation

$$y = f(x)$$

meets the straight line that has for its equation

$$y = b$$

one or more times in the interval contained between the ordinates that correspond to the abscissas x_0 and X. Now it is evident under the given hypothesis that this is what happens. Indeed, because the function $f(x)$ is continuous between the limits $x = x_0$ and $x = X$, the curve which has $y = f(x)$ as its equation and which passes [51] 1° through the point corresponding to the coordinates x_0, $f(x_0)$, and 2° through the point corresponding to the coordinates X and $f(X)$, is continuous between these two points. Because the constant ordinate b of the straight line which has $y = b$ as its equation is found between the ordinates $f(x_0)$ and $f(X)$ of the two points being considered, the straight line necessarily will pass between these two points, which it could not do without meeting the above-mentioned curve in the interval.

Furthermore, as we will do in Note III, we can prove theorem IV by a direct and purely analytic method, which also has the advantage of providing the numerical solution to the equation

$$f(x) = b.$$

2.3 On singular values of functions in various particular cases.

When a function of one or several variables admits but a single value for a system of values attributed to the variables which it contains, this unique value is ordinarily deduced from the definition itself of the function. If a particular case arises in which the given definition cannot immediately give the value of the function under consideration, we seek the limit or limits towards which this function converges as the variables approach indefinitely the particular values assigned to them. If there exist one or more limits of this kind, they are regarded as the values of the function under

[17] This is the Intermediate Value Theorem. Cauchy gives a rigorous proof of this theorem in Note III [Cauchy 1821, pp. 460–462, Cauchy 1897, pp. 378–380].

the given hypothesis, however many there may be. We call *singular values* of the proposed function those values determined as we have just described. For example, such values are those which we obtain by attributing infinite values to the variables, and also those values which correspond to the solutions of continuity.[18] Research on singular values of functions is one of the most important and most delicate questions of Analysis: it offers more or less [52] difficulty depending on the nature of the functions and the number of variables which they contain.

If we first consider simple functions of a single variable, we find that it is easy to determine their singular values. These values always correspond to one of the three cases

$$x = -\infty, \quad x = 0 \quad \text{or} \quad x = \infty,$$

and are, respectively,

for the functions				
$a+x$	a arbitrary	$a+(-\infty) = -\infty$	....	$a+\infty = \infty$
$a-x$	a arbitrary	$a-(-\infty) = \infty$	....	$a-\infty = -\infty$
ax	a positive	$a\times(-\infty) = -\infty$	....	$a\times\infty = \infty$
	a negative	$a\times(-\infty) = \infty$	....	$a\times\infty = -\infty$
$\frac{a}{x}$	a positive	$\frac{a}{-\infty} = 0$	$\frac{a}{0} = \pm\infty$	$\frac{a}{\infty} = 0$
	a negative	$\frac{a}{-\infty} = 0$	$\frac{a}{0} = \mp\infty$	$\frac{a}{\infty} = 0$
x^a	a positive	....	$0^a = 0$	$\infty^a = \infty$
	a negative	....	$0^a = \infty$	$\infty^a = 0$
A^x	$A > 1$	$A^{-\infty} = 0$	$A^0 = 1$	$A^\infty = \infty$
	$A < 1$	$A^{-\infty} = \infty$	$A^0 = 1$	$A^\infty = 0$
$\log(x)$	base > 1	....	$\log(0) = -\infty$	$\log(0) = \infty$
	base < 1	....	$\log(0) = \infty$	$\log(0) = -\infty$
$\sin x$	....	$\sin(-\infty) = M((-1,+1))$	....	$\sin(\infty) = M((-1,+1))$
$\cos x$	....	$\cos(-\infty) = M((-1,+1))$	....	$\cos(\infty) = M((-1,+1))$

Here, as in the preliminaries, the notation $M((-1,+1))$ denotes one of the average quantities between the two limits

$$-1 \quad \text{and} \quad +1.$$

[18] Recall [Cauchy 1821, p. 35, Cauchy 1897, p. 43] that a "solution of continuity" is a point where continuity dissolves, what we would call a point of discontinuity.

It is worth observing that, in the case where we suppose that $a = \pm m$, [53] where m is an integer number, the simple function

$$x^a$$

always admits three singular values, namely:

$$\begin{array}{ll} \text{when} & \left\{ \begin{array}{llll} m \text{ being even} & (-\infty)^m = \infty, & 0^m = 0, & \infty^m = \infty, \\ m \text{ being odd} & (-\infty)^m = -\infty, & 0^m = 0, & \infty^m = \infty, \end{array} \right. \\ a = +m & \end{array}$$

$$\begin{array}{ll} \text{when} & \left\{ \begin{array}{llll} m \text{ being even} & (-\infty)^{-m} = 0, & 0^{-m} = \infty, & \infty^{-m} = 0, \\ m \text{ being odd} & (-\infty)^{-m} = 0, & ((0))^{-m} = \pm\infty, & \infty^{-m} = 0. \end{array} \right. \\ a = -m & \end{array}$$

Now let us consider functions composed of a single variable x. Sometimes it is easy to find their singular values. Thus, for example, if we denote by k any integer number, we recognize without trouble that the composite function

$$\tan x = \frac{\sin x}{\cos x}$$

has its singular values contained in the three formulas

$$\begin{aligned} &\tan((\infty)) = M((-\infty, \infty)), \\ &\tan\left((2k\pi \pm \tfrac{\pi}{2})\right) = \pm\infty \quad \text{and} \\ &\tan((-\infty)) = M((-\infty, \infty)), \end{aligned}$$

while the singular values of the inverse function

$$\arctan x = \arcsin \frac{x}{\sqrt{1+x^2}}$$

are, respectively,

$$\arctan(-\infty) = -\frac{\pi}{2} \quad \text{and} \quad \arctan(\infty) = \frac{\pi}{2}.$$

Often such questions also present true difficulties. For example, we do not immediately see how to determine the singular value of the function

$$x^x,$$

[54] when we suppose that $x = 0$, or that of the function

$$x^{\frac{1}{x}},$$

when we take $x = \infty$. To give an idea of the methods which lead to the solution of questions of this kind, I am going to establish here two theorems by the aid of which we can, in a great number of cases, determine the singular values which the two functions

$$\frac{f(x)}{x} \quad \text{and} \quad [f(x)]^{\frac{1}{x}}$$

take when we suppose that $x = \infty$.

Theorem I. — *If the difference*

$$f(x+1) - f(x)$$

converges towards a certain limit k, for increasing values of x, then the fraction

$$\frac{f(x)}{x}$$

converges at the same time towards the same limit.

Proof. — First suppose that the quantity k has a finite value, and denote by ε a number as small as we wish. Because the increasing values of x make the difference

$$f(x+1) - f(x)$$

converge towards the limit k, we can give the number h a value large enough that, when x is equal to or greater than h, the difference in question is always contained between the limits

$$k - \varepsilon \quad \text{and} \quad k + \varepsilon.$$

Given this, if we denote by n any integer number, each [55] of the quantities

$$\begin{gathered} f(h+1) - f(h), \\ f(h+2) - f(h+1), \\ \ldots\ldots\ldots\ldots\ldots\ldots\ldots, \\ f(h+n) - f(h+n-1), \end{gathered}$$

and consequently their arithmetic mean, namely

$$\frac{f(h+n) - f(h)}{n},$$

is contained between the limits $k - \varepsilon$ and $k + \varepsilon$. Thus we have

$$\frac{f(h+n) - f(h)}{n} = k + \alpha,$$

where α is a quantity contained between the limits $-\varepsilon$ and $+\varepsilon$. Now let

$$h + n = x.$$

The preceding equation becomes

$$\frac{f(x)-f(h)}{x-h}=k+\alpha, \tag{1}$$

and we thus conclude

$$f(x)=f(h)+(x-h)(k+\alpha),$$

$$\frac{f(x)}{x}=\frac{f(h)}{x}+\left(1-\frac{h}{x}\right)(k+\alpha). \tag{2}$$

Moreover, to make the value of x increase indefinitely, it suffices to make the integer number n increase indefinitely without changing the value of h. Consequently, let us suppose that in equation (2) we consider h as a constant quantity and x as a variable quantity which converges towards the limit ∞. The quantities

$$\frac{f(h)}{x} \quad \text{and} \quad \frac{h}{x},$$

contained in the right-hand side, converge towards the limit zero, [56] and the right-hand side itself converges towards a limit of the form

$$k+\alpha,$$

where α is always contained between $-\varepsilon$ and $+\varepsilon$. Thus the ratio

$$\frac{f(x)}{x}$$

has for its limit a quantity contained between $k-\varepsilon$ and $k+\varepsilon$. This conclusion remains true however small the number ε may be, and as a result the limit in question is precisely the quantity k. In other words, we have

$$\lim\frac{f(x)}{x}=k=\lim[f(x+1)-f(x)]. \tag{3}$$

Second, let us suppose that $k=\infty$. Denoting by H a number however large we may wish, we can always find a number h so large that, for x equal to or greater than h, the difference

$$f(x+h)-f(x),$$

which converges towards the limit ∞, becomes always greater than H. Reasoning as above, we establish the formula

$$\frac{f(h+n)-f(h)}{n}>H.$$

Now, if we set $h+n=x$, we find the following formula instead of equation (2),

$$\frac{f(x)}{x} > \frac{f(h)}{x} + H\left(1 - \frac{h}{x}\right),$$

from which we conclude that

$$\lim \frac{f(x)}{x} > H$$

by making x converge towards the limit ∞. The limit of the ratio

$$\frac{f(x)}{x}$$

[57] is thus greater than the number H, however great it may be. This limit, larger than any assignable number, cannot be anything but positive infinity.

Finally, let us suppose that $k = -\infty$. To reduce this last case to the preceding one, it suffices to observe that because the difference

$$f(x+1) - f(x)$$

has as its limit $-\infty$, the following

$$[-f(x+1)] - [-f(x)]$$

has for its limit $+\infty$. We then conclude that the limit of $\frac{-f(x)}{x}$ is equal to $+\infty$, and consequently the limit of $\frac{f(x)}{x}$ equals $-\infty$.

Corollary I. — To give an application of the preceding theorem, let us suppose that

$$f(x) = \log(x),$$

where log is the characteristic of logarithms in a system for which the base is greater than 1. We find that

$$f(x+1) - f(x) = \log(x+1) - \log(x) = \log\left(1 + \frac{1}{x}\right),$$

and consequently

$$k = \log\left(1 + \frac{1}{\infty}\right) = \log(1) = 0.$$

We can thus affirm that as x grows indefinitely, the ratio

$$\frac{\log(x)}{x}$$

converges towards the limit zero, and it follows that *in a system for which the base is greater than* 1, *the logarithms of numbers grow much less rapidly than the numbers themselves.*

Corollary II. — Suppose, on the other hand, that

$$f(x) = A^x,$$

[58] where A denotes a number greater than 1. We find that

$$f(x+1) - f(x) = A^{x+1} - A^x = A^x (A-1),$$

and consequently

$$k = A^\infty (A-1) = \infty.$$

We can thus affirm that when x grows indefinitely, the ratio

$$\frac{A^x}{x}$$

converges towards the limit ∞, and it follows that *the exponential* A^x, *when the number* A *is greater than* 1, *eventually grows more rapidly than the variable* x.

Corollary III. — We ought to observe, moreover, that it is not necessary to use theorem I to find the value of the ratio

$$\frac{f(x)}{x}$$

corresponding to $x = \infty$ except in the case where the function $f(x)$ becomes infinite along with the variable x. If this function remains finite for $x = \infty$, the ratio $\frac{f(x)}{x}$ evidently has zero as its limit.

I pass to a theorem which serves to determine in many cases the value of

$$[f(x)]^{\frac{1}{x}}$$

for $x = \infty$. It consists of this:

Theorem II. — *If the function* $f(x)$ *is positive for very large values of* x *and the ratio*

$$\frac{f(x+1)}{f(x)}$$

converges towards the limit k *when* x *grows indefinitely, then the expression*

$$[f(x)]^{\frac{1}{x}}$$

converges at the same time to the same limit.

[59] *Proof.* — First suppose that the quantity k, necessarily positive, has a finite value, and denote by ε a number as small as we wish. Because increasing values of x make the ratio

$$\frac{f(x+1)}{f(x)}$$

converge towards the limit k, we can give the number h a value large enough that when x is equal to or greater than h, the ratio in question is always contained between the limits

$$k-\varepsilon \quad \text{and} \quad k+\varepsilon.$$

Given this, if we denote by n any integer number, each of the quantities

$$\frac{f(h+1)}{f(h)}, \quad \frac{f(h+2)}{f(h+1)}, \quad \ldots, \quad \frac{f(h+n)}{f(h+n-1)}$$

and consequently their geometric mean, namely

$$\left[\frac{f(x+h)}{f(x)}\right]^{\frac{1}{n}},$$

is contained between the limits $k+\varepsilon$ and $k-\varepsilon$. Thus we have

$$\left[\frac{f(h+n)}{f(h)}\right]^{\frac{1}{n}} = k+\alpha,$$

where α is a quantity contained between the limits $-\varepsilon$ and $+\varepsilon$. Now let

$$h+n=x.$$

The preceding equation becomes

$$\left[\frac{f(x)}{f(h)}\right]^{\frac{1}{x-h}} = k+\alpha, \tag{4}$$

and we thus conclude

$$f(x) = f(h)\,(k+\alpha)^{x-h},$$
$$[f(x)]^{\frac{1}{x}} = [f(h)]^{\frac{1}{x}}\,(k+\alpha)^{1-\frac{x}{h}}. \tag{5}$$

[60] Moreover, to make the value of x increase indefinitely, it suffices to make the integer number n increase indefinitely without changing the value of h. Consequently, let us suppose that in equation (5) we consider h as a constant quantity and x as a variable quantity which converges towards the limit ∞. The quantities

$$[f(h)]^{\frac{1}{x}} \quad \text{and} \quad 1-\frac{h}{x},$$

contained in the right-hand side, converge towards the limit 1, and the right-hand side itself converges towards a limit of the form

$$k+\alpha,$$

where α is always contained between $-\varepsilon$ and $+\varepsilon$. Thus the expression

$$[f(h)]^{\frac{1}{x}}$$

has for its limit a quantity contained between $k-\varepsilon$ and $k+\varepsilon$. This conclusion remains true however small the number ε, and as a result the limit in question is precisely the quantity k. In other words, we have

$$\lim [f(x)]^{\frac{1}{x}} = k = \lim \frac{f(x+1)}{f(x)}. \tag{6}$$

On the other hand, let us suppose that the quantity k is infinite, that is to say, because this quantity is positive, that $k=\infty$. Then, denoting by H a number as large as we wish, we can always find a number h so large that when x is equal to or greater than h, the ratio

$$\frac{f(x+1)}{f(x)},$$

which converges towards the limit ∞, becomes always greater than H. Reasoning as above, we establish the formula

$$\left[\frac{f(h+n)}{f(h)}\right]^{\frac{1}{n}} > H.$$

[61] Now, if we set $h+n=x$, we find the following formula instead of formula (5)

$$[f(x)]^{\frac{1}{x}} > [f(h)]^{\frac{1}{x}} H^{1-\frac{h}{x}},$$

from which we conclude that

$$\lim [f(x)]^{\frac{1}{x}} > H,$$

by making x converge towards the limit ∞. The limit of the expression

$$[f(x)]^{\frac{1}{x}}$$

is thus greater than the number H, however great it may be. This limit, larger than any assignable number, cannot be anything but positive infinity.

Note. — We can easily prove equation (6) by using theorem I to find the limit towards which the logarithm

$$\log [f(x)]^{\frac{1}{x}} = \frac{\log [f(x)]}{x}$$

converges and then returning from logarithms to numbers.

Corollary I. — To give an application of theorem II, let us suppose that

$$f(x) = x.$$

We have

$$\frac{f(x+1)}{f(x)} = \frac{x+1}{x} = 1 + \frac{1}{x},$$

and consequently, by passing to the limits,

$$k = 1.$$

Then, if we make the variable x grow indefinitely, the function

$$x^{\frac{1}{x}}$$

converges towards the limit 1.

[62] *Corollary II.* — On the other hand, let

$$f(x) = ax^n + bx^{n-1} + cx^{n-2} + \ldots = P,$$

so that P denotes a polynomial in x of degree n. We find that

$$\frac{f(x+1)}{f(x)} = \frac{a\left(1+\frac{1}{x}\right)^n + \frac{b}{x}\left(1+\frac{1}{x}\right)^{n-1} + \frac{c}{x^2}\left(1+\frac{1}{x}\right)^{n-2} + \ldots}{a + \frac{b}{x} + \frac{c}{x^2} + \ldots}$$

and, by passing to the limits,

$$k = \frac{a}{a} = 1.$$

Thus, if P represents any integer polynomial, then $P^{\frac{1}{x}}$ has 1 as its limit.

Corollary III. — Finally let

$$f(x) = \log(x).$$

We find that

$$\frac{f(x+1)}{f(x)} = \frac{\log(x+1)}{\log(x)} = \frac{\log(x) + \log\left(1+\frac{1}{x}\right)}{\log(x)} = 1 + \frac{\log\left(1+\frac{1}{x}\right)}{\log(x)},$$

and passing to the limits,

$$k = 1.$$

Consequently, $[\log(x)]^{\frac{1}{x}}$ also has 1 as its limit.

Theorems I and II evidently remain true in the case where the variable x takes only integer values. Indeed, to make the proofs that we have given to these two theorems apply in this particular case, it suffices to suppose that the quantity denoted by h in each of these proofs is a very large integer number. If in the same case we represent the successive values of the function $f(x)$ corresponding to the various integer values of x, namely

$$f(1),\quad f(2),\quad f(3),\quad \ldots,\quad f(n),$$

by

$$A_1,\quad A_2,\quad A_3,\quad \ldots,\quad A_n,$$

[63] we obtain the following propositions instead of theorems I and II:

Theorem III. — *If the sequence of quantities*

$$A_1,\quad A_2,\quad A_3,\quad \ldots,\quad A_n,\quad \ldots$$

is such that the difference between two consecutive terms of this sequence, namely

$$A_{n+1} - A_n,$$

converges constantly towards a fixed limit A for increasing values of n, then the ratio

$$\frac{A_n}{n}$$

converges at the same time towards the same limit.

Theorem IV. — *If the sequence of numbers*

$$A_1,\quad A_2,\quad A_3,\quad \ldots,\quad A_n,$$

is such that the ratio between two consecutive terms, namely

$$\frac{A_{n+1}}{A_n},$$

converges constantly towards a fixed limit A for increasing values of n, then the expression

$$(A_n)^{\frac{1}{n}}$$

converges at the same time towards the same limit.

To give an application of this last theorem, let us suppose that

$$A_n = 1 \cdot 2 \cdot 3 \ldots n.$$

The sequence $A_1, A_2, \ldots$ becomes

$$1,\quad 1 \cdot 2,\quad 1 \cdot 2 \cdot 3,\quad \ldots,\quad 1 \cdot 2 \cdot 3 \ldots (n-1)n,\quad \ldots,$$

and the ratio between two consecutive terms of this same series, namely

$$\frac{A_{n+1}}{A_n} = \frac{1\cdot 2\cdot 3\ldots n\,(n+1)}{1\cdot 2\cdot 3\ldots n} = n+1,$$

[64] evidently converges towards the limit ∞ for increasing values of n. Consequently, the expression

$$(A_n)^{\frac{1}{n}} = (1\cdot 2\cdot 3\ldots n)^{\frac{1}{n}}$$

converges towards the same limit.

On the other hand, we find that the expression

$$\left(\frac{1}{1\cdot 2\cdot 3\ldots n}\right)^{\frac{1}{n}}$$

converges, for increasing values of n, towards the limit zero.

Often, with the aid of theorems I and II, we can determine the singular value of a composite function of the variable x when this variable vanishes. Thus, for example, if we wish to obtain the singular value of x^x corresponding to $x = 0$, it suffices to look for the limit towards which the expression $\left(\frac{1}{x}\right)^{\frac{1}{x}} = \frac{1}{x^{\frac{1}{x}}}$ converges for increasing values of x. This limit, by virtue of theorem II (corollary I), is equal to 1.

Likewise, we conclude from theorem I (corollary I) that the function

$$x\log(x)$$

vanishes with the variable x.

When the two terms of a fraction are infinitely small quantities, the numerical values of which decrease indefinitely with that of the variable α, the singular value of this fraction for $\alpha = 0$ is sometimes finite, sometimes zero or infinite. Indeed, let us denote by k and k' two finite constants that are not zero, and by ε and ε' two variable numbers which converge with α towards the limit zero. Two infinitely small quantities, one of order n, the other of order n', can be represented, respectively, by

$$k\alpha^n(1\pm\varepsilon) \quad \text{and} \quad k'\alpha^{n'}(1\pm\varepsilon'),$$

[65] and their ratio, namely

$$\frac{k'\alpha^{n'}(1\pm\varepsilon')}{k\alpha^n(1\pm\varepsilon)} = \frac{k'}{k}\frac{1\pm\varepsilon'}{1\pm\varepsilon}\alpha^{n'-n} = \frac{k'}{k}\frac{1\pm\varepsilon'}{1\pm\varepsilon}\frac{1}{\alpha^{n-n'}},$$

evidently has as its limit

$$\begin{aligned} &\tfrac{k'}{k}, \text{ if we suppose that } n' = n,\\ &0, \text{ if we suppose that } n' > m, \quad \text{and}\\ &\pm\infty, \text{ if we suppose that } n' < n. \end{aligned}$$

Likewise, we can prove that *the limit towards which the ratio of two infinitely large quantities converges when their numerical values increase indefinitely with that of a variable x can be zero, finite or infinite.* But this limit has a determined sign, constantly equal to the product of the signs of the two quantities being considered.

Among the fractions for which the two terms converge with the variable α towards the limit zero, we ought to include the following[19]

$$\frac{f(x+\alpha)-f(x)}{\alpha},$$

always attributing to the variable x a value in the neighborhood of which the function $f(x)$ remains continuous.[20] Indeed, under this hypothesis, the difference

$$f(x+\alpha)-f(x)$$

is an infinitely small quantity. We might also remark that in general it is an infinitely small quantity of the first order, so that the ratio

$$\frac{f(x+\alpha)-f(x)}{\alpha}$$

ordinarily converges towards a finite limit different from zero as the numerical value of α diminishes. This limit is, for example,

$$2x, \quad \text{if we take} \quad f(x)=x^2$$

and

$$-\frac{a}{x^2}, \quad \text{if we take} \quad f(x)=\frac{a}{x}.$$

[66] In the particular case where we suppose that $x=0$, the ratio

$$\frac{f(x+\alpha)-f(x)}{\alpha}$$

reduces to

$$\frac{f(\alpha)-f(0)}{\alpha}.$$

Among the ratios of this last kind, we will restrict ourselves to considering the following

$$\frac{\sin\alpha}{\alpha}.$$

Because it can be put into the form

---

[19] This, and what follows over the next few pages, are as close as Cauchy gets to using the derivative in the *Cours d'analyse*. It highlights the fact that the book is about the foundations of calculus, and not about calculus itself. It is not until the third lesson of his *Résumé* [Cauchy 1823] that he takes the next step and defines the derivative as the limit of the difference quotient.

[20] Note that Cauchy does not seem to consider the necessary and sufficient conditions for a function $f(x)$ to be differentiable at a point x.

$$\frac{\sin(-\alpha)}{-\alpha},$$

its limit will remain the same, whatever the sign of α may be. Given this, suppose that the arc α takes a very small positive value. Because the chord[21] of the double arc 2α is represented by $2\sin\alpha$, we evidently have $2\alpha > 2\sin\alpha$, and as a consequence,

$$\alpha > \sin\alpha.$$

Moreover, the sum of the tangents taken at the endpoints of the arc 2α is represented by $2\tan\alpha$, and, by forming a portion of a polygon which encloses this arc, we now have $2\tan\alpha > 2\alpha$,[22] and consequently

$$\tan\alpha > \alpha.$$

By combining the two formulas which we have just established, we find that[23]

$$\sin\alpha < \alpha < \tan\alpha,$$

then by replacing $\tan\alpha$ with its value

$$\sin\alpha < \alpha < \frac{\sin\alpha}{\cos\alpha},$$

and consequently we have

$$1 < \frac{\alpha}{\sin\alpha} < \frac{1}{\cos\alpha} \quad \text{and}$$

$$1 > \frac{\sin\alpha}{\alpha} > \cos\alpha.$$

[67] Now, when α decreases, $\cos\alpha$ converges towards the limit 1. Thus, *a fortiori* the ratio $\frac{\sin\alpha}{\alpha}$ is always contained between 1 and $\cos\alpha$, and consequently we have[24]

$$\lim\frac{\sin\alpha}{\alpha} = 1. \tag{7}$$

Because the study of the limits towards which the ratios $\frac{f(x+\alpha)-f(x)}{\alpha}$ and $\frac{f(\alpha)-f(0)}{\alpha}$ converge is one of the principal objects of the infinitesimal Calculus, there is no need to dwell any further on this.

[21] The chord is an obsolete trigonometric function; see p. 10 or [Cauchy 1821, p. 11, Cauchy 1897, p. 24] for others. The chord of x is $2\sin\left(\frac{x}{2}\right)$.

[22] Following Lagrange, Cauchy does not supply diagrams in his text. Presumably, he expected the reader to supply any diagrams necessary for following the argument.

[23] [Cauchy 1897, p. 66] has "sin$a$" instead of "sinα." This error is not in [Cauchy 1821, p. 63]. (tr.)

[24] Cauchy is using what we call the Squeeze Theorem here. He considers it evident and sees no need either to state the theorem explicitly or to prove it.

It remains for us to examine the singular values of functions of several variables. Sometimes these values are completely determined and independent of the relations which we may establish among the variables. Thus, for example, if we denote by

$$\alpha, \quad \beta, \quad x \quad \text{and} \quad y$$

four positive variables of which the first two converge towards the limit zero and the last two towards the limit ∞, we recognize without trouble that the expressions

$$\alpha\beta, \quad xy, \quad \frac{\alpha}{x}, \quad \frac{y}{\beta}, \quad \alpha^y \quad \text{and} \quad x^y$$

have for their respective limits

$$0, \quad \infty, \quad 0, \quad \infty, \quad 0 \quad \text{and} \quad \infty.$$

But more often the singular value of a function of several variables cannot be entirely determined except in the particular case where, in making these variables converge towards their respective limits, we establish certain relations among them, and when these relations are not fixed, the singular value in question is a quantity either totally indeterminate, or only required to remain contained between known limits. Thus, as we have remarked above, the singular value to which the ratio of two infinitely small variables is reduced in the case where each of its variables vanishes can be any quantity, either finite, zero or infinite. [68] In other words, this singular value is completely indeterminate. If instead of two infinitely small variables we consider two infinitely large variables, we find that the ratio of these last ones, when their numerical values increase indefinitely, converge again towards an arbitrary limit, which may be positive or negative according to whether the two variables are of the same sign or of opposite signs. It is equally easy to assure ourselves that the product of an infinitely small variable by an infinitely large one has for its limit a quantity that is completely indeterminate.

In order to present a final application of the principles which we have just established, let us look for the values that must be attributed to variables x and y in order that the value of the function

$$y^{\frac{1}{x}}$$

become indeterminate. If A denotes a number greater than 1 and if log is the characteristic of logarithms in the system for which the base is A, we evidently have

$$y = A^{\log(y)},$$

and consequently

$$y^{\frac{1}{x}} = A^{\frac{\log(y)}{x}}.$$

Now, it is clear that the expression

$$A^{\frac{\log(y)}{x}}$$

converges towards an indeterminate limit whenever the ratio

$$\frac{\log(y)}{x}$$

itself converges towards such a limit. This may arise in two different cases, namely: 1° when $\log(y)$ and x are two infinitely large quantities, that is to say when x and y have for their respective limits 0 and 1; and 2° when $\log(y)$ and x are two infinitely large quantities, that is to say when x has an infinite limit and y has [69] 0 or ∞ as its limit. In either case, it is worth observing that the indeterminate limit of the expression

$$A^{\frac{\log(y)}{x}} = y^{\frac{1}{x}}$$

is necessarily positive. It may even happen that this limit must remain contained between the extreme values of 0 and 1, or else between 1 and ∞. Suppose, for example, that each of the variables x and y converges towards the limit ∞. In this case, because the limit of the ratio

$$\frac{\log(y)}{x}$$

can be any positive quantity, the limit of $y^{\frac{1}{x}} = A^{\frac{\log(y)}{x}}$ must be an average quantity between 1 and ∞. Moreover, this average is indeterminate as long as we do not establish a particular relation between the infinitely large variables x and y. But if we suppose that

$$y = f(x),$$

where $f(x)$[25] denotes a function which increases indefinitely with the variable x, then the average value in question, which is none other than the limit of

$$[f(x)]^{\frac{1}{x}},$$

takes a determinate value, which we can always calculate with the aid of theorem II.

If, in place of the function $y^{\frac{1}{x}}$, we consider the following

$$y^x,$$

we find that this last one becomes indeterminate: 1° when the variable y converges towards the limit 1 and the variable x towards $-\infty$ or $+\infty$, and 2° when the variable x has zero for its limit and y converges towards zero or positive infinity.

In calculation, we sometimes encounter singular expressions which cannot be considered except as limits towards which functions of several variables converge, as these same [70] variables become infinitely small or infinitely large, or even more generally, converge towards fixed limits. Examples of such expressions are

[25] This was incorrectly written as $f(y)$ in [Cauchy 1897, p. 69], but was correctly given as $f(x)$ in [Cauchy 1821, p. 67]. (tr.)

$$0\times 0,\quad \frac{0}{0},\quad \infty\times\infty,\quad \frac{\infty}{\infty},\quad 0\times\infty,\quad 0^0,\quad 1^\infty,\quad \ldots,$$

among which we ought to consider the first two as the limits towards which the product and the ratios of two infinitely small variables converge, the next two as the limits of the product and of the ratio of two infinitely large positive variables, etc. In particular, if we consider the singular expressions which the functions

$$x+y,\quad xy,\quad \frac{x}{y},\quad y^x \quad\text{and}\quad y^{\frac{1}{x}}$$

produce, we find that when the variables remain independent, the values of these same expressions can be easily determined by that which precedes. The equations which serve to determine these values are, respectively,

For the functions

$$\begin{array}{lll}
x+y & \infty+\infty=\infty, & \infty-\infty=M((-\infty,+\infty)); \\[2ex]
xy & \left\{\begin{array}{l} 0\times 0=0, \\ \infty\times\infty=-\infty\times-\infty=\infty, \end{array}\right. & \begin{array}{l} 0\times\infty=0\times-\infty=M((-\infty,+\infty)), \\ \infty\times-\infty=-\infty; \end{array} \\[2ex]
\frac{x}{y} & \left\{\begin{array}{l} \frac{0}{0}=M((-\infty,+\infty)), \\ \frac{\infty}{\infty}=\frac{-\infty}{-\infty}=M((0,\infty)), \end{array}\right. & \begin{array}{l} \frac{0}{\infty}=\frac{0}{-\infty}=0,\ \frac{\infty}{0}=\frac{-\infty}{0}=\pm\infty, \\ \frac{\infty}{-\infty}=\frac{-\infty}{\infty}=M((-\infty,0)); \end{array} \\[2ex]
y^x & \left\{\begin{array}{l} 0^0=\infty^0=M((0,\infty)), \\ 0^{-\infty}=\infty^\infty=\infty, \end{array}\right. & \begin{array}{l} 0^\infty=\infty^{-\infty}=0, \\ 1^\infty=1^{-\infty}=M((0,\infty)); \end{array} \\[2ex]
y^{\frac{1}{x}} & \left\{\begin{array}{l} 0^{\frac{1}{0}}=\infty^{\frac{1}{0}}=0 \text{ or } \infty, \\ 0^{\frac{1}{-\infty}}=\infty^{\frac{1}{\infty}}=M((1,\infty)), \end{array}\right. & \begin{array}{l} 0^{\frac{1}{\infty}}=\infty^{\frac{1}{-\infty}}=M((0,1)), \\ 1^{\frac{1}{0}}=M((0,\infty)) \end{array}
\end{array}$$

Chapter 3
On symmetric functions and alternating functions. The use of these functions for the solution of equations of the first degree in any number of unknowns. On homogeneous functions.

3.1 On symmetric functions.

[71] A *symmetric* function of several quantities is one which conserves the same value and the same sign after any exchange made among its quantities. Thus, for example, each of the functions

$$x+y, \quad x^y+y^x, \quad xyz, \quad \sin x+\sin y+\sin z, \quad \ldots$$

is symmetric with respect to the variables which it contains, while

$$x-y, \quad x^y, \quad \ldots$$

are not symmetric functions of the variables x and y. Likewise,

$$b+c, \quad b^2+c^2, \quad bc, \quad \ldots$$

are symmetric functions of the two quantities b and c, and

$$b+c+d, \quad b^2+c^2+d^2, \quad bc+bd+cd \quad \text{and} \quad bcd$$

are symmetric functions of the three quantities b, c and d, etc.

Among the symmetric functions of several quantities b, c, ..., g and h, we ought to distinguish those which serve as the coefficients of the various powers of a in the expansion of the product

$$(a-b)(a-c)\ldots(a-g)(a-h),$$

and whose properties lead to a very elegant solution to several [72] equations of the first degree among n variables x, y, z, ..., u, v, when the equations are of the form

R.E. Bradley, C.E. Sandifer, *Cauchy's Cours d'analyse*, Sources and Studies in the History of Mathematics and Physical Sciences, DOI 10.1007/978-1-4419-0549-9_3,

$$
(1)\qquad \begin{cases}
x+ \quad y+ \quad z+\ldots+ \quad u+ \quad v = k_0, \\
ax+ \quad by+ \quad cz+\ldots+ \quad gu+ \quad hv = k_1, \\
a^2x+ \quad b^2y+ \quad c^2z+\ldots+ \quad g^2u+ \quad h^2v = k_2, \\
\ldots\ldots\ldots\ldots\ldots\ldots\ldots\ldots\ldots\ldots\ldots, \\
a^{n-1}x+ b^{n-1}y+ c^{n-1}z+\ldots+ g^{n-1}u+ h^{n-1}v = k_{n-1}.
\end{cases}
$$

Indeed, let

$$
\begin{aligned}
&A_{n-2} = -(b+c+\ldots+g+h), \\
&A_{n-3} = bc+\ldots+bg+bh+\ldots+cg+ch+\ldots+gh, \\
&\ldots\ldots\ldots\ldots\ldots\ldots\ldots\ldots\ldots\ldots\ldots\ldots, \\
&A_0 \quad = \pm bc\ldots gh
\end{aligned}
$$

be the symmetric functions in question, so that we have

$$
a^{n-1}+A_{n-2}a^{n-2}+\ldots+A_1a+A_0 = (a-b)(a-c)(a-d)\ldots.
$$

If, in this last formula, we replace a successively by b, by c, …, by g, and by h, we have

$$
\begin{aligned}
b^{n-1}+A_{n-2}b^{n-2}+\ldots+A_1b+a_0 &= 0, \\
c^{n-1}+A_{n-2}c^{n-2}+\ldots+A_1c+a_0 &= 0, \\
\ldots\ldots\ldots\ldots\ldots\ldots\ldots\ldots&\ldots, \\
g^{n-1}+A_{n-2}g^{n-2}+\ldots+A_1g+a_0 &= 0, \\
h^{n-1}+A_{n-2}h^{n-2}+\ldots+A_1h+a_0 &= 0.
\end{aligned}
$$

Then, if we add equations (1) term by term, after multiplying the first one by A_0, the second by A_1, …, the next-to-last by A_{n-2}, and the last by one, we obtain the following,

$$
\begin{aligned}
&\left(a^{n-1}+A_{n-2}a^{n-2}+\ldots+A_1a+A_0\right)x \\
&\qquad = k_{n-1}+A_{n-2}k_{n-2}+\ldots+A_1k_1+A_0k_0,
\end{aligned}
$$

and we conclude that

$$
(2)\qquad x = \frac{\begin{pmatrix} k_{n-1}-(b+c+\ldots+g+h)\,k_{n-2} \\ +(bc+\ldots+bg+bh+\ldots+cg+ch+\ldots+gh)\,k_{n-3} \\ -\ldots\pm bc\ldots gh\cdot k_0 \end{pmatrix}}{(a-b)(a-c)\ldots(a-g)(a-h)}.
$$

[73] By an analogous process, we can determine the values of the other unknowns $y, z, \ldots, u, v$.

When we substitute for the constants

$$
k_0, \quad k_1, \quad k_2, \quad \ldots, \quad k_{n-1}
$$

in equations (1), the successive integer powers of a particular quantity k, namely

$$
k^0 = 1, \quad k, \quad k^2, \quad \ldots, \quad k^{n-1},
$$

the value found for x reduces to

$$x = \frac{(k-b)(k-c)\dots(k-g)(k-h)}{(a-b)(a-c)\dots(a-g)(a-h)}. \tag{3}$$

3.2 On alternating functions.

An *alternating* function of several quantities is one which changes sign, but keeps the same value next to the sign, when we interchange two of these quantities. Consequently, by a series of such exchanges, the function becomes alternatingly positive and negative. According to this definition,

$$x-y, \quad xy^2-x^2y, \quad \log\frac{x}{y}, \quad \sin x - \sin y, \quad \dots$$

are alternating functions of the two variables x and y,

$$(x-y)(x-z)(y-z)$$

is an alternating function of the three variables x, y and z, and so forth.

Among the alternating functions of several variables

$$x, \quad y, \quad z, \quad \dots, \quad u, \quad v,$$

we ought to distinguish those which are rational and integer with respect to each of these same variables. Suppose that such a function [74] is expanded and put into the form of a polynomial. One of its terms, taken at random, has the form

$$kx^py^qz^r\dots u^sv^t,$$

where $p, q, r, \dots, s, t$ denote integer numbers and k denotes any coefficient whatsoever. Moreover, because the function ought to change sign, but keep the same value next to the sign after interchanging the variables x and y, it is necessary that there correspond to the term in question another term of contrary sign,

$$-kx^qy^pz^r\dots u^sv^t,$$

derived from the first by virtue of this exchange. Thus the function is composed of terms, alternately positive and negative, which, combined two by two, produce binomials of the form

$$kx^py^qz^r\dots u^sv^t - kx^qy^pz^r\dots u^sv^t = k(x^py^q - x^qy^p)z^r\dots u^sv^t.$$

In each binomial of this kind, p and q will necessarily be two integer numbers, distinct from each other. Because the difference

$$x^p y^q - x^q y^p$$

is evidently divisible by $y-x$, or what amounts to the same thing, by $x-y$, it follows that each binomial, and consequently the sum of the binomials, or the given function, is divisible by

$$\pm(y-x).$$

Moreover, by the reasoning above, we can substitute any two other variables x and z, or y and z, ..., for the two variables x and y. Consequently, we definitively obtain the following conclusions:

1° An alternating but integer function of several variables $x, y, z, \ldots, u, v$, is composed of terms alternately positive and negative, in each of which the various variables all have different exponents; [75]

2° Such a function is divisible by the product of the differences

$$(1)\qquad \begin{cases} \pm(y-x),\ \pm(z-x),\ \ldots,\ \pm(u-x),\ \pm(v-x), \\ \qquad\qquad \pm(z-y),\ \ldots,\ \pm(u-y),\ \pm(v-y), \\ \qquad\qquad\qquad\quad \ldots,\ \pm(u-z),\ \pm(v-z), \\ \qquad\qquad\qquad\qquad\quad \ldots\ldots\ldots,\ \ldots\ldots\ldots, \\ \qquad\qquad\qquad\qquad\qquad\qquad \pm(v-u), \end{cases}$$

each taken with whichever sign we please.

The product in question here, as we can easily recognize, is itself an alternating function of the variables which we are considering. To prove this, it suffices to observe that this product changes sign, but keeps the same value next to the sign, after interchanging two variables, x and y for example. But indeed, according to whether we adopt for each difference the sign $+$ or the sign $-$, this product is found to be equal either to $+\varphi$ or to $-\varphi$, the value of φ being determined by the equation

$$(2)\qquad \begin{aligned} \varphi = {} & (y-x)(z-x)\ldots(u-x)(v-x) \\ & \times(z-y)\ldots(u-y)(v-y)\times\ldots\times(v-u). \end{aligned}$$

Because it is evident that this value of φ changes only its sign by virtue of interchanging the variables x and y, we can conclude that it will be the same for a function equivalent either to $+\varphi$ or to $-\varphi$.

In order to fix these ideas, imagine that we take each of the differences (1) with the sign +. The product of all these differences will be the function φ determined by equation (2), or what amounts to the same thing, by the following

$$(3)\qquad \begin{aligned} \varphi = {} & (y-x)\times(z-x)(z-y)\times\ldots \\ & \times(v-x)(v-y)(v-z)\ldots(v-u). \end{aligned}$$

If additionally we let n be the number of variables $x, y, z, \ldots, u, v$, then $n-1$ is evidently the number of differences which contain a particular variable. Consequently, in each term of the function φ expanded and put into the form of a polynomial, the exponent of any variable [76] cannot surpass $n-1$. Finally, because in any particular term, the different variables ought to have different exponents, it is clear that these exponents will be respectively equal to the numbers

$$0, \quad 1, \quad 2, \quad 3, \quad \ldots, \quad n-1.$$

Each term, disregarding the sign and the numerical coefficient, is thus equivalent to the product of the various variables arranged in some order, and respectively raised to powers $0, 1, 2, 3, \ldots, n-1$. We ought to add that each product of this kind is found only once, sometimes with the sign $+$, sometimes with the sign $-$, in the expansion of the function φ. For example, the product

$$x^0 y^1 z^2 \ldots u^{n-2} v^{n-1}$$

cannot be formed except by the multiplication of the first letters of the binomial factors which compose the right-hand side of equation (3).

With the aid of the principles that we have just established, it is easy to construct in its entirety the expansion of the function φ and to demonstrate its various properties (on this subject see Note IV). We are now going to show how one is led, by the consideration of such an expansion, to the solution of general equations of the first degree of several variables.

Let

$$(4) \qquad \begin{cases} a_0x \ +b_0y \ +c_0z \ + \ldots + g_0u \ +h_0v \ = k_0, \\ a_1x \ +b_1y \ +c_1z \ + \ldots + g_1u \ +h_1v \ = k_1, \\ a_2x \ +b_2y \ +c_2z \ + \ldots + g_2u \ +h_2v \ = k_2, \\ \ldots\ldots\ldots\ldots\ldots\ldots\ldots\ldots\ldots\ldots\ldots\ldots, \\ a_{n-1}x + b_{n-1}y + c_{n-1}z + \ldots + g_{n-1}u + h_{n-1}v = k_{n-1} \end{cases}$$

be n linear equations among the n variables or unknowns

$$x, \quad y, \quad z, \quad \ldots, \quad u, \quad v,$$

[77] and the constants

$$\begin{array}{lllllll} a_0, & b_0, & c_0, & \ldots, g_0, & h_0, & k_0, \\ a_1, & b_1, & c_1, & \ldots, g_1, & h_1, & k_1, \\ a_2, & b_2, & c_2, & \ldots, g_2, & h_2, & k_2, \\ \ldots, & \ldots, & \ldots, & \ldots, \ldots, & \ldots, & \ldots, \\ a_{n-1}, & b_{n-1}, & c_{n-1}, & \ldots, g_{n-1}, & h_{n-1}, & k_{n-1}, \end{array}$$

chosen arbitrarily. Moreover, let P represent the result of replacing the variables

$$x, \quad y, \quad z, \quad \ldots, \quad u, \quad v$$

in the function φ by the letters

$$a, \quad b, \quad c, \quad \ldots, \quad g, \quad h,$$

considered as new quantities. Consequently we have

$$(5)\qquad P=(b-a)\times(c-a)(c-b)\times\ldots\\ \times(h-a)(h-b)(h-c)\ldots(h-g).$$

The product P is the simplest alternating function of the quantities a, b, c, ..., g, h, and if we expand this function by algebraic multiplication of these binomial factors, each term of the expansion will be equivalent, except for the sign, to the product of these same quantites arranged in a certain order, and respectively raised to the powers 0, 1, 2, 3, ..., $n-1$. Given this, imagine that in each term we replace the exponents with letters for their indices, by writing, for example,

$$a_0b_1c_2\ldots g_{n-2}h_{n-1}$$

in place of the term

$$a^0b^1c^2\ldots g^{n-2}h^{n-1},$$

and denote by D the expansion of the product P. The quantity D, just like the product P, evidently has the property of changing its sign whenever we interchange two [78] of the given letters, for example, the letters a and b. From this, it is easy to conclude that the value of D is reduced to zero if in all of its terms we write the letter b in place of the letter a without writing at the same time a in place of b. It is the same if everywhere we write one of the letters c, ..., g, h in place of the letter a. Consequently, suppose that in the polynomial D we denote the sum of all the terms that have a_0 as their common factor by A_0a_0, the sum of the terms which contain the factor a_1 by A_1a_1, ..., and finally the sum of the terms that have the factor a_{n-1} by $A_{n-1}a_{n-1}$, so that the value of D is given by the equation

$$(6)\qquad D=A_0a_0+A_1a_1+A_2a_2+\ldots+A_{n-1}a_{n-1}.$$

Then we find, by writing successively in the right-hand side of this equation the letters b, c, ..., g, h in place of the letter a,

$$(7)\qquad \begin{cases} 0=A_0b_0+A_1b_1+A_2b_2+\ldots+A_{n-1}b_{n-1}, \\ 0=A_0c_0+A_1c_1+A_2c_2+\ldots+A_{n-1}c_{n-1}, \\ \ldots\ldots\ldots\ldots\ldots\ldots\ldots\ldots\ldots\ldots\ldots\ldots, \\ 0=A_0g_0+A_1g_1+A_2g_2+\ldots+A_{n-1}g_{n-1}, \\ 0=A_0h_0+A_1h_1+A_2h_2+\ldots+A_{n-1}h_{n-1}. \end{cases}$$

Now suppose that we add equations (4) together term by term, after multiplying the first by A_0, the second by A_1, the third by A_2, ..., the last by A_{n-1}. In this sum, we see that the coefficients of the unknowns y, z, ..., u, v disappear by virtue of formulas (7), and we obtain definitively the equation

$$Dx=A_0k_0+A_1k_1+A_2k_2+\ldots+A_{n-1}k_{n-1},$$

from which we conclude

$$(8)\qquad x=\frac{A_0k_0+A_1k_1+A_2k_2+\ldots+A_{n-1}k_{n-1}}{D}.$$

Moreover, of the two quantities

$$D \quad \text{and} \quad A_0k_0 + A_1k_1 + A_2k_2 + \ldots + A_{n-1}k_{n-1},$$

[79] the first is what arises from the expansion of the product

$$(b-a)\times(c-a)(c-b)\times\ldots\times(h-a)(h-b)(h-c)\ldots(h-g),$$

when we replace the exponents of the letters in this expansion with the indices, and the second is what becomes of the quantity D, equivalent to the right-hand side of formula (6), when we substitute the letter k for the letter a. Consequently, we can consider the value of x to be determined by the equation

$$(9) \qquad x = \frac{(b-k)\times(c-k)(c-b)\times\ldots\times(h-k)(h-b)(h-c)\ldots(h-g)}{(b-a)\times(c-a)(c-b)\times\ldots\times(h-a)(h-b)(h-c)\ldots(h-g)},$$

provided that we agree to expand the two terms of the fraction that forms the right-hand side and to replace in each expansion the exponents of the letters by their indices. Taken literally, the value which equation (9) seems to give to the unknown x is not exact and not capable of being made exact without the stated modifications. This is what we call a *symbolic value* of this unknown.

The method which has led us to the symbolic value of x furnishes equally the symbolic values of the other unknowns. To give an application of this method, suppose that we wish to solve the linear equations

$$(10) \qquad \begin{cases} a_0x + b_0y + c_0z = k_0, \\ a_1x + b_1y + c_1z = k_1, \\ a_2x + b_2y + c_2z = k_2. \end{cases}$$

Under this hypothesis, we find the symbolic value of the unknown x to be,[1]

$$(11) \qquad \begin{cases} x = \dfrac{(b-k)(c-k)(c-b)}{(b-a)(c-a)(c-b)} \\ \quad = \dfrac{k^0b^1c^2 - k^0b^2c^1 + k^1b^2c^0 - k^1b^0c^2 + k^2b^0c^1 - k^2b^1c^0}{a^0b^1c^2 - a^0b^2c^1 + a^1b^2c^0 - a^1b^0c^2 + a^2b^0c^1 - a^2b^1c^0}, \end{cases}$$

[80] and consequently, the true value of the unknown is[2]

$$(12) \qquad x = \frac{k_0b_1c_2 - k_0b_2c_1 + k_1b_2c_0 - k_1b_0c_2 + k_2b_0c_1 - k_2b_1c_0}{a_0b_1c_2 - a_0b_2c_1 + a_1b_2c_0 - a_1b_0c_2 + a_2b_0c_1 - a_2b_1c_0}.$$

Note. — When, in equations (4), we replace the indices of the letters a, b, c, ..., g, h, k by the exponents, the symbolic value of x given by equation (9) evidently

[1] In [Cauchy 1897, p. 79], there are typographical errors in the second line of (11), with c_0 written in place of c^0 in two instances. These errors were not present in [Cauchy 1821, p. 81]. (tr.)

[2] We recognize this as Cramer's Rule, named for Gabriel Cramer (1704–1752); see [Cramer 1750].

becomes the true value, and coincides, as we ought to expect, with that furnished by formula (3) of § 1.

3.3 On homogeneous functions.

A function of several variables x, y, z, $\ldots$ is *homogeneous* when changing x to tx, y to ty, z to tz, $\ldots$, where t is a new variable independent of the others, makes this function vary in the ratio of 1 to some fixed power of t. The exponent of this power is called the *degree* of the homogeneous function. In other words,

$$f(x,y,z,\ldots)$$

is a homogeneous function of degree a with respect to the variables x, y, z, $\ldots$, if for any t, we have

(1) $$f(tx,ty,tz,\ldots) = t^a f(x,y,z,\ldots).$$

Thus, for example,

$$x^2+xy+y^2, \quad \sqrt{xy} \quad \text{and} \quad \ln x - \ln y$$

are three homogeneous functions of the variables x and y, the first of the second degree, the second of the first degree and the third of degree zero. An integer function of the variables $x, y, z, \ldots$ composed of terms chosen so that the sum of the exponents of the various [81] variables is the same in all the terms is evidently homogeneous.

If we let $t = \frac{1}{x}$ in formula (1), we conclude that

(2) $$f(x,y,z,\ldots) = x^a f\left(1,\frac{y}{x},\frac{z}{x},\ldots\right).$$

This last equation establishes a property of homogeneous functions that we can state in the following manner:

Whenever a function of several variables x, y, z, $\ldots$ is homogeneous, it is equivalent to a product of any one of the variables raised to a certain power by a function of the ratios of these same variables combined in pairs.

We can add that this property applies exclusively to homogeneous functions. And, indeed, suppose that $f(x,y,z,\ldots)$ is equivalent to the product of x^a by a function of the ratios among the variables x, y, z, $\ldots$ combined in pairs. Because we can express each of these ratios by means of those which have x for their denominators by writing, for example, in place of $\frac{z}{y}$,

$$\frac{\left(\frac{z}{x}\right)}{\left(\frac{y}{x}\right)},$$

it follows that the value of $f(x,y,z,\ldots)$ is given by an equation of the form

$$f(x,y,z,\ldots) = x^a \varphi\left(\frac{y}{x}, \frac{z}{x}, \ldots\right).$$

This equation remains true, whatever the values of x, y, z, ... may be, and if we replace

$$x \text{ by } tx, \quad y \text{ by } ty, \quad z \text{ by } tz, \quad \ldots,$$

it becomes

$$f(tx,ty,tz,\ldots) = t^a x^a \varphi\left(\frac{y}{x}, \frac{z}{x}, \ldots\right).$$

[82] Consequently, under the given hypothesis, we have,

$$f(tx,ty,tz,\ldots) = t^a f(x,y,z,\ldots),$$

whatever t may be. In other words,

$$f(x,y,z,\ldots)$$

will be a homogeneous function of degree a with respect to the variables x, y, z, ...

Chapter 4
Determination of integer functions, when a certain number of particular values are known. Applications.

4.1 Research on integer functions of a single variable for which a certain number of particular values are known.

[83] To determine a function when a certain number of particular values are taken to be known is what we call to *interpolate*. When it is a matter of a function of one or two variables, this function can be considered as the ordinates of a curve or of a surface, and the problem of *interpolation* consists of fixing the general value of this ordinate given a certain number of particular values, that is to say, to make the curve or the surface pass through a certain number of points. This question can be solved in an infinity of ways, and in general the problem of interpolation is indeterminate. However, the indeterminacy will cease if, to the knowledge of the particular values of the desired function, we add the expressed condition that this function be integer, and of a degree such that the number of its terms becomes precisely equal to the number of particular values given.

To fix these ideas, suppose that we consider first the integer functions of a single variable x. We establish easily in this regard the following propositions:

Theorem I. — *If an integer function of the variable x vanishes for* [84] *a particular value of this variable, for example for $x = x_0$, it is algebraically divisible by $x - x_0$.*

Theorem II. — *If an integer function of the variable x vanishes for each of the values of x contained in the series*

$$x_0, \quad x_1, \quad x_2, \quad \ldots, \quad x_{n-1},$$

where n denotes any integer, it will necessarily be divisible by the product

$$(x - x_0)\,(x - x_1)\,(x - x_2)\ldots(x - x_{n-1})\,.$$

R.E. Bradley, C.E. Sandifer, *Cauchy's Cours d'analyse*, Sources and Studies in the History of Mathematics and Physical Sciences, DOI 10.1007/978-1-4419-0549-9_4,

Now let $\varphi(x)$ and $\psi(x)$ be two integer functions of the variable x, both of degree $n-1$, and which become equal to each other for each of the n particular values of x contained in the series $x_0, x_1, x_2, \ldots, x_{n-1}$. I say that these two functions are identically equal, that is to say that we have,

$$\psi(x) = \varphi(x),$$

whatever x may be. Indeed, if this equality did not occur, we would find in the difference

$$\psi(x) = \varphi(x),$$

an integer polynomial for which the degree does not surpass $n-1$ but which vanishes for each of the values of x mentioned above, and is still divisible by the product

$$(x-x_0)(x-x_1)(x-x_2)\ldots(x-x_{n-1}),$$

that is to say by a polynomial of degree n, which is absurd. We are assured *a fortiori* of the absolute equality of the two functions $\varphi(x)$ and $\psi(x)$ if we know that they become equal to each other for a number of values of x greater than n. We can thus state the following theorem:

Theorem III. — *If two integer functions of the variable x become* [85] *equal for a number of values of this variable greater than the degree of each of these two functions, they are identically equal, whatever x may be.*

We thereby deduce as a corollary this other theorem:

Theorem IV. — *Two integer functions of the variable x are identically equal whenever they become equal for all integer values of that variable, or even for all integer values which surpass a given limit.*

Indeed, in this case the number of values of x for which the two functions become equal is indefinite.

It follows from theorem III that an integer function u of degree $n-1$ is completely determined if we know its particular values

$$u_0, \quad u_1, \quad u_2, \quad \ldots, \quad u_{n-1}$$

corresponding to the values

$$x_0, \quad x_1, \quad x_2, \quad \ldots, \quad x_{n-1}$$

of the variable x. Under this hypothesis, we look for the general value of the function u.[1] If we suppose first that the particular values $u_0, u_1, \ldots, u_{n-1}$ all reduce to zero

[1] The interpolation technique that Cauchy is about to describe is known as Lagrange interpolation. See, for example, [Burden and Faires 2001, pp. 107–118].

with the exception, u_0, then the function u ought to vanish for $x = x_1$, for $x = x_2$, ..., and finally for $x = x_{n-1}$, and it is divisible by the product

$$(x - x_1)(x - x_2)\dots(x - x_{n-1}),$$

and consequently it is of the form

$$u = k(x - x_1)(x - x_2)\dots(x - x_{n-1}),$$

where k must be a constant quantity. Moreover, because u must reduce to u_0 for $x = x_0$, we conclude that

$$u_0 = k(x - x_1)(x - x_2)\dots(x - x_{n-1})$$

[86] and consequently

$$u = u_0 \frac{(x - x_1)(x - x_2)\dots(x - x_{n-1})}{(x_0 - x_1)(x_0 - x_2)\dots(x_0 - x_{n-1})}.$$

Likewise, if the particular values u_0, u_1, u_2, ..., u_{n-1} all reduce to zero with the exception of the second one, u_1, we find that

$$u = u_1 \frac{(x - x_0)(x - x_2)\dots(x - x_{n-1})}{(x_1 - x_0)(x_1 - x_2)\dots(x_1 - x_{n-1})},$$

....

Finally, if they all reduce to zero with the exception of the last one, u_{n-1}, we find

$$u = u_{n-1} \frac{(x - x_0)(x - x_1)\dots(x - x_{n-2})}{(x_{n-1} - x_0)(x_{n-1} - x_1)\dots(x_{n-1} - x_{n-2})}.$$

In adding together these various values of u corresponding to the various hypotheses that we have just made, we obtain for the sum a polynomial in x of degree $n-1$ which evidently has the property that it reduces to u_0 when $x = x_0$, to u_1 when $x = x_1$, ..., and to u_{n-1} when $x = x_{n-1}$. Thus this polynomial is the general value of u which solves the given question, so that this value is found to be determined by the formula

$$(1) \quad \begin{cases} u = & u_0 \dfrac{(x - x_1)(x - x_2)\dots(x - x_{n-1})}{(x_0 - x_1)(x_0 - x_2)\dots(x_0 - x_{n-1})} \\ & + u_1 \dfrac{(x - x_0)(x - x_2)\dots(x - x_{n-1})}{(x_1 - x_0)(x_1 - x_2)\dots(x_1 - x_{n-1})} \\ & + \dots\dots\dots\dots \\ & + u_{n-1} \dfrac{(x - x_0)(x - x_1)\dots(x - x_{n-2})}{(x_{n-1} - x_0)(x_{n-1} - x_1)\dots(x_{n-1} - x_{n-2})}. \end{cases}$$

We could have deduced the same formula directly from the method which we employed above (Chap. III, § I) to solve linear equations of several variables in a particular case (on this subject, see Note V).

Denoting by a a constant quantity, if we replace in formula (1) the function u by the function $u-a$, which evidently is [87] of the same degree, and the particular values of u by the particular values of $u-a$, we obtain the equation[2]

$$(2)\quad \left\{\begin{aligned} u-a = {} & (u_0-a)\frac{(x-x_1)(x-x_2)\dots(x-x_{n-1})}{(x_0-x_1)(x_0-x_2)\dots(x_0-x_{n-1})} \\ & + (u_1-a)\frac{(x-x_0)(x-x_2)\dots(x-x_{n-1})}{(x_1-x_0)(x_1-x_2)\dots(x_1-x_{n-1})} \\ & + \dots\dots\dots\dots\dots\dots \\ & + (u_{n-1}-a)\frac{(x-x_0)(x-x_1)\dots(x-x_{n-2})}{(x_{n-1}-x_0)(x_{n-1}-x_1)\dots(x_{n-1}-x_{n-2})}, \end{aligned}\right.$$

and by comparing this equation to formula (1), we find the following

$$(3)\quad \left\{\begin{aligned} 1 = {} & \frac{(x-x_1)(x-x_2)\dots(x-x_{n-1})}{(x_0-x_1)(x_0-x_2)\dots(x_0-x_{n-1})} \\ & + \frac{(x-x_0)(x-x_2)\dots(x-x_{n-1})}{(x_1-x_0)(x_1-x_2)\dots(x_1-x_{n-1})} \\ & + \dots\dots\dots\dots\dots\dots \\ & + \frac{(x-x_0)(x-x_1)\dots(x-x_{n-2})}{(x_{n-1}-x_0)(x_{n-1}-x_1)\dots(x_{n-1}-x_{n-2})}. \end{aligned}\right.$$

This last equation is an identity and remains true whatever x may be.

Equations (1) and (2) can both serve to solve the problem of interpolation for integer functions, but in general it is advisable to prefer equation (2), considering that we can make one of the terms of the right-hand side disappear by taking the constant a to be equal to one of the quantities

$$u_0, \quad u_1, \quad u_2, \quad \dots, \quad u_{n-1}.$$

Suppose, for example, that we are trying to make a straight line pass through two given points. Denote by x_0 and y_0 the rectangular coordinates of the first point, by x_1 and y_1 the those of the second, and by y the ordinate variable of the straight line. By replacing the letter u in formula (2) by the letter y, then making $n=1$ and $a=y_0$, we find the equation of the line to be

$$(4)\quad y-y_0 = (y_1-y_0)\frac{x-x_0}{x_1-x_0}.$$

[2] In both [Cauchy 1821, p. 91] and [Cauchy 1897, p. 87], there is a typographical error in the last line of formula (2), in which the denominator contains an x_2 where it should be x_1. (tr.)

[88] On the other hand, suppose that we are trying to make a parabola whose axis is parallel to the y axis pass through three given points. Let

$$x_1 \text{ and } y_1, \quad x_2 \text{ and } y_2, \quad \text{and} \quad x_3 \text{ and } y_3$$

be the rectangular coordinates of the three points. Also, let y be the ordinate variable of the parabola. By replacing the letter u in formula (2) by the letter y, then making $n = 2$ and $a = y_0$, we find the equation of the parabola to be

$$(5) \qquad \left\{ \begin{aligned} y - y_1 = \quad & (y_0 - y_1)\frac{(x-x_1)(x-x_2)}{(x_0-x_1)(x_0-x_2)} \\ & + (y_2 - y_1)\frac{(x-x_0)(x-x_1)}{(x_2-x_0)(x_2-x_1)}, \end{aligned} \right.$$

or what amounts to the same thing,

$$(6) \qquad y - y_1 = \frac{x-x_1}{x_2-x_0}\left[(y_0-y_1)\frac{x-x_2}{x_1-x_0} + (y_2-y_1)\frac{x-x_0}{x_2-x_1}\right].$$

When in equation (1) we take $u = x^m$ (m denoting an integer number less than n), the particular values of u represented by

$$u_0, \quad u_1, \quad u_2, \quad \ldots, \quad u_{n-1}$$

evidently reduce to

$$x_0^m, \quad x_1^m, \quad x_2^m, \quad \ldots, \quad x_{n-1}^m.$$

Thus we have, for integer values of m which do not surpass $n-1$,[3]

$$(7) \qquad \left\{ \begin{aligned} x^m = \quad & x_0^m \frac{(x-x_1)(x-x_2)\ldots(x-x_{n-1})}{(x_0-x_1)(x_0-x_2)\ldots(x_0-x_{n-1})} \\ & + x_1^m \frac{(x-x_0)(x-x_2)\ldots(x-x_{n-1})}{(x_1-x_0)(x_1-x_2)\ldots(x_1-x_{n-1})} \\ & + \ldots\ldots\ldots\ldots\ldots\ldots\ldots\ldots\ldots \\ & + x_{n-1}^m \frac{(x-x_0)(x-x_1)\ldots(x-x_{n-2})}{(x_{n-1}-x_0)(x_{n-1}-x_1)\ldots(x_{n-1}-x_{n-2})}. \end{aligned} \right.$$

This last formula contains equation (3) as a particular case. Moreover, if we observe that each power of x, and in particular [89] the power x^{n-1}, ought necessarily to have the same coefficient on both sides of formula (7), we find:

1° by supposing that $m < n-1$,

[3] [Cauchy 1897, p. 88] has an unbalanced parenthesis in the last denominator of this formula. This typographical error is not in [Cauchy 1821, p. 92]. (tr.)

$$
(8) \quad \begin{cases} 0 = & \dfrac{x_0^m}{(x_0 - x_1)(x_0 - x_2)\dots(x_0 - x_{n-1})} \\ & + \dfrac{x_1^m}{(x_1 - x_0)(x_1 - x_2)\dots(x_1 - x_{n-1})} \\ & + \dots\dots\dots\dots\dots\dots\dots\dots\dots\dots \\ & + \dfrac{x_{n-1}^m}{(x_{n-1} - x_0)(x_{n-1} - x_1)\dots(x_{n-1} - x_{n-2})}; \end{cases}
$$

2° By supposing that $m = n - 1$,

$$
(9) \quad \begin{cases} 1 = & \dfrac{x_0^{n-1}}{(x_0 - x_1)(x_0 - x_2)\dots(x_0 - x_{n-1})} \\ & + \dfrac{x_1^{n-1}}{(x_1 - x_0)(x_1 - x_2)\dots(x_1 - x_{n-1})} \\ & + \dots\dots\dots\dots\dots\dots\dots\dots\dots\dots \\ & + \dfrac{(x_{n-1})^{n-1}}{(x_{n-1} - x_0)(x_{n-1} - x_1)\dots(x_{n-1} - x_{n-2})}. \end{cases}
$$

It is worth remarking that formula (8) remains true in the case where we suppose that $m = 0$ and then it becomes[4]

$$
(10) \quad \begin{cases} 0 = & \dfrac{1}{(x_0 - x_1)(x_0 - x_2)\dots(x_0 - x_{n-1})} \\ & + \dfrac{1}{(x_1 - x_0)(x_1 - x_2)\dots(x_1 - x_{n-1})} \\ & + \dots\dots\dots\dots\dots\dots\dots\dots\dots\dots \\ & + \dfrac{1}{(x_{n-1} - x_0)(x_{n-1} - x_1)\dots(x_{n-1} - x_{n-2})}. \end{cases}
$$

4.2 Determination of integer functions of several variables, when a certain number of particular values are assumed to be known.

The methods by which we determine functions of one variable when a certain number of particular values are assumed to be [90] known can be easily extended, as we are going to see, to functions of several variables.

To fix these ideas, let us first consider functions of two variables, x and y. Let $\varphi(x,y)$ and $\psi(x,y)$ be two such functions, both of degree $n-1$ with respect to each of the variables, and which become equal to each other whenever, by attributing to

[4] This result is due to Euler [Euler 1769, vol. 2, § 1169]. See also [Sandifer 2007, pp. 133–137].

the variable x one of the particular values

$$x_0, \quad x_1, \quad x_2, \quad \ldots, \quad x_{n-1}$$

at the samc time we attribute to the variable y one of the following

$$y_0, \quad y_1, \quad y_2, \quad \ldots, \quad y_{n-1}.$$

Then $\varphi(x_0, y)$ and $\psi(x_0, y)$ are two functions of the single variable y, which ought to be equal to each other for n particular values of this variable. Consequently (by virtue of theorem III, § I), these two functions are constantly equal, whatever y may be. Then we have identically

$$\varphi(x_0, y) = \psi(x_0, y).$$

Likewise we find

$$\begin{aligned} &\varphi(x_1, y) = \psi(x_1, y), \\ &\varphi(x_2, y) = \psi(x_2, y), \\ &\ldots\ldots\ldots\ldots\ldots\ldots\ldots, \\ &\varphi(x_{n-1}, y) = \psi(x_{n-1}, y). \end{aligned}$$

Moreover, the left-hand sides of the preceding n equations are particular values of the function $\varphi(x, y)$ in the case where we consider just x as the variable, and the right-hand sides represent the corresponding particular values of the function $\psi(x, y)$. The two functions

$$\varphi(x, y) \quad \text{and} \quad \psi(x, y),$$

when we attribute to y a constant value chosen arbitrarily, thus become equal for n particular values of x, and because they are both of degree $n-1$ with respect to x, it follows [91] that they remain equal, not only for any value attributed to the variable y but also for any value of x. We are assured, *a fortiori*, of the absolute equality of the two functions $\varphi(x, y)$ and $\psi(x, y)$ if we know that they become equal whenever the values of x and y are respectively taken in two series each composed of more than n different terms. Thus we can state the following proposition:

Theorem I. — *If two integer functions of the variables x and y become equal whenever the values of these two variables are respectively taken from two series both of which contain a number of terms greater than the highest exponents of x and y in these same functions, then they are identically equal.*

We thereby deduce as a corollary this other theorem:

Theorem II. — *Two integer functions of the variables x and y are identically equal whenever they become equal for all integer values of these variables, or even for all integer values which surpass a given limit.*

Indeed, in this case the number of values of x and y for which the two functions become equal is indefinite.

It follows from theorem I that, if we suppose that the function $\varphi(x,y)$ is integer and of degree $n-1$ with respect to each of the variables x and y, this function is completely determined when we know the particular values which it receives when, in taking for the values of x one of the quantities

$$x_0, \quad x_1, \quad x_2, \quad \ldots, \quad x_{n-1},$$

we take at the same time for the value of y one of the following

$$y_0, \quad y_1, \quad y_2, \quad \ldots, \quad y_{n-1}.$$

Under the same hypothesis, the general value of the function can [92] be easily deduced from formula (1) of the preceding section.[5] Indeed, if we replace u by $\varphi(x,y)$ in this formula, we get

$$(1)\quad \left\{ \begin{aligned} \varphi(x,y) = {} & \frac{(x-x_1)(x-x_2)\ldots(x-x_{n-1})}{(x_0-x_1)(x_0-x_2)\ldots(x_0-x_{n-1})}\varphi(x_0,y) \\ & + \frac{(x-x_0)(x-x_2)\ldots(x-x_{n-1})}{(x_1-x_0)(x_1-x_2)\ldots(x_1-x_{n-1})}\varphi(x_1,y) \\ & + \ldots\ldots\ldots\ldots\ldots\ldots\ldots\ldots\ldots\ldots \\ & + \frac{(x-x_0)(x-x_1)\ldots(x-x_{n-2})}{(x_{n-1}-x_0)(x_{n-1}-x_1)\ldots(x_{n-1}-x_{n-2})}\varphi(x_{n-1},y), \end{aligned} \right.$$

and we have, moreover, denoting by m one of the integer numbers $1, 2, 3, \ldots, n-1$,

$$(2)\quad \left\{ \begin{aligned} \varphi(x_m,y) = {} & \frac{(y-y_1)(y-y_2)\ldots(y-y_{n-1})}{(y_0-y_1)(y_0-y_2)\ldots(y_0-y_{n-1})}\varphi(x_m,y_0) \\ & + \frac{(y-y_0)(y-y_2)\ldots(y-y_{n-1})}{(y_1-y_0)(y_1-y_2)\ldots(y_1-y_{n-1})}\varphi(x_m,y_1) \\ & + \ldots\ldots\ldots\ldots\ldots\ldots\ldots\ldots\ldots\ldots \\ & + \frac{(y-y_0)(y-y_1)\ldots(y-y_{n-2})}{(y_{n-1}-y_0)(y_{n-1}-y_1)\ldots(y_{n-1}-y_{n-2})}\varphi(x_m,y_{n-1}). \end{aligned} \right.$$

We draw the general value of $\varphi(x,y)$ immediately from the two preceding equations. For example, by supposing that $n=2$, we find

[5] Cauchy used the word *paragraphe*, which we will consistently translate as "section." (tr.)

$$(3)\qquad \begin{cases} \varphi(x,y) = & \dfrac{x-x_1}{x_0-x_1}\dfrac{y-y_1}{y_0-y_1}\varphi(x_0,y_0) \\ & + \dfrac{x-x_0}{x_1-x_0}\dfrac{y-y_1}{y_0-y_1}\varphi(x_1,y_0) \\ & + \dfrac{x-x_1}{x_0-x_1}\dfrac{y-y_0}{y_1-y_0}\varphi(x_0,y_1) \\ & + \dfrac{x-x_0}{x_1-x_0}\dfrac{y-y_0}{y_1-y_0}\varphi(x_1,y_1). \end{cases}$$

If we consider functions of three or more variables, we obtain results entirely similar to those which we have just found for functions of [93] only two variables. We find, for example, in place of theorem II the following proposition:

Theorem III. — *Two integer functions of several variables x, y, z, ... are identically equal to each other whenever they become equal for all integer values of these variables, or even for all integer variables which surpass a given limit.*

4.3 Applications.

To apply the principles established in the preceding sections, let us consider in particular products formed by the multiplication of successive factors for which each surpasses the following one by one, the first factor being one of the variables x, y, z, ... By means of these kinds of products, we seek to express the very similar product that we would obtain by taking for the first factor to be the sum of the given variables, namely

$$x+y+z+\ldots.$$

If we reduce the number of variables to two, the problem at hand can be stated as follows:

Problem I. — *To express the product*

$$(1)\qquad (x+y)(x+y-1)(x+y-2)\ldots(x+y-n+1),$$

in which n denotes any integer number, by means of the following products

$$x(x-1)(x-2)\ldots(x-n+1) \quad \textit{and}$$
$$y(y-1)(y-2)\ldots(y-n+1)$$

and all such products which arise by changing the value of n.

Solution. — To solve the preceding question more easily, let us first suppose that x and y are integer numbers greater than or equal to n. Then the product (1) is nothing other than the numerator [94] of the fraction that expresses the number of possible combinations of $x+y$ letters taken n at a time. This number is precisely

$$\frac{(x+y)\,(x+y-1)\,(x+y-2)\ldots(x+y-n+1)}{1\cdot 2\cdot 3\ldots n}.$$

Given this, imagine that

$$a,\quad b,\quad c,\quad \ldots,\quad p,\quad q,\quad r,\quad \ldots$$

are $x+y$ letters, and that we divide them into two groups so that there are x letters, a, b, c, …, in the first group and y letters, p, q, r, …, in the second group. Among the combinations formed with these different letters, some contain only letters taken from the first group. The number of combinations of this kind is

$$\frac{x\,(x-1)\,(x-2)\ldots(x-n+1)}{1\cdot 2\cdot 3\ldots n}.$$

Others contain $n-1$ letters taken from the first group and one letter taken from the second. We easily determine the number of combinations of this second kind and we see that it is equal to

$$\frac{x\,(x-1)\,(x-2)\ldots(x-n+2)}{1\cdot 2\cdot 3\ldots(n-1)}\,\frac{y}{1}.$$

Likewise, we find that the number of combinations which contain $n-2$ letters taken from the first group and two letters from the second group is

$$\frac{x\,(x-1)\,(x-2)\ldots(x-n+3)}{1\cdot 2\cdot 3\ldots(n-2)}\,\frac{y\,(y-1)}{1\cdot 2},$$

etc. Finally, the number of combinations which contain only letters taken from the second group is

$$\frac{y\,(y-1)\,(y-2)\ldots(y-n+1)}{1\cdot 2\cdot 3\ldots n}.$$

The sum of the numbers of combinations of each kind ought [95] to produce the total number of combinations of $x+y$ given letters taken n at a time. We conclude that[6]

$$(2)\quad \left\{\begin{aligned} &\frac{(x+y)\,(x+y-1)\ldots(x+y-n+1)}{1\cdot 2\cdot 3\ldots n}\\ &\quad=\frac{x\,(x-1)\ldots(x-n+1)}{1\cdot 2\cdot 3\ldots n}+\frac{x\,(x-1)\ldots(x-n+2)}{1\cdot 2\cdot 3\ldots(n-1)}\,\frac{y}{1}\\ &\quad+\frac{x\,(x-1)\ldots(x-n+3)}{1\cdot 2\cdot 3\ldots(n-2)}\,\frac{y\,(y-1)}{1\cdot 2}+\ldots\\ &\quad+\frac{x}{1}\,\frac{y\,(y-1)\ldots(y-n+2)}{1\cdot 2\cdot 3\ldots(n-1)}+\frac{y\,(y-1)\ldots(y-n+1)}{1\cdot 2\cdot 3\ldots n}.\end{aligned}\right.$$

[6] The numeral 3 was missing from the denominator in the third line of equation (2) in [Cauchy 1897, p. 95], but present in [Cauchy 1821, p. 100]. (tr.)

The preceding equation, being thus proved in the case where the variables x and y take integer values greater than n,[7] remains true, by virtue of theorem II (§ II), for all values of these variables, and the value of product (1) derived from the same equation is

$$(3)\quad \begin{cases} (x+y)(x+y-1)\dots(x+y-n+1) \\ \qquad = x(x-1)\dots(x-n+1) \\ \qquad\quad +\frac{n}{1}x(x-1)\dots(x-n+2)y \\ \qquad\quad +\frac{n(n-1)}{1\cdot 2}x(x-1)\dots(x-n+3)y(y-1)+\dots \\ \qquad\quad +\frac{n}{1}xy(y-1)\dots(y-n+2) \\ \qquad\quad +y(y-1)\dots(y-n+1). \end{cases}$$

Corollary I. — If we replace x by $-x$ and y by $-y$, in equation (2) we obtain the following:[8]

$$(4)\quad \begin{cases} \frac{(x+y)(x+y+1)\dots(x+y+n-1)}{1\cdot 2\cdot 3\dots n} \\ \qquad = \frac{x(x+1)\dots(x+n-1)}{1\cdot 2\cdot 3\dots n} + \frac{x(x+1)\dots(x+n-2)}{1\cdot 2\cdot 3\dots(n-1)}\frac{y}{1} \\ \qquad\quad + \frac{x(x+1)\dots(x+n-3)}{1\cdot 2\cdot 3\dots(n-2)}\frac{y(y+1)}{1\cdot 2}+\dots \\ \qquad\quad + \frac{x}{1}\frac{y(y+1)\dots(y+n-2)}{1\cdot 2\cdot 3\dots(n-1)} + \frac{y(y+1)\dots(y+n-1)}{1\cdot 2\cdot 3\dots n}. \end{cases}$$

Corollary II. — If we replace x by $\frac{x}{2}$ and y in equation (2) [96] by $\frac{y}{2}$, we find

$$(5)\quad \begin{cases} \frac{(x+y)(x+y-2)\dots(x+y-2n+2)}{2\cdot 4\cdot 6\dots(2n)} \\ \qquad = \frac{x(x-2)\dots(x-2n+2)}{2\cdot 4\cdot 6\dots(2n)} + \frac{x(x-2)\dots(x-2n+4)}{2\cdot 4\cdot 6\dots(2n-2)}\frac{y}{2} \\ \qquad\quad + \dots\dots\dots\dots\dots\dots\dots\dots \\ \qquad\quad + \frac{x}{2}\frac{y(y-2)\dots(y-2n+4)}{2\cdot 4\cdot 6\dots(2n-2)} + \frac{y(y-2)\dots(y-2n+2)}{2\cdot 4\cdot 6\dots(2n)}. \end{cases}$$

---

[7] Cauchy has modified, perhaps inadvertently, the condition "greater than or equal to" stated at the beginning of this solution.

[8] We have restored parentheses to the second line of equation (4) that were missing in [Cauchy 1897, p. 95]. They had been present in [Cauchy 1821, p. 100]. (tr.)

Corollary III. — By expanding both sides of equation (2) and keeping on each side only the terms in which the sum of the exponents of the variables is equal to n, we obtain the formula

$$(6)\qquad \begin{cases} \dfrac{(x+y)^n}{1\cdot 2\cdot 3\dots n} = \dfrac{x^n}{1\cdot 2\cdot 3\dots n} + \dfrac{x^{n-1}}{1\cdot 2\cdot 3\dots (n-1)}\dfrac{y}{1} \\ \qquad + \dfrac{x^{n-2}}{1\cdot 2\cdot 3\dots (n-2)}\dfrac{y^2}{1\cdot 2} + \dots \\ \qquad + \dfrac{x}{1}\dfrac{y^{n-1}}{1\cdot 2\cdot 3\dots (n-1)} + \dfrac{y^n}{1\cdot 2\cdot 3\dots n}. \end{cases}$$

The value of $(x+y)^n$ taken from this last formula is precisely that given by the *Newton binomial.*

The formulas that we have just derived can easily be extended to the case where we consider more than two variables, and the method which has brought us to the solution of problem I is equally applicable to the following question:

Problem II. — *With x, y, z, ... denoting any number of variables, to express the product*

$$(x+y+z+\dots)(x+y+z+\dots-1)(x+y+z+\dots-2)\dots(x+y+z+\dots-n+1)$$

as a function of the following ones

$$\begin{aligned} &x(x-1)(x-2)\dots(x-n+1),\\ &y(y-1)(y-2)\dots(y-n+1),\\ &z(z-1)(z-2)\dots(z-n+1),\\ &\dots\dots\dots\dots\dots\dots\dots\dots, \end{aligned}$$

and all such products which arise by changing the value of n.

[97] We begin by solving the problem in the case where x, y, z, ... denote integer numbers greater than n, and on the basis of this principle, the fraction

$$\frac{(x+y+z+\dots)(x+y+z+\dots-1)(x+y+z+\dots-2)\dots(x+y+z+\dots-n+1)}{1\cdot 2\cdot 3\dots n}$$

is equal to the number of combinations that we can form with $x+y+z+\dots$ letters taken n at a time. Then we pass to the case where the variables x, y, z, ... become any quantities based on theorem III of § II. When we have thus proved the formula which solves the given question, we deduce without trouble the value of the power

$$(x+y+z+\dots)^n.$$

We then solve the problem, indeed, by expanding both sides of the formula we found, and keeping on each side only the terms in which the combined exponents of the variables x, y, z, ... form a sum equal to n.

Chapter 5
Determination of continuous functions of a single variable that satisfy certain conditions.

5.1 Research on a continuous function formed so that if two such functions are added or multiplied together, their sum or product is the same function of the sum or product of the same variables.

[98] When, instead of integer functions we imagine any functions, so that we leave the form entirely arbitrary, we can no longer successfully determine them given a certain number of particular values, however large that number might be, but we can sometimes do so in the case where we assume certain general properties of these functions. For example, a continuous function of x, represented by $\varphi(x)$, can be completely determined when it is required to satisfy, for all possible values of the variables x and y, one of the equations

$$(1) \qquad \varphi(x+y) = \varphi(x) + \varphi(y) \quad \text{or}$$
$$(2) \qquad \varphi(x+y) = \varphi(x) \times \varphi(y),$$

as well as when, for all positive real values of the same variables, one of the following equations:

$$(3) \qquad \varphi(xy) = \varphi(x) + \varphi(y) \quad \text{or}$$
$$(4) \qquad \varphi(xy) = \varphi(x) \times \varphi(y).$$

The solution of these four equations presents four different problems, which we will treat one after another.

[99] **Problem I.** — *To determine the function $\varphi(x)$ in such a manner that it remains continuous between any two real limits of the variable x and so that for all real values of the variables x and y, we have*

$$(1) \qquad \varphi(x+y) = \varphi(x) + \varphi(y).$$

R.E. Bradley, C.E. Sandifer, *Cauchy's Cours d'analyse*, Sources and Studies in the History of Mathematics and Physical Sciences, DOI 10.1007/978-1-4419-0549-9_5,

Solution. — If in equation (1) we successively replace y by $y+z$, z by $z+u$, ..., we get

$$\varphi\,(x+y+z+u+\ldots) = \varphi\,(x)+\varphi\,(y)+\varphi\,(z)+\varphi\,(u)+\ldots,$$

however many variables x, y, z, u, ... there may be. Also, if we denote this number of variables by m and a positive constant by α, and then we make

$$x = y = z = u = \ldots = \alpha,$$

then the formula which we have just found becomes

$$\varphi\,(m\alpha) = m\varphi\,(\alpha)\,.$$

To extend this last equation to the case where the integer number m is replaced by a fractional number $\frac{m}{n}$, or even by an arbitrary number μ, we set, in the first case,

$$\beta = \frac{m}{n}\alpha,$$

where m and n denote integer numbers, and we conclude that

$$n\beta = m\alpha,$$
$$n\varphi\,(\beta) = m\varphi\,(\alpha) \quad \text{and}$$
$$\varphi\,(\beta) = \varphi\left(\tfrac{m}{n}\alpha\right) = \tfrac{m}{n}\varphi\,(\alpha)\,.$$

Then, by supposing that the fraction $\frac{m}{n}$ varies in such a way as to converge towards any number μ, and passing to the limit, we find that

$$\varphi\,(\mu\alpha) = \mu\varphi\,(\alpha)\,.$$

[100] If we now take $\alpha - 1$, then we have, for all positive values of μ,

(5) $$\varphi\,(\mu) = \mu\varphi\,(1)\,,$$

and consequently, by making μ converge towards the limit zero,

$$\varphi\,(0) = 0.$$

Moreover, if in equation (1) we set $x = \mu$ and $y = -\mu$, we conclude that

$$\varphi\,(-\mu) = \varphi\,(0) - \varphi\,(\mu) = -\mu\varphi\,(1)\,.$$

Thus, equation (5) remains true when we change μ to $-\mu$. In other words, we have, for any values, positive or negative, of the variable x,

(6) $$\varphi\,(x) = x\varphi\,(1)\,.$$

It follows from formula (6) that any function $\varphi(x)$ which remains continuous between any limits of the variable and satisfies equation (1) is necessarily of the form

$$\varphi(x) = ax, \tag{7}$$

where a denotes a constant quantity. I add that the function ax enjoys the stated properties whatever the value of the constant a may be. Indeed, between any limits of the variable x, the product ax is a continuous function of that variable, and what's more, the assumption that $\varphi(x) = ax$ changes equation (1) into this other equation,

$$a(x+y) = ax + ay,$$

which is evidently always an identity. Thus formula (7) gives a solution to the proposed question, whatever value is attributed to the constant a. Because we have the ability to choose this constant arbitrarily, we call it an *arbitrary constant.*

Problem II. — *To determine the function $\varphi(x)$ in such a manner that it remains continuous between any two real limits of the variable x and so that* [101] *for all real values of the variables x and y, we have*

$$\varphi(x+y) = \varphi(x)\varphi(y). \tag{2}$$

Solution. — First, it is easy to assure ourselves that the function $\varphi(x)$ required to satisfy equation (2) will admit only positive values. Indeed, if we make $y = x$ in equation (2), we find that

$$\varphi(2x) = [\varphi(x)]^2,$$

and then, writing $\frac{1}{2}x$ in place of x, we conclude that

$$\varphi(x) = [\varphi(\tfrac{1}{2}x)]^2.$$

Thus the function $\varphi(x)$ is always equal to a square, and consequently it is always positive. Given this, suppose that in equation (2) we successively replace y by $y+z$, z by $z+u$, ... We then get

$$\varphi(x+y+z+u+\ldots) = \varphi(x)\varphi(y)\varphi(z)\varphi(u)\ldots,$$

however many variables $x, y, z, u, \ldots$ there may be. Also, if we denote this number of variables by m, and a positive constant by α, and then we make

$$x = y = z = u = \ldots = \alpha,$$

then the formula we have just found becomes

$$\varphi(m\alpha) = [\varphi(\alpha)]^m.$$

To extend this last formula to the case where the integer number m is replaced by a fractional number $\frac{m}{n}$, or even by an arbitrary number μ, we set, in the first case,

$$\beta = \frac{m}{n}\alpha,$$

where m and n denote two integer numbers, and we conclude that

$$n\beta = m\alpha,$$
$$[\varphi(\beta)]^n = [\varphi(\alpha)]^m \quad \text{and}$$
$$\varphi(\beta) = \varphi\left(\tfrac{m}{n}\alpha\right) = [\varphi(\alpha)]^{\frac{m}{n}}.$$

[102] Then, by supposing that the fraction $\frac{m}{n}$ varies in such a way as to converge towards any number μ and passing to the limit, we find that

$$\varphi(\mu\alpha) = [\varphi(\alpha)]^{\mu}.$$

Now if we take $\alpha = 1$, we have for all positive values of μ

(8) $$\varphi(\mu) = [\varphi(1)]^{\mu},$$

and consequently, by making μ converge towards the limit zero,

$$\varphi(0) = 1.$$

Moreover, if in equation (2) we set $x = \mu$ and $y = -\mu$, we conclude that

$$\varphi(-\mu) = \frac{\varphi(0)}{\varphi(\mu)} = [\varphi(1)]^{-\mu}.$$

Thus, equation (8) remains true when we change μ to $-\mu$. In other words, we have, for any values, positive or negative, of the variable x,

(9) $$\varphi(x) = [\varphi(1)]^{x}.$$

It follows from equation (9) that any function $\varphi(x)$ that solves the second problem is necessarily of the form

(10) $$\varphi(x) = A^x,$$

where A denotes a positive constant. I add that we can attribute to this constant any value between the limits 0 and ∞. Indeed, for any positive value of A, the function A^x remains continuous from $x = -\infty$ to $x = +\infty$, and the equation

$$A^{x+y} = A^x A^y$$

is an identity. The quantity A is thus an arbitrary constant that admits only positive values.

[103] *Note.* — We can get equation (9) very simply in the following manner. If we take logarithms of both sides of equation (2) in any system, we find that

$$\log \varphi(x+y) = \log \varphi(x) + \log \varphi(y),$$

and we conclude (see problem I) that

$$\log \varphi(x) = x \log \varphi(1),$$

then, by passing again from logarithms to numbers,

$$\varphi(x) = [\varphi(1)]^x.$$

Problem III. — *To determine the function $\varphi(x)$ in such a manner that it remains continuous between any two positive limits of the variable x and so that for all positive values of the variables x and y we have*

(3) $$\varphi(xy) = \varphi(x) + \varphi(y).$$

Solution. — It would be easy to apply a method similar to the one we used to solve the first problem to the solution of problem III. However, we will arrive more promptly at the solution we seek by putting equation (3) into a form analogous to that of equation (1), as we are going to do.

If A denotes any number and log denotes the characteristic of logarithms in the system for which the base is A, then for all positive values of the variables x and y we have

$$x = A^{\log x} \quad \text{and} \quad y = A^{\log y},$$

so that equation (3) becomes

$$\varphi\left(A^{\log x + \log y}\right) = \varphi\left(A^{\log x}\right) + \varphi\left(A^{\log y}\right).$$

Because in this last formula the variable quantities $\log x$ and $\log y$ admit any values, positive or negative, it follows [104] that we have, for all possible real values of x and y,

$$\varphi\left(A^{x+y}\right) = \varphi(A^x) + \varphi(A^y).$$

We conclude that [see problem I, eqn. (6)]

$$\varphi(A^x) = x\varphi\left(A^1\right) = x\varphi(A),$$

and consequently

$$\varphi\left(A^{\log x}\right) = \varphi(A) \log x,$$

or what amounts to the same thing

(11) $$\varphi(x) = \varphi(A)\log x.$$

It follows from formula (11) that every function $\varphi(x)$ that solves problem III is necessarily of the form

(12) $$\varphi(x) = a\log(x),$$

where a denotes a constant. Moreover, it is easy to assure ourselves: 1° that the constant a remains entirely arbitrary; and 2° that by choosing the number A suitably, which is itself arbitrary, we can reduce the constant a to one.

Problem IV. — *To determine the function $\varphi(x)$ in such a manner that it remains continuous between any two positive limits of the variable x and so that for all positive values of the variables x and y we have*

(4) $$\varphi(xy) = \varphi(x)\,\varphi(y).$$

Solution. — It would be easy to apply a method similar to that which we used to solve the second problem to the solution of problem IV. However, we will arrive more promptly at the solution we seek if we observe that, by denoting by log the characteristic of logarithms in the system for which the base is A, we can put equation (4) into the form

$$\varphi\left(A^{\log x+\log y}\right) = \varphi\left(A^{\log x}\right)\varphi\left(A^{\log y}\right).$$

Because in this last equation the variable quantities $\log x$ [105] and $\log y$ admit any values, positive or negative, it follows that we have, for all possible real values of the variables x and y,

$$\varphi\left(A^{x+y}\right) = \varphi\left(A^{x}\right)\varphi\left(A^{y}\right).$$

We conclude that [see problem II, eqn. (9)]

$$\varphi\left(A^{x}\right) = [\varphi(A)]^{x}$$

and consequently

$$\varphi\left(A^{\log x}\right) = [\varphi(A)]^{\log x} = x^{\log\varphi(A)},$$

or what amounts to the same thing,

(13) $$\varphi(x) = x^{\log\varphi(A)}.$$

It follows from equation (13) that any function $\varphi(x)$ that solves problem IV is necessarily of the form

(14) $$\varphi(x) = x^{a},$$

where a denotes a constant. Moreover it is easy to assure ourselves that this constant ought to remain entirely arbitrary.

The four values of $\varphi(x)$ which respectively satisfy equations (1), (2), (3) and (4), namely

$$ax, \quad A^x, \quad a\log x \quad \text{and} \quad x^a,$$

have this much in common, that each of them contains an arbitrary constant, a or A. Thus we ought to conclude that there is a great difference between the questions where it is a matter of calculating the unknown values of certain quantities and the questions in which we propose to discover the unknown nature of certain functions that have given properties. Indeed, in the first case, the values of unknown quantities are ultimately expressed by means of other known and determined quantities, while in the second case the unknown functions can, as we have seen here, admit arbitrary constants into their expression.

5.2 Research on a continuous function formed so that if we multiply two such functions together and then double the product, the result equals that function of the sum of the variables added to the same function of the difference of the variables.

[106] In each of the problems of the preceding section, the equation to be solved contained, along with the unknown function $\varphi(x)$, two other similar functions, namely $\varphi(y)$ and $\varphi(x+y)$ or $\varphi(xy)$. Now we are going to propose a new problem of the same kind, but in which the equation of the condition that the function $\varphi(x)$ must satisfy contains four such functions in place of three. It consists of the following:

Problem. — *To determine the function $\varphi(x)$ in such a manner that it remains continuous between any two real limits of the variable x and so that for all real values of the variables x and y we have*

(1) $$\varphi(y+x)+\varphi(y-x)=2\varphi(x)\varphi(y).$$

Solution. — If we make $x=0$ in equation (1), we get[1]

$$\varphi(0)=1.$$

The function $\varphi(x)$ thus reduces to 1 for the particular value $x=0$, and because we suppose that it is continuous between any limits, it is clear that, in the neighborhood of this particular value, it is only very slightly different from 1, and consequently is positive. Thus, by denoting a very small number by α, we can choose this number in such a way that the function $\varphi(x)$ remains constantly positive between the limits

[1] Cauchy does not mention the trivial solutions $\varphi(x)\equiv 0$ and $\varphi(x)\equiv 1$.

$$x=0 \quad \text{and} \quad x=\alpha.$$

Given this, two things could happen: either the positive value of $\varphi(\alpha)$ will be contained between the limits 0 and 1, or this value will be [107] greater than 1. We will examine successively these two hypotheses.

Now suppose that $\varphi(\alpha)$ has a value contained between the limits 0 and 1. We can represent this value by the cosine of a certain arc θ contained between the limits 0 and $\frac{\pi}{2}$, and as a consequence we can set

$$\varphi(\alpha)=\cos\theta.$$

Moreover, if equation (1) is put into the form

$$\varphi(y+x)=2\varphi(x)\varphi(y)-\varphi(y-x),$$

and we successively make

$$\begin{aligned} x&=\alpha \text{ and } y=\alpha,\\ x&=\alpha \text{ and } y=2\alpha,\\ x&=\alpha \text{ and } y=3\alpha,\\ &\ldots\ldots \qquad \ldots\ldots, \end{aligned}$$

then we deduce the formulas

$$\begin{aligned} \varphi(2\alpha)&=2\cos^2\theta-1=\cos 2\theta,\\ \varphi(3\alpha)&=2\cos\theta\cos 2\theta-\cos\theta=\cos 3\theta,\\ \varphi(4\alpha)&=2\cos\theta\cos 3\theta-\cos 2\theta=\cos 4\theta, \end{aligned}$$

one after another and in general,

$$\varphi(m\alpha)=2\cos\theta\cos(m-1)\theta-\cos(m-2)\theta=\cos m\theta,$$

where m denotes any integer number. I add that the formula

$$\varphi(m\alpha)=\cos m\theta$$

remains true even if we replace the integer number m by a fraction or even by any number μ. We will prove this easily as follows.

If we make $x=\frac{1}{2}\alpha$ and $y=\frac{1}{2}\alpha$ in equation (1), then we get

$$\left[\varphi\left(\frac{1}{2}\alpha\right)\right]^2=\frac{\varphi(0)+\varphi(\alpha)}{2}=\frac{1+\cos\theta}{2}=\left(\cos\frac{1}{2}\theta\right)^2.$$

Then, by taking the positive roots of both sides and [108] observing that the two functions $\varphi(x)$ and $\cos x$ remain positive, the first between the limits $x=0$ and $x=\alpha$ and the second between the limits $x=0$ and $x=\theta$, we find

$$\varphi\left(\frac{1}{2}\alpha\right) = \cos\frac{1}{2}\theta.$$

Likewise, if we make

$$x = \frac{1}{4}\alpha \quad \text{and} \quad y = \frac{1}{4}\theta$$

in equation (1), then we get[2]

$$\left[\varphi\left(\frac{1}{4}\alpha\right)\right]^2 = \frac{\varphi(0) + \varphi\left(\frac{1}{2}\alpha\right)}{2} = \frac{1 + \cos\frac{1}{2}\theta}{2} = \left(\cos\frac{1}{4}\theta\right)^2.$$

Then, by extracting the positive roots of the first and last parts, we get

$$\varphi\left(\frac{1}{4}\alpha\right) = \cos\frac{1}{4}\theta.$$

By similar reasoning, we successively obtain the formulas

$$\begin{aligned} \varphi\left(\frac{1}{8}\alpha\right) &= \cos\frac{1}{8}\theta, \\ \varphi\left(\frac{1}{16}\alpha\right) &= \cos\frac{1}{16}\theta, \\ \cdots\cdots\cdots & \cdots\cdots\cdots\cdots, \end{aligned}$$

and in general

$$\varphi\left(\frac{1}{2^n}\alpha\right) = \cos\frac{1}{2^n}\theta,$$

where n denotes any integer number. If we operate on the preceding expression for $\varphi\left(\frac{1}{2^n}\alpha\right)$ to deduce that for $\varphi\left(\frac{m}{2^n}\alpha\right)$ as we operated on the expression for $\varphi(\alpha)$ to deduce that for $\varphi(m\alpha)$, then we find

$$\varphi\left(\frac{m}{2^n}\alpha\right) = \cos\frac{m}{2^n}\theta.$$

Then, by supposing that the fraction $\frac{m}{2^n}$ varies in such a way as to approach [109]
indefinitely the number μ, and passing to the limit, we obtain the equation

$$\text{(2)} \qquad \varphi(\mu\alpha) = \cos\mu\theta.$$

Moreover, if we make[3]

$$x = \mu\alpha \quad \text{and} \quad y = 0$$

in formula (1), then we conclude that

---

[2] In [Cauchy 1897, p. 108], the numerator of the second part contains the expression $\varphi\left(\frac{1}{2}\right)\alpha$ in place of $\varphi\left(\frac{1}{2}\alpha\right)$. This error did not appear in [Cauchy 1821, p. 116]. (tr.)

[3] In [Cauchy 1897, p. 109], this reads $x = \mu a$. It is correctly written $x = \mu\alpha$ in [Cauchy 1821, p. 117]. (tr.)

$$\varphi(-\mu\alpha) = [2\varphi(0) - 1]\,\varphi(\mu\alpha) = \cos\mu\theta = \cos(-\mu\theta).$$

Thus, equation (2) remains true when we replace μ by $-\mu$. In other words, we have, for any values, positive or negative, of the variable x,

$$\varphi(\alpha x) = \cos\theta x. \tag{3}$$

If we change x to $\frac{x}{\alpha}$ in this last formula, we get[4]

$$\varphi(x) = \cos\frac{\theta}{\alpha}x = \cos\left(-\frac{\theta}{\alpha}x\right). \tag{4}$$

The preceding value of $\varphi(x)$ corresponds to the case where the positive quantity $\varphi(\alpha)$ remains contained between the limits 0 and 1.

Now let us suppose that this same quantity is greater than 1. It is easy to see that under this second hypothesis we can find a positive value of r that satisfies the equation

$$\varphi(\alpha) = \frac{1}{2}\left(r + \frac{1}{r}\right).$$

Indeed, it suffices to take

$$r = \varphi(\alpha) + \left\{[\varphi(\alpha)]^2 - 1\right\}^{\frac{1}{2}}.$$

Given this, if we successively make

$$\begin{aligned} x &= \alpha \text{ and } y = \alpha, \\ x &= \alpha \text{ and } y = 2\alpha, \\ x &= \alpha \text{ and } y = 3\alpha, \\ &\cdots\cdots \qquad \cdots\cdots, \end{aligned}$$

in equation (1), [110] then we deduce, one after another, the formulas

$$\begin{aligned} \varphi(2\alpha) &= \tfrac{1}{2}\left(r + \tfrac{1}{r}\right)^2 - 1 = \tfrac{1}{2}\left(r^2 + \tfrac{1}{r^2}\right), \\ \varphi(3\alpha) &= \tfrac{1}{2}\left(r + \tfrac{1}{r}\right)\left(r^2 + \tfrac{1}{r^2}\right) - \tfrac{1}{2}\left(r + \tfrac{1}{r}\right) = \tfrac{1}{2}\left(r^3 + \tfrac{1}{r^3}\right), \\ \varphi(4\alpha) &= \tfrac{1}{2}\left(r + \tfrac{1}{r}\right)\left(r^3 + \tfrac{1}{r^3}\right) - \tfrac{1}{2}\left(r^2 + \tfrac{1}{r^2}\right) = \tfrac{1}{2}\left(r^4 + \tfrac{1}{r^4}\right), \\ &\cdots\cdots\cdots\cdots\cdots\cdots\cdots\cdots\cdots\cdots\cdots\cdots, \end{aligned}$$

In general,

$$\varphi(m\alpha) = \frac{1}{2}\left(r + \frac{1}{r}\right)\left(r^{m-1} + \frac{1}{r^{m-1}}\right) - \frac{1}{2}\left(r^{m-2} + \frac{1}{r^{m-2}}\right)$$

[4] The trivial solution $\varphi(x) \equiv 1$ is included in equation (4) as the case $\theta = 0$.

$$= \frac{1}{2}\left(r^m + \frac{1}{r^m}\right),$$

where m denotes any integer number. I add that the formula

$$\varphi(m\alpha) = \frac{1}{2}\left(r^m + \frac{1}{r^m}\right)$$

remains true even if we replace the integer number m by a fraction or even by any number μ. We will prove this easily as follows.

If we make $x = \frac{1}{2}\alpha$ and $y = \frac{1}{2}\alpha$ in equation (1), we get

$$\left[\varphi\left(\frac{1}{2}\alpha\right)\right]^2 = \frac{\varphi(0) + \varphi(\alpha)}{2} = \frac{1 + \frac{1}{2}\left(r + \frac{1}{r}\right)}{2} = \frac{1}{4}\left(r^{\frac{1}{2}} + r^{-\frac{1}{2}}\right)^2.$$

Then, by taking the positive roots of both sides and observing that the function $\varphi(x)$ remains positive between the limits $x = 0$ and $x = \alpha$, we find

$$\varphi\left(\frac{1}{2}\alpha\right) = \frac{1}{2}\left(r^{\frac{1}{2}} + r^{-\frac{1}{2}}\right).$$

Likewise, if we make

$$x = \frac{1}{4}\alpha \quad \text{and} \quad y = \frac{1}{4}\alpha$$

in equation (1), then we get[5]

$$\left[\varphi\left(\frac{1}{4}\alpha\right)\right]^2 = \frac{\varphi(0) + \varphi\left(\frac{1}{2}\alpha\right)}{2}$$

$$= \frac{1 + \frac{1}{2}\left(r^{\frac{1}{2}} + r^{-\frac{1}{2}}\right)}{2} = \frac{1}{4}\left(r^{\frac{1}{4}} + r^{-\frac{1}{4}}\right)^2.$$

[111] Then, by taking the positive roots of the first and the last parts, we get

$$\varphi\left(\frac{1}{4}\alpha\right) = \frac{1}{2}\left(r^{\frac{1}{4}} + r^{-\frac{1}{4}}\right).$$

By similar reasoning, we successively obtain the formulas

$$\varphi\left(\frac{1}{8}\alpha\right) = \frac{1}{2}\left(r^{\frac{1}{8}} + r^{-\frac{1}{8}}\right),$$
$$\varphi\left(\frac{1}{16}\alpha\right) = \frac{1}{2}\left(r^{\frac{1}{16}} + r^{-\frac{1}{16}}\right),$$
$$\cdots\cdots\cdots\cdots\cdots\cdots,$$

[5] The negative signs were missing from the exponents $-\frac{1}{2}$ and $-\frac{1}{4}$ in [Cauchy 1897, p. 110]. They were present in [Cauchy 1821, p. 119]. (tr.)

and in general

$$\varphi\left(\frac{1}{2^n}\alpha\right) = \frac{1}{2}\left(r^{\frac{1}{2^n}} + r^{-\frac{1}{2^n}}\right),$$

where n denotes any integer number. If we operate on the preceding expression for $\varphi\left(\frac{1}{2^n}\alpha\right)$ to deduce that for $\varphi\left(\frac{m}{2^n}\alpha\right)$ as we operated on the expression for $\varphi(\alpha)$ to deduce that for[6] $\varphi(m\alpha)$, then we find

$$\varphi\left(\frac{m}{2^n}\alpha\right) = \frac{1}{2}\left(r^{\frac{m}{2^n}} + r^{-\frac{m}{2^n}}\right).$$

Then, by supposing that the fraction $\frac{m}{2^n}$ varies in such a way as to approach indefinitely the number μ, and passing to the limit, we obtain the equation

(5) $$\varphi(\mu\alpha) = \frac{1}{2}\left(r^{\mu} + r^{-\mu}\right).$$

Moreover, if we make

$$x = \mu\alpha \quad \text{and} \quad y = 0$$

in formula (1), then we conclude that

$$\varphi(-\mu\alpha) = [2\varphi(0) - 1]\,\varphi(\mu\alpha) = \frac{1}{2}\left(r^{-\mu} + r^{\mu}\right).$$

[112] Thus, equation (5) remains true when we replace μ by $-\mu$. In other words, we have, for all values, positive or negative, of the variable x,

(6) $$\varphi(\alpha x) = \frac{1}{2}\left(r^{x} + r^{-x}\right).$$

If we change x to $\frac{x}{\alpha}$ in this last formula, we get

(7) $$\varphi(x) = \frac{1}{2}\left(r^{\frac{x}{\alpha}} + r^{-\frac{x}{\alpha}}\right).$$

When we make $\pm\frac{\theta}{\alpha} = a$ in equation (4) and $r^{\pm\frac{1}{\alpha}} = A$ in equation (7), these equations give, respectively, the following forms:

(8) $$\varphi(x) = \cos ax \quad \text{and}$$

(9) $$\varphi(x) = \frac{1}{2}\left(A^{x} + A^{-x}\right).$$

Thus, if we denote a constant quantity by a and a constant number by A, then any function $\varphi(x)$ that remains continuous between any limits of the variable and that satisfies equation (1) is necessarily contained in one of the two forms that we have just described. Moreover, it is easy to assure ourselves that the values of $\varphi(x)$ given

[6] This word "for" is the translation of the word "de," which was present in [Cauchy 1821, p. 120], but absent from [Cauchy 1897, p. 111]. (tr.)

by equations (8) and (9) solve the proposed question, whatever values may be attributed to the quantity a and the number A. This number and this quantity are two arbitrary constants, of which one admits only positive quantities.

From what we have just said, the two functions[7]

$$\cos ax \quad \text{and} \quad \frac{1}{2}\left(A^x + A^{-x}\right)$$

have the common property of satisfying equation (1), and this establishes a remarkable analogy between them. Both of these two [113] functions still reduce to one for $x = 0$. But one essential difference between the first and the second is that the numerical value of the first is constantly less than the limit 1, whenever it does not reach this limit, while, under the same hypothesis, the numerical value of the second is constantly above the limit 1.

[7] Using modern notation, we observe that the second solution may be written as $\cosh ax$, where $a = \ln A$. Lambert (1728–1777) was the first to note such parallels between the trigonometric and hyperbolic functions in [Lambert 1768].

Chapter 6
On convergent and divergent series. Rules for the convergence of series. The summation of several convergent series.

6.1 General considerations on series.

[114][1] We call a *series* an indefinite sequence of quantities,

$$u_0,\ u_1,\ u_2,\ u_3,\ \ldots,$$

which follow from one to another according to a determined law. These quantities themselves are the various *terms* of the series under consideration. Let

$$s_n = u_0 + u_1 + u_2 + \ldots + u_{n-1}$$

be the sum of the first n terms, where n denotes any integer number. If, for ever increasing values of n, the sum s_n indefinitely approaches a certain limit s, the series is said to be *convergent*, and the limit in question is called the *sum* of the series. On the contrary, if the sum s_n does not approach any fixed limit as n increases indefinitely, the series is *divergent*, and does not have a sum. In either case, the term which corresponds to the index n, that is u_n, is what we call the *general term*. For the series to be completely determined, it is enough that we give this general term as a function of the index n.

One of the simplest series is the geometric progression,

$$1,\ x,\ x^2,\ x^3,\ \ldots,$$

which has x^n for its general term, that is to say the nth power of the quantity [115] x. If we form the sum of the first n terms of this series, then we find

$$1 + x + x^2 + \ldots + x^{n-1} = \frac{1}{1-x} - \frac{x^n}{1-x}.$$

[1] Both [Cauchy 1821, p. ix] and [Cauchy 1897, p. 473] use the title "On convergent and divergent (real) series. ..." in the table of contents. (tr.)

R.E. Bradley, C.E. Sandifer, *Cauchy's Cours d'analyse*, Sources and Studies in the History of Mathematics and Physical Sciences, DOI 10.1007/978-1-4419-0549-9_6,

As the values of n increase, the numerical value of the fraction $\frac{x^n}{1-x}$ converges towards the limit zero, or increases beyond all limits, according to whether we suppose that the numerical value of x is less than or greater than 1. Under the first hypothesis, we ought to conclude that the progression

$$1,\ x,\ x^2,\ x^3,\ \ldots$$

is a convergent series which has $\frac{1}{1-x}$ as its sum, whereas, under the second hypothesis, the same progression is a divergent series which does not have a sum.

Following the principles established above, in order that the series

$$u_0,\ u_1,\ u_2,\ \ldots,\ u_n,\ u_{n+1},\ \ldots \tag{1}$$

be convergent, it is necessary and it suffices that increasing values of n make the sum

$$s_n = u_0 + u_1 + u_2 + \ldots + u_{n-1}$$

converge indefinitely towards a fixed limit s. In other words, it is necessary and it suffices that, for infinitely large values of the number n, the sums

$$s_n,\ s_{n+1},\ s_{n+2},\ \ldots$$

differ from the limit s, and consequently from one another, by infinitely small quantities. Moreover, the successive differences between the first sum s_n and each of the following sums are determined, respectively, by the equations

$$\begin{aligned}
s_{n+1} - s_n &= u_n,\\
s_{n+2} - s_n &= u_n + u_{n+1},\\
s_{n+3} - s_n &= u_n + u_{n+1} + u_{n+2},\\
&\ldots\ldots\ldots\ldots\ldots\ldots
\end{aligned}$$

Hence, in order for series (1) to be convergent, it is first of all necessary [116] that the general term u_n decrease indefinitely as n increases. But this condition does not suffice, and it is also necessary that, for increasing values of n, the different sums,

$$\begin{aligned}
&u_n + u_{n+1},\\
&u_n + u_{n+1} + u_{n+2},\\
&\ldots\ldots\ldots\ldots\ldots\ldots,
\end{aligned}$$

that is to say, the sums of as many of the quantities

$$u_n,\ u_{n+1},\ u_{n+2},\ \ldots,$$

as we may wish, beginning with the first one, eventually constantly assume numerical values less than any assignable limit. Conversely, whenever these various conditions are fulfilled, the convergence of the series is guaranteed.[2]

Let us take, for example, the geometric progression

$$1,\ x,\ x^2,\ x^3,\ \ldots. \tag{2}$$

If the numerical value of x is greater than 1, that of the general term x^n increases indefinitely with n, and this remark alone suffices to establish the divergence of the series. The series is still divergent if we let $x = \pm 1$, because the numerical value of the general term x^n, which is 1, does not decrease indefinitely for increasing values of n. However, if the numerical value of x is less than 1, then the sums of any number of terms of the series, beginning with x^n, namely:

$$\begin{aligned} &x^n, \\ x^n + x^{n+1} &= x^n \frac{1-x^2}{1-x}, \\ x^n + x^{n+1} + x^{n+2} &= x^n \frac{1-x^3}{1-x}, \\ \ldots\ldots\ldots\ldots\ldots\ldots &\ldots\ldots\ldots\ldots, \end{aligned}$$

are all contained between the limits

$$x^n \quad \text{and} \quad \frac{x^n}{1-x},$$

[117] each of which becomes infinitely small for infinitely large values of n. Consequently, the series is convergent, as we already knew.

As a second example, let us take the numerical series

$$1,\ \frac{1}{2},\ \frac{1}{3},\ \frac{1}{4},\ \ldots,\ \frac{1}{n},\ \frac{1}{n+1},\ \ldots. \tag{3}$$

The general term of this series, namely $\frac{1}{n+1}$, decreases indefinitely as n increases. Nevertheless, the series is not convergent, because the sum of the terms from $\frac{1}{n+1}$ up to $\frac{1}{2n}$ inclusive, namely

$$\frac{1}{n+1} + \frac{1}{n+2} + \ldots + \frac{1}{2n-1} + \frac{1}{2n},$$

is always greater than the product

$$n\frac{1}{2n} = \frac{1}{2},$$

[2] This is the Cauchy Convergence Criterion. It is still one of the few necessary and sufficient conditions for convergence of series.

whatever the value of n. As a consequence, this sum does not decrease indefinitely with increasing values of n, as would be the case if the series were convergent. Let us add that, if we denote the sum of the first n terms of series (3) by s_n and the highest power of 2 bounded by $n+1$ by 2^m, then we have

$$\begin{aligned} s_n &= 1+\frac{1}{2}+\frac{1}{3}+\ldots+\frac{1}{n+1} \\ &> 1+\frac{1}{2}+\left(\frac{1}{3}+\frac{1}{4}\right)+\left(\frac{1}{5}+\frac{1}{6}+\frac{1}{7}+\frac{1}{8}\right)+\ldots \\ &\quad +\left(\frac{1}{2^{m-1}+1}+\frac{1}{2^{m-1}+2}+\ldots+\frac{1}{2^m}\right), \end{aligned}$$

and, *a fortiori*,

$$s_n > 1+\frac{1}{2}+\frac{1}{2}+\frac{1}{2}+\ldots+\frac{1}{2} = 1+\frac{m}{2}.$$

We conclude from this that the sum s_n increases indefinitely with the integer number m, and consequently with n, which is a new proof of the divergence of the series.[3]

[118] Let us further consider the numerical series

$$(4) \qquad 1, \frac{1}{1}, \frac{1}{1\cdot 2}, \frac{1}{1\cdot 2\cdot 3}, \ldots, \frac{1}{1\cdot 2\cdot 3\ldots n}, \ldots.$$

The terms of this series with index greater than n, namely

$$\frac{1}{1\cdot 2\cdot 3\ldots n}, \frac{1}{1\cdot 2\cdot 3\ldots n(n+1)}, \frac{1}{1\cdot 2\cdot 3\ldots n(n+1)(n+2)}, \ldots,$$

are, respectively, less than the corresponding terms of the geometric progression

$$\frac{1}{1\cdot 2\cdot 3\ldots n}, \frac{1}{1\cdot 2\cdot 3\ldots n}\frac{1}{n}, \frac{1}{1\cdot 2\cdot 3\ldots n}\frac{1}{n^2}, \ldots.$$

As a consequence, the sum of however many of the initial terms as we may wish is always less than the sum of the corresponding terms of the geometric progression, which is a convergent series, and so *a fortiori*,[4] it is less than the sum of this series, which is to say

$$\frac{1}{1\cdot 2\cdot 3\ldots n}\frac{1}{1-\frac{1}{n}} = \frac{1}{1\cdot 2\cdot 3\ldots (n-1)}\frac{1}{n-1}.$$

Because this last sum decreases indefinitely as n increases, it follows that series (4) is itself convergent. It is conventional to denote the sum of this series by the letter e. By adding together the first n terms, we obtain an approximate value of the number e,

[3] Cauchy may not be claiming originality for this "new" proof. It was first given by Oresme (see, for example, [Dunham 1990, pp. 202–203]), but Cauchy was probably not aware of it.

[4] This is an implicit use of the Comparison Test. Cauchy never states this test explicitly.

$$1+\frac{1}{1}+\frac{1}{1\cdot 2}+\frac{1}{1\cdot 2\cdot 3}+\ldots+\frac{1}{1\cdot 2\cdot 3\ldots(n-1)}.$$

According to what we have just said, the error made will be smaller than the product of the nth term by $\frac{1}{n-1}$. Therefore, for example, if we let $n=11$, we find as the approximate value of e

(5) $$e=2.7182818\ldots,$$

and the error made in this case is less than the product [119] of the fraction $\frac{1}{1\cdot2\cdot3\cdot4\cdot5\cdot6\cdot7\cdot8\cdot9\cdot10}$ by $\frac{1}{10}$, that is $\frac{1}{36{,}288{,}000}$, so that it does not affect the seventh decimal place.

The number e, determined as we have just said, is often used in the summation of series and in the infinitesimal Calculus. Logarithms taken in the system with this number as its base are called *Napierian*, for *Napier*, the inventor of logarithms, or *hyperbolic*, because they measure the various parts of the area between the equilateral hyperbola and its asymptotes.[5]

In general, we denote the sum of a convergent series by the sum of the first terms, followed by an ellipsis. Thus, when the series

$$u_0,\ u_1,\ u_2,\ u_3,\ \ldots$$

is convergent, the sum of this series is denoted

$$u_0+u_1+u_2+u_3+\ldots.$$

By virtue of this convention, the value of the number e is determined by the equation

(6) $$e=1+\frac{1}{1}+\frac{1}{1\cdot 2}+\frac{1}{1\cdot 2\cdot 3}+\frac{1}{1\cdot 2\cdot 3\cdot 4}+\ldots,$$

and, if one considers the geometric progression

$$1,\ x,\ x^2,\ x^3,\ \ldots,$$

we have, for numerical values of x less than 1,

(7) $$1+x+x^2+x^3+\ldots=\frac{1}{1-x}.$$

Denoting the sum of the convergent series

$$u_0,\ u_1,\ u_2,\ u_3,\ \ldots$$

by s and the sum of the first n terms by s_n, we have

$$\begin{aligned} s &= u_0+u_1+u_2+\ldots+u_{n-1}+u_n+u_{n+1}+\ldots \\ &= s_n+u_n+u_{n+1}+\ldots, \end{aligned}$$

[5] I.e., the area under the curve $y=\frac{1}{x}$.

and, as a consequence,

$$s - s_n = u_n + u_{n+1} + \dots.$$

[120] From this last equation, it follows that the quantities

$$u_n,\ u_{n+1},\ u_{n+2},\ \dots$$

form a new convergent series, the sum of which is equal to $s - s_n$. If we represent this sum by r_n, we have

$$s = s_n + r_n,$$

and r_n is called the *remainder* of series (1) beginning from the nth term.

Suppose the terms of series (1) involve some variable x. If the series is convergent and its various terms are continuous functions of x in a neighborhood of some particular value of this variable, then

$$s_n,\ r_n \text{ and } s$$

are also three functions of the variable x, the first of which is obviously continuous with repect to x in a neighborhood of the particular value in question. Given this, let us consider the increments in these three functions when we increase x by an infinitely small quantity α. For all possible values of n, the increment in s_n is an infinitely small quantity. The increment of r_n, as well as r_n itself, becomes infinitely small for very large values of n. Consequently, the increment in the function s must be infinitely small.[6] From this remark, we immediately deduce the following proposition:

Theorem I. — *When the various terms of series* (1) *are functions of the same variable x, continuous with respect to this variable in the neighborhood of a particular value for which the series converges, the sum s of the series is also a continuous function of x in the neighborhood of this particular value.*[7]

By virtue of this theorem, the sum of series (2) must be a continuous function of the variable x between the limits $x = -1$ and $x = 1$, [121] as we may verify by considering the values of s given by the equation

$$s = \frac{1}{1-x}.$$

6.2 On series for which all the terms are positive.

Whenever all the terms of the series

[6] This passage is quoted in [Lützen 2003, p. 168].

[7] This theorem as stated is incorrect. If we impose the additional condition of uniform convergence on the functions s_n, then it does hold. This theorem is controversial. Some have argued that Cauchy really had uniform convergence in mind. See [Lützen 2003, pp. 168–169] for further discussion.

(1) $$u_0,\ u_1,\ u_2,\ \ldots,\ u_n,\ \ldots$$

are positive, we may usually decide whether it is convergent or divergent by using the following theorem:

Theorem I.[8] — *Consider the limit or limits towards which the expression* $(u_n)^{\frac{1}{n}}$ *converges as n increases indefinitely, and let k denote the largest of these limits, or in other words, the limit of the largest values of the expression in question. Series* (1) *converges whenever* $k < 1$ *and diverges whenever* $k > 1$.

Proof. — First of all, suppose that $k < 1$ and choose an arbitrary third number U between the two numbers 1 and k, so that we have

$$k < U < 1.$$

As n increases beyond assignable limit, the largest values of $(u_n)^{\frac{1}{n}}$ cannot approach indefinitely the limit k without eventually being constantly less than U. Consequently, it is possible to assign an integer value to n large enough so that when n is greater than or equal to this value, we constantly have[9]

$$(u_n)^{\frac{1}{n}} < U, \quad \text{or} \quad u_n < U^n.$$

It follows that the terms of the series

$$u_0,\ u_1,\ u_2,\ \ldots,\ u_{n+1},\ u_{n+2},\ \ldots$$

[122] are eventually always smaller than the corresponding terms of the geometric progression

$$1,\ U,\ U^2,\ \ldots,\ U^n,\ U^{n+1},\ U^{n+2},\ \ldots.$$

As this progression is convergent (because $U < 1$) we may, by the previous remark, conclude *a fortiori* the convergence of series (1).

On the other hand, suppose that $k > 1$ and again pick a third number U between the two numbers 1 and k, so that we have

$$k > U > 1.$$

As n increases without limit, the largest values of $(u_n)^{\frac{1}{n}}$ in approaching k indefinitely eventually become greater than U. We may therefore satisfy the condition

$$(u_n)^{\frac{1}{n}} > U$$

or, what amounts to the same thing, the following condition

---

[8] This theorem is now known as the Root Test. It is cited as the definition of upper and lower limits in [DSB Cauchy, p. 136].

[9] In [Cauchy 1897, p. 121], the subscript is missing in the term $(u_n)^{\frac{1}{n}}$. It is present in [Cauchy 1821, p. 133]. (tr.)

$$u_n > U^n,$$

for values of n as large as we might wish. As a consequence, we find in the series

$$u_0,\ u_1,\ u_2,\ \ldots,\ u_n,\ u_{n+1},\ u_{n+2},\ \ldots$$

an indefinite number of terms greater than the corresponding terms of the geometric progression

$$1,\ U,\ U^2,\ \ldots,\ U^n,\ U^{n+1},\ U^{n+2},\ \ldots.$$

As this progression is divergent (because $U > 1$) and, as a consequence its various terms increase to infinity, the remark that we have just made suffices to establish the divergence of series (1).

In a great number of cases we may determine the values of the quantity k with the assistance of theorem IV (Chap. II, § III). Indeed, [123] by virtue of this theorem, any time the ratio $\frac{u_{n+1}}{u_n}$ converges towards a fixed limit, that limit is precisely the value of k. We may therefore state the following proposition:

Theorem II.[10] — *If, for increasing values of n, the ratio*

$$\frac{u_{n+1}}{u_n}$$

converges towards a fixed limit k, series (1) *converges whenever* $k < 1$ *and diverges whenever* $k > 1$.

For example, if we consider the series

$$1,\ \frac{1}{1},\ \frac{1}{1\cdot 2},\ \frac{1}{1\cdot 2\cdot 3},\ \ldots,\ \frac{1}{1\cdot 2\cdot 3\ldots n},\ \ldots,$$

then we find

$$\frac{u_{n+1}}{u_n} = \frac{1\cdot 2\cdot 3\ldots n}{1\cdot 2\cdot 3\ldots n(n+1)} = \frac{1}{n+1}, \quad \text{so } k = \frac{1}{\infty} = 0,$$

and consequently the series is convergent, as we already knew.

The first of the two theorems that we have just established leaves no doubt about the convergence or divergence of a series whose terms are positive, except in the particular case where the quantity k becomes equal to one. In this particular case, it is not always easy to answer the question of convergence. However, we will now prove two new propositions, which frequently help us to decide the issue.

Theorem III.[11] — *Whenever each term of series* (1) *is smaller than the one preceding it, that series and the following one*

---

[10] This is the Ratio Test; see [DSB Cauchy, p. 136].

[11] This is known as the Cauchy Condensation Test.

(2) $$u_0,\ 2u_1,\ 4u_3,\ 8u_7,\ 16u_{15},\ldots$$

are either both convergent or both divergent.

Proof. — First of all, suppose that series (1) is convergent and [124] let s denote its sum. Then

$$\begin{aligned} u_0 &= u_0, \\ 2u_1 &= 2u_1, \\ 4u_3 &< 2u_2+2u_3, \\ 8u_7 &< 2u_4+2u_5+2u_6+2u_7, \\ \ldots\ldots &\ldots\ldots\ldots\ldots\ldots\ldots\ldots, \end{aligned}$$

and consequently, the sum of as many of the terms of series (2) as we may wish is smaller than

$$u_0+2u_1+2u_2+2u_3+2u_4+\ldots=2s-u_0.$$

It follows that series (2) converges.

On the other hand, suppose that series (1) diverges. The sum of its terms, taken in great number, eventually surpasses any assignable limit. Because we have

$$\begin{aligned} u_0 &= u_0, \\ 2u_1 &> u_1+u_2, \\ 4u_3 &> u_3+u_4+u_5+u_6, \\ 8u_7 &> u_7+u_8+u_9+u_{10}+u_{11}+u_{12}+u_{13}+u_{14}, \\ \ldots\ldots &\ldots\ldots\ldots\ldots\ldots\ldots\ldots\ldots\ldots\ldots\ldots\ldots, \end{aligned}$$

we must conclude that the sum of the quantities

$$u_0,\ 2u_1,\ 4u_3,\ 8u_7,\ \ldots,$$

taken in great number, is itself eventually greater than any given quantity. Series (2) is therefore divergent, conforming to the stated theorem.

Corollary. — Let μ be any quantity. If series (1) is

(3) $$1,\ \frac{1}{2^\mu},\ \frac{1}{3^\mu},\ \frac{1}{4^\mu},\ \ldots,$$

then series (2) becomes

$$1,\ 2^{1-\mu},\ 4^{1-\mu},\ 8^{1-\mu},\ \ldots.$$

[125] This last series is a geometric progression, convergent whenever we have $\mu > 1$ and divergent in the opposite case. As a consequence, series (3) is itself convergent if μ is a number greater than 1, and divergent if $\mu = 1$ or $\mu < 1$. For example, of the three series

(4) $$1, \frac{1}{2^2}, \frac{1}{3^2}, \frac{1}{4^2}, \ldots,$$

(5) $$1, \frac{1}{2}, \frac{1}{3}, \frac{1}{4}, \ldots,$$

(6) $$1, \frac{1}{2^{\frac{1}{2}}}, \frac{1}{3^{\frac{1}{2}}}, \frac{1}{4^{\frac{1}{2}}}, \ldots,$$

the first is convergent and the other two divergent.

Theorem IV.[12] — *Suppose that* log *denotes the characteristic of the logarithm in any system and that the ratio*

$$\frac{\log(u_n)}{\log\left(\frac{1}{n}\right)}$$

converges towards a finite limit h for increasing values of n. Series (1) *is convergent if $h > 1$ and divergent if $h < 1$.*

Proof. — First of all, suppose $h > 1$ and choose any third quantity a between the two quantities 1 and h, so that we have

$$h > a > 1.$$

The ratio $\frac{\log(u_n)}{\log\left(\frac{1}{n}\right)}$, or its equivalent

$$\frac{\log\left(\frac{1}{u_n}\right)}{\log(n)},$$

eventually, for very large values of n, is constantly greater than a. In other words, if n increases beyond [126] a certain limit, we always have

$$\frac{\log\left(\frac{1}{u_n}\right)}{\log(n)} > a,$$

or what amounts to the same thing,

$$\log\left(\frac{1}{u_n}\right) > a\log(n),$$

and, as a consequence,

$$\frac{1}{u_n} > n^a, \quad \text{so} \quad u_n < \frac{1}{n^a}.$$

It follows that the terms of series (1) eventually are constantly smaller than the corresponding terms of the following series

[12] This is the Logarithmic Convergence Test.

$$1, \frac{1}{2^a}, \frac{1}{3^a}, \frac{1}{4^a}, \ldots, \frac{1}{n^a}, \frac{1}{(n+1)^a}, \ldots.$$

As this last series is convergent (because $a > 1$), we may, by the previous remark, conclude *a fortiori* the convergence of series (1).

On the other hand, suppose that $h < 1$, and again pick a third quantity a between 1 and h, so that we have

$$h < a < 1.$$

Eventually, for very large values of n, we constantly have

$$\frac{\log\left(\frac{1}{u_n}\right)}{\log(n)} < a,$$

or what amounts to the same thing,

$$\log\left(\frac{1}{u_n}\right) < a\log(n),$$

and, as a consequence,

$$\frac{1}{u_n} < n^a, \quad \text{so} \quad u_n > \frac{1}{n^a}.$$

It follows that the terms of series (1) eventually are constantly [127] greater than the corresponding terms of the following series

$$1, \frac{1}{2^a}, \frac{1}{3^a}, \frac{1}{4^a}, \ldots, \frac{1}{n^a}, \frac{1}{(n+1)^a}, \ldots.$$

As this last series is convergent (because $a < 1$), we may, by the remark we have just made, conclude *a fortiori* the divergence of series (1).

Given two convergent series, the terms of which are positive, we may, by adding or multiplying these same terms, form a new series, the sum of which results from the addition or the multiplication of the sums of the first two. On this subject, we establish the two following theorems:

Theorem V. — *Let*

$$(7) \qquad \begin{cases} u_0, \ u_1, \ u_2, \ \ldots, \ u_n, \ \ldots, \\ v_0, \ v_1, \ v_2, \ \ldots, \ v_n, \ \ldots \end{cases}$$

be two convergent series composed only of positive terms, having s and s', respectively, as sums. Then

$$(8) \qquad u_0+v_0, \ u_1+v_1, \ u_2+v_2, \ \ldots, \ u_n+v_n, \ \ldots$$

is a new convergent series, which has $s+s'$ as its sum.

Proof. — If we let

$$\begin{aligned} s_n &= u_0 + u_1 + u_2 + \ldots + u_{n-1} \quad \text{and} \\ s'_n &= v_0 + v_1 + v_2 + \ldots + v_{n-1}, \end{aligned}$$

then s_n and s'_n converge, for increasing values of n, towards the limits s and s', respectively. As a consequence, $s_n + s'_n$, that is the sum of the first n terms of series (8), converges towards the limit $s + s'$, which suffices to establish the stated theorem.

Theorem VI. — *Under the same hypotheses as the previous theorem,*

$$(9) \qquad \begin{cases} u_0 v_0,\ u_0 v_1 + u_1 v_0,\ u_0 v_2 + u_1 v_1 + u_2 v_0,\ \ldots \\ \ldots,\ u_0 v_n + u_1 v_{n-1} + \ldots + u_{n-1} v_1 + u_n v_0,\ \ldots \end{cases}$$

is a new convergent series, which has ss' as its sum.

[128] *Proof.* — Once again, let s_n and s'_n be the sums of the first n terms of the two series (7), and additionally denote the sum of the first n terms of series (9) by s''_n. If we denote by m the greatest integer included in $\frac{n-1}{2}$, that is to say $\frac{n-1}{2}$ when n is odd and $\frac{n-2}{2}$ otherwise, we clearly have[13]

$$\begin{aligned} u_0 v_0 + (u_0 v_1 + u_1 v_0) + \ldots + (u_0 v_{n-1} + u_1 v_{n-2} + \ldots + u_{n-2} v_1 + u_{n-1} v_0) \\ < (u_0 + u_1 + \ldots + u_{n-1})(v_0 + v_1 + \ldots + v_{n-1}) \end{aligned}$$

and

$$> (u_0 + u_1 + \ldots + u_m)(v_0 + v_1 + \ldots + v_m).$$

In other words,

$$s''_n < s_n s'_n \quad \text{and} \quad > s_{m+1} s'_{m+1}.$$

Now suppose that we make n increase beyond all limit. The number

$$m = \frac{n - \frac{3}{2} \pm \frac{1}{2}}{2}$$

itself increases indefinitely, and the two sums s_n and s_{m+1} converge towards the limit s, while s'_n and s'_{m+1} converge towards the limit s'. As a consequence, the two products $s_n s'_n$ and $s_{m+1} s'_{m+1}$, as well as the sum s''_n contained between these two products, converge towards the limit ss', which suffices to establish theorem VI.[14]

6.3 On series which contain positive terms and negative terms.

Suppose that the series

[13] The left-hand side of this inequality contained some subscripting errors in [Cauchy 1821, p. 141], which were not included in the *Errata* of that edition. These were corrected in [Cauchy 1897, p. 128].

[14] This is another implicit application of the Squeeze Theorem.

(1) $$u_0, u_1, u_2, \ldots, u_n, \ldots$$

is composed of terms that are sometimes positive and sometimes negative, and let

(2) $$\rho_0, \rho_1, \rho_2, \ldots, \rho_n, \ldots$$

[129] be, respectively, the numerical values of these same terms, so that we have

$$u_0 = \pm\rho_0,\ u_1 = \pm\rho_1,\ u_2 = \pm\rho_2,\ \ldots,\ u_n = \pm\rho_n,\ \ldots.$$

The numerical value of the sum

$$u_0 + u_1 + u_2 + \ldots + u_{n-1}$$

will never surpass[15]

$$\rho_0 + \rho_1 + \rho_2 + \ldots + \rho_{n-1},$$

so it follows that the convergence of series (2) always entails that of series (1).[16] We ought to add that series (1) is divergent if some terms of series (2) eventually increase beyond all assignable limit. This latter case occurs whenever the greatest values of $(\rho_n)^{\frac{1}{n}}$ converge towards a limit greater than 1, for increasing values of n. On the other hand, whenever this limit is less than 1, series (2) is always convergent. As a consequence, we may state the following theorem:

Theorem I.[17] — *Let ρ_n be the numerical value of the general term u_n of series* (1), *and let k denote the limit towards which the largest values of the expression $(\rho_n)^{\frac{1}{n}}$ converge as n increases indefinitely. Series* (1) *is convergent if we have $k < 1$ and divergent if we have $k > 1$.*

Whenever the fraction $\frac{\rho_{n+1}}{\rho_n}$, that is, the numerical value of the ratio $\frac{u_{n+1}}{u_n}$, converges towards a fixed limit, then by virtue of theorem IV (Chap. II, § III), this limit is the desired value of k. This remark brings us to the proposition which I will now state:

Theorem II.[18] — *If the numerical value of the ratio*

$$\frac{u_{n+1}}{u_n}$$

converges towards a fixed limit k for increasing values of n, then series 1 *is convergent whenever we have $k < 1$ and divergent whenever we have $k > 1$.*

[130] For example, if we consider the series

---

[15] Here Cauchy makes an implicit use of the generalized triangle inequality.

[16] Cauchy does not define absolute convergence, but has essentially shown here that absolute convergence implies convergence.

[17] This is another application of the Root Test.

[18] This is another application of the Ratio Test.

$$1,\ -\frac{1}{1},\ +\frac{1}{1\cdot 2},\ -\frac{1}{1\cdot 2\cdot 3},\ +\ldots,$$

we find that

$$\frac{u_{n+1}}{u_n} = -\frac{1}{n+1}, \quad \text{so that} \quad k = \frac{1}{\infty} = 0,$$

from which it follows that the series is convergent.

The first of the two theorems we have just established leaves no doubt about the convergence or divergence of a particular series, except in the particular case where the quantity denoted by k becomes equal to one. In this particular case, we may often establish the convergence of the given series either by verifying that the numerical values of the various terms form a convergent series or by means of the following theorem:

Theorem III.[19] — *If the numerical value of the general term u_n in series* (1) *decreases constantly and indefinitely for increasing values of n, and if further the different terms are alternately positive and negative, then the series converges.*

For example, consider the series

$$(3) \qquad 1,\ -\frac{1}{2},\ +\frac{1}{3},\ -\frac{1}{4},\ +\ldots\pm\frac{1}{n},\ \mp\frac{1}{n+1},\ \ldots.$$

The sum of the terms whose index is greater than n, if we suppose them to be m in number, is

$$\pm\left(\frac{1}{n+1} - \frac{1}{n+2} + \frac{1}{n+3} - \frac{1}{n+4} + \ldots \pm \frac{1}{n+m}\right).$$

Now the numerical value of this sum, namely

$$\begin{aligned}
&\frac{1}{n+1} - \frac{1}{n+2} + \frac{1}{n+3} - \frac{1}{n+4} + \ldots \pm \frac{1}{n+m} \\
&\quad = \frac{1}{n+1} - \left(\frac{1}{n+2} - \frac{1}{n+3}\right) - \left(\frac{1}{n+4} - \frac{1}{n+5}\right) - \ldots \\
&\quad = \left(\frac{1}{n+1} - \frac{1}{n+2}\right) + \left(\frac{1}{n+3} - \frac{1}{n+4}\right) \\
&\qquad + \left(\frac{1}{n+5} - \frac{1}{n+6}\right) + \ldots,
\end{aligned}$$

[131] because it is obviously contained between

$$\frac{1}{n+1} \quad \text{and} \quad \frac{1}{n+1} - \frac{1}{n+2},$$

decreases indefinitely for increasing values of n, whatever the value of m, which suffices to establish the convergence of the given series. The same arguments may

[19] This is the Alternating Series Test.

obviously be applied to any series of this kind. I will cite, for example, the following

$$1,\ -\frac{1}{2^\mu},\ +\frac{1}{3^\mu},\ -\frac{1}{4^\mu},\ \ldots, \tag{4}$$

which remains convergent for all positive values of μ, by virtue of theorem III.

If we suppress the – sign preceding each term of even index in series (4), we obtain series (3) of section II, which is divergent whenever we have $\mu = 1$ or $\mu < 1$. As a consequence, to transform a convergent series into a divergent series, or vice versa, it sometimes suffices to change the sign of certain terms. Moreover, this remark applies exclusively to series for which the quantity denoted by k in theorem II reduces to 1.

Given a convergent series, the terms of which are positive, we can only augment the convergence by diminishing the numerical values of the same terms and changing the signs of some of them. It is worth noting that we produce this double effect if we multiply each term by a sine or by a cosine, and this observation suffices to establish the following proposition:

Theorem IV.[20] — *When the series*

$$\rho_0,\ \rho_1,\ \rho_2,\ \ldots,\ \rho_n,\ \ldots, \tag{2}$$

made up entirely of positive terms, is convergent, then each of the following

$$\begin{cases} \rho_0\cos\theta_0,\ \rho_1\cos\theta_1,\ \rho_2\cos\theta_2,\ \ldots,\ \rho_n\cos\theta_n,\ \ldots, \\ \rho_0\sin\theta_0,\ \rho_1\sin\theta_1,\ \rho_2\sin\theta_2,\ \ldots,\ \rho_n\sin\theta_n,\ \ldots \end{cases} \tag{5}$$

[132] *is also convergent, whatever the values of the arcs* $\theta_0,\ \theta_1,\ \theta_2,\ \ldots,\ \theta_n,\ \ldots$.

Corollary. — If we suppose in general that

$$\theta_n = n\theta,$$

where θ denotes an arbitrary arc, then the two series in (5) become, respectively,

$$\begin{cases} \rho_0,\ \rho_1\cos\theta,\ \rho_2\cos 2\theta,\ \ldots,\ \rho_n\cos n\theta,\ \ldots, \\ \qquad \rho_1\sin\theta,\ \rho_2\sin 2\theta,\ \ldots,\ \rho_n\sin n\theta,\ \ldots. \end{cases} \tag{6}$$

These last two series will therefore always be convergent whenever series (2) is convergent.

If we consider two series at the same time, both of which include positive terms and negative terms, we easily prove theorems V and VI of §II about them, as we will now see.

Theorem V. — *Let*

[20] This is another implicit application of the Comparison Test and Cauchy's notion of absolute convergence.

(7)
$$\begin{cases} u_0, \ u_1, \ u_2, \ \ldots, \ u_n, \ \ldots, \\ v_0, \ v_1, \ v_2, \ \ldots, \ v_n, \ \ldots \end{cases}$$

be two convergent series having s and s', respectively, as sums. Then

(8)
$$u_0+v_0, \ u_1+v_1, \ u_2+v_2, \ \ldots, \ u_n+v_n, \ \ldots$$

is a new convergent series, having $s+s'$ as its sum.

Proof. — If we let

$$s_n = u_0+u_1+u_2+\ldots+u_{n-1} \quad \text{and}$$
$$s'_n = v_0+v_1+v_2+\ldots+v_{n-1},$$

then, for increasing values of n, s_n and s'_n converge towards the limits s and s', respectively. As a consequence, $s_n+s'_n$, that is the sum of the first n terms of series (8), converges towards the limit $s+s'$, which suffices to establish the stated theorem.

Theorem VI.[21] — *Under the same hypotheses as the previous* [133] *theorem, if each of series* (7) *remains convergent when we replace its various terms with their numerical values, then*

(9)
$$\begin{cases} u_0v_0, \\ u_0v_1+u_1v_0, \\ u_0v_2+u_1v_1+u_2v_0, \\ \ldots\ldots\ldots\ldots\ldots\ldots, \\ u_0v_n+u_1v_{n-1}+\ldots+u_{n-1}v_1+u_nv_0, \\ \ldots\ldots\ldots\ldots\ldots\ldots\ldots\ldots\ldots\ldots \end{cases}$$

is a new convergent series having ss' as its sum.

Proof. — Once again, let s_n and s'_n be the sums of the first n terms of the two series (7), and additionally denote the sum of the first n terms of series (9) by s''_n. Then we have

$$s_ns'_n - s''_n = u_{n-1}v_{n-1} + (u_{n-1}v_{n-2}+u_{n-2}v_{n-1}) + \ldots$$
$$+ (u_{n-1}v_1+u_{n-2}v_2+\ldots+u_2v_{n-2}+u_1v_{n-1}).$$

Furthermore, theorem VI was proved in the second section in the case where series (7) consists only of positive terms. It is a consequence of this hypothesis that each the quantities $s_ns'_n$ and s''_n converges towards the limit ss', for increasing values of n. Consequently, the difference $s_ns'_n - s''_n$, or what amounts to the same thing, the sum

$$u_{n-1}v_{n-1} + (u_{n-1}v_{n-2}+u_{n-2}v_{n-1}) + \ldots$$
$$+ (u_{n-1}v_1+u_{n-2}v_2+\ldots+u_2v_{n-2}+u_1v_{n-1}),$$

[21] This is sometimes known as Mertens' Theorem.

converges towards the limit zero.

Now, if some of the terms of series (7) are positive and the others are negative, suppose that we denote the numerical values of the various terms by

$$(10)\qquad \begin{cases} \rho_0,\ \rho_1,\ \rho_2,\ \ldots,\ \rho_n,\ \ldots, \\ \rho'_0,\ \rho'_1,\ \rho'_2,\ \ldots,\ \rho'_n,\ \ldots \end{cases}$$

respectively. Suppose further, as in the statement of the theorem, that series (10), composed [134] of these same numerical values, are both convergent. By virtue of the remark we have just made, the sum

$$\begin{aligned} \rho_{n-1}\rho'_{n-1} &+ (\rho_{n-1}\rho'_{n-2}+\rho_{n-1}\rho'_{n-1})+\ldots \\ &+ (\rho_{n-1}\rho'_1+\rho_{n-2}\rho'_2+\ldots+\rho_2\rho'_{n-2}+\rho_1\rho'_{n-1}) \end{aligned}$$

converges towards the limit zero for increasing values of n. Because the numerical value of that sum is evidently greater than that of the following

$$\begin{aligned} u_{n-1}v_{n-1} &+ (u_{n-1}v_{n-2}+u_{n-2}v_{n-1})+\ldots \\ &+ (u_{n-1}v_1+u_{n-2}v_2+\ldots+u_2v_{n-2}+u_1v_{n-1}), \end{aligned}$$

it follows that this latter, or what amounts to the same thing, the difference $s_ns'_n - s''_n$ itself converges towards the limit zero. Consequently, ss', which is the limit of the product $s_ns'_n$, is also that of s''_n. In other words, series (9) is convergent and has as its sum the product ss'.

Scholium. — The previous theorem could not remain true if series (7), assumed to be convergent, ceased to be so after the reduction of each term to its numerical value. Suppose, for example, that we take both of series (7) to be

$$(11)\qquad 1,\ -\frac{1}{2^{\frac{1}{2}}},\ +\frac{1}{3^{\frac{1}{2}}},\ -\frac{1}{4^{\frac{1}{2}}},\ +\frac{1}{5^{\frac{1}{2}}},\ -\ldots.$$

Series (9) becomes

$$(12)\qquad \begin{cases} 1, \\ -\left(\frac{1}{\sqrt{2}}+\frac{1}{\sqrt{2}}\right), \\ +\left(\frac{1}{\sqrt{3}}+\frac{1}{\sqrt{2\cdot 2}}+\frac{1}{\sqrt{3}}\right), \\ -\left(\frac{1}{\sqrt{4}}+\frac{1}{\sqrt{3\cdot 2}}+\frac{1}{\sqrt{2\cdot 3}}+\frac{1}{\sqrt{4}}\right), \\ +\ldots\ldots\ldots\ldots\ldots\ldots\ldots\ldots\ldots \end{cases}$$

[135] This last series is divergent because its general term, namely

$$\pm\left(\frac{1}{\sqrt{n}}+\frac{1}{\sqrt{(n-1)2}}+\frac{1}{\sqrt{(n-2)3}}+\ldots+\frac{1}{\sqrt{2(n-1)}}+\frac{1}{\sqrt{n}}\right),$$

has a numerical value clearly greater than

$$\frac{n}{\left[\frac{n}{2}\left(\frac{n}{2}+1\right)\right]^{\frac{1}{2}}} = \left(\frac{4n}{n+2}\right)^{\frac{1}{2}}$$

when n is even, and greater than

$$\frac{n}{\left[\left(\frac{n+1}{2}\right)^2\right]^{\frac{1}{2}}} = \frac{2n}{n+1}$$

when n is odd. That is, in every possible case, it has a numerical value greater than 1. Nevertheless, series (11) is convergent. However, we ought to observe that it ceases to be convergent when we replace each term with its numerical value, because it then changes to series (6) of § II.

6.4 On series ordered according to the ascending integer powers of a single variable.

Let

(1) $$a_0,\ a_1x,\ a_2x^2,\ \ldots,\ a_nx^n,\ \ldots$$

be a series ordered according to the ascending integer powers of the variable x,[22] where

(2) $$a_0,\ a_1,\ a_2,\ \ldots,\ a_n,\ \ldots$$

denote constant coefficients, positive or negative. Furthermore, let A be the quantity that corresponds to the quantity k of the previous section (see § III, theorem II), with respect to series (2).[23] The same quantity, when calculated for series (1), is the numerical value of the product

$$Ax.$$

[136] As a consequence, series (1) is convergent if this numerical value is less than 1, which is to say in other words, if the numerical value of the variable x is less than $\frac{1}{A}$.[24] On the other hand, series (1) is divergent if the numerical value of x is greater than $\frac{1}{A}$. We may therefore state the following proposition:

Theorem I. — *Let A be the limit towards which the nth root of the largest numerical values of a_n converge, for increasing values of n. Series* (1) *is convergent for all values of x contained between the limits*

---

[22] Such series had not yet been given the modern name *power series*.

[23] Theorem II of § III is the Ratio Test, so Cauchy is saying that $A = \lim \frac{a_{n+1}}{a_n}$, when this limit exists. However, his statements of theorems I and II below and the discussion in between suggest that he means $A = \limsup \sqrt[n]{|a_n|}$.

[24] This number $\frac{1}{A}$ had not yet been given the modern name *radius of convergence*.

$$x = -\frac{1}{A} \quad and \quad x = +\frac{1}{A},$$

and divergent for all values of x situated outside of these same limits.

Whenever the numerical value of the ratio $\frac{a_{n+1}}{a_n}$ converges towards a fixed limit, this limit is the desired value of A (by virtue of theorem IV, Chap. II, § III). This remark brings us to a new proposition that I will write:

Theorem II. — *If the numerical value of the ratio*

$$\frac{a_{n+1}}{a_n}$$

converges towards the limit A for increasing values of n, series (1) *is convergent for all values of x contained between the limits*

$$x = -\frac{1}{A} \quad and \quad x = +\frac{1}{A},$$

and divergent for all values of x situated outside of these same limits.

Corollary I. — For an example, take the series

(3) $$1,\ 2x,\ 3x^2,\ 4x^3,\ \ldots,\ (n+1)x^n,\ \ldots.$$

Because under this hypothesis we find that

$$\frac{a_{n+1}}{a_n} = \frac{n+2}{n+1} = 1 + \frac{1}{n+1}$$

[137] and as a consequence,

$$A = 1,$$

we thereby conclude that series (3) is convergent for all values of x contained between the limits

$$x = -1 \quad \text{and} \quad x = +1,$$

and divergent for all values of x situated outside of these limits.

Corollary II. — For a second example, take the series

(4) $$\frac{x}{1},\ \frac{x^2}{2},\ \frac{x^3}{3},\ \ldots,\ \frac{x^n}{n},\ \ldots,$$

in which the constant term is understood to be zero. Under this hypothesis, we find that

$$\frac{a_{n+1}}{a_n} = \frac{n}{n+1} = \frac{1}{1+\frac{1}{n}}$$

and as a consequence, $A = 1$. Series (4) is therefore again convergent or divergent according to whether the numerical value of x is less or greater than 1.

Corollary III. — If we take the following for series (1)

$$(5)\ 1,\quad \frac{\mu}{1}x,\quad \frac{\mu(\mu-1)}{1\cdot 2}x^2,\quad \ldots,\quad \frac{\mu(\mu-1)(\mu-2)\ldots(\mu-n+1)}{1\cdot 2\cdot 3\ldots n}x^n,\quad \ldots,$$

where μ denotes any quantity, then we find that

$$\frac{a_{n+1}}{a_n} = \frac{\mu-n}{n+1} = -\frac{1-\frac{\mu}{n}}{1+\frac{1}{n}}$$

and as a consequence,

$$A = \lim \frac{1-\frac{\mu}{n}}{1+\frac{1}{n}} = \frac{1-\frac{1}{\infty}}{1+\frac{1}{\infty}} = 1.$$

We thereby conclude that series (5) is, like series (3) and (4), convergent [138] or divergent, according to whether we assign a numerical value less or greater than 1 to the variable x.

Corollary IV. — Now consider the series

$$(6)\qquad 1,\quad \frac{x}{1},\quad \frac{x^2}{1\cdot 2},\quad \frac{x^3}{1\cdot 2\cdot 3},\quad \ldots,\quad \frac{x^n}{1\cdot 2\cdot 3\ldots n},\quad \ldots.$$

Because in this case we have

$$\frac{a_{n+1}}{a_n} = \frac{1}{n+1}$$

and, as a consequence

$$A = \frac{1}{\infty} = 0,$$

we thereby conclude that the series is convergent between the limits

$$x = -\frac{1}{0} = -\infty \quad \text{and} \quad x = +\frac{1}{0} = +\infty,$$

that is, for all possible real values of the variable x.

Corollary V. — Finally, consider the series

$$(7)\qquad 1,\ 1\cdot x,\ 1\cdot 2\cdot x^2,\ 1\cdot 2\cdot 3\cdot x^3,\ \ldots,\ 1\cdot 2\cdot 3\ldots n\cdot x^n,\ \ldots.$$

In applying theorem II to this series, we find

$$\frac{a_{n+1}}{a_n} = n+1 \quad \text{and} \quad A = \infty$$

and consequently, we have

$$\frac{1}{A} = 0.$$

We thereby conclude that series (7) is always divergent, except when we suppose that $x = 0$, in which case it reduces to its first term 1.

By examining the results which we have just obtained, we recognize immediately that, among series ordered according to increasing integer powers of the variable x, some are either [139] convergent or divergent according to the value assigned to this variable, while others are always convergent, no matter x might be, and others are always divergent, except for $x = 0$. We may add that theorem I leaves no uncertainty about the convergence of such a series, except in the case where the numerical value of x becomes equal to the positive constant given by $\frac{1}{A}$, that is, when we suppose

$$x = \pm\frac{1}{A}.$$

In this particular case, the series is sometimes convergent, sometimes divergent, and the convergence sometimes depends on the sign of the variable x. For example, if in series (4), for which $A = 1$, we successively let

$$x = 1 \quad \text{and} \quad x = -1,$$

we obtain the following

$$(8) \qquad 1, \ \tfrac{1}{2}, \ \tfrac{1}{3}, \ \tfrac{1}{4}, \ \ldots, \ \tfrac{1}{n}, \ \ldots,$$

$$(9) \qquad -1, \ +\tfrac{1}{2}, \ -\tfrac{1}{3}, \ +\tfrac{1}{4}, \ \ldots, \ \pm\tfrac{1}{n}, \ \ldots,$$

of which the first is divergent (see the corollary to theorem III in § II) and the second is convergent, as follows from theorem III (§ III).

It is also essential to remark that, as follows from theorem I, whenever a series ordered according to the ascending integer powers of a variable x is convergent for a numerical value of x different from zero, it remains convergent if we diminish that numerical value, or even let it decrease indefinitely.

Whenever two series ordered according to the ascending integer powers of the variable x are convergent for the same value of the variable, we may apply theorems V and VI of § III to them. [140] This remark suffices to establish two propositions, which I will state:

Theorem III. — *Suppose that the two series*

$$(10) \qquad \begin{cases} a_0, \ a_1 x, \ a_2 x^2, \ \ldots, \ a_n x^n, \ \ldots, \\ b_0, \ b_1 x, \ b_2 x^2, \ \ldots, \ b_n x^n, \ \ldots \end{cases}$$

are both convergent when we assign a particular value to the variable x, and have s and s', repectively, as their sums. Then

$$(11) \qquad a_0 + b_0, \ (a_1 + b_1)x, \ (a_2 + b_2)x^2, \ \ldots, \ (a_n + b_n)x^n, \ \ldots$$

is a new convergent series, having $s+s'$ as its sum.

Corollary. — We easily extend this theorem to as many series as we might wish. For example, if the three series

$$\begin{array}{l} a_0,\ a_1x,\ a_2x^2,\ \ldots, \\ b_0,\ b_1x,\ b_2x^2,\ \ldots, \\ c_0,\ c_1x,\ c_2x^2,\ \ldots \end{array}$$

are convergent for the same value assigned to the variable x, and if we denote their respective sums by s, s' and s'', then

$$a_0+b_0+c_0,\ (a_1+b_1+c_1)x,\ (a_2+b_2+c_2)x^2,\ \ldots$$

is a new convergent series, which has $s+s'+s''$ as its sum.

Theorem IV. — *Under the same hypotheses as the previous theorem, if each of series* (10) *remains convergent when we replace its various terms with their numerical values, then*

$$(12)\qquad \begin{cases} a_0b_0,\ (a_0b_1+a_1b_0)x,\ (a_0b_2+a_1b_1+a_2b_0)x^2,\ldots, \\ \ldots,\ (a_0b_n+a_1b_{n-1}+\ldots+a_{n-1}b_1+a_nb_0)x^n,\ \ldots \end{cases}$$

is a new convergent series, having ss' as its sum.

[141] *Corollary I.* — The previous theorem is found contained in the formula

$$(13)\qquad \begin{cases} (a_0+a_1x+a_2x^2+\ldots(b_0+b_1x+b_2x^2+\ldots \\ \quad = a_0b_0+(a_0b_1+a_1b_0)x+(a_0b_2+a_1b_1+a_2b_0)x^2+\ldots, \end{cases}$$

which remains true in the case where each of series (10) remains convergent when we replace its various terms with their numerical values. Under this hypothesis, formula (13) may be used to expand the product of the sums of the two series into a new series of the same form.

Corollary II. — We may multiply together three or more series similar to (10), each of which remains convergent when we replace its various terms with their numerical values, by repeating the operation indicated in equation (13) several times. The product thus obtained is the sum of a new convergent series, ordered according to the increasing integer powers of the variable x.

Corollary III. — In the two preceding corollaries, suppose that all the series whose sums we multiply are equal. Then the product we obtain is the integer power of the sum of each of these, and this last sum is also represented by the sum of a series of the same kind. For example, if we let $a_0=b_0$, $a_1=b_1$, $a_2=b_2$, ..., in equation (13) we get

(14) $$(a_0+a_1x+a_2x^2+\ldots)^2=a_0^2+2a_0a_1x+(2a_0a_2+a_1^2)x^2+\ldots.$$

Corollary IV. — If we take

$$\frac{\mu(\mu-1)(\mu-2)\ldots(\mu-n+1)}{1\cdot2\cdot3\ldots n}x^n$$

and

$$\frac{\mu'(\mu'-1)(\mu'-2)\ldots(\mu'-n+1)}{1\cdot2\cdot3\ldots n}x^n$$

as the general terms of series (10), where μ and μ' denote any two quantities, and if the variable x is contained between the limits $x=-1$ and $x=+1$, then each of series (10) [142] is convergent, even if we replace the various terms with their numerical values, and the general term of series (12) is

$$\begin{aligned}&\left[\frac{\mu(\mu-1)\ldots(\mu-n+1)}{1\cdot2\cdot3\ldots n}+\frac{\mu(\mu-1)\ldots(\mu-n+2)}{1\cdot2\cdot3\ldots(n-1)}\frac{\mu'}{1}+\ldots\right.\\&\quad\left.+\frac{\mu}{1}\frac{\mu'(\mu'-1)\ldots(\mu'-n+2)}{1\cdot2\cdot3\ldots(n-1)}+\frac{\mu'(\mu'-1)\ldots(\mu'-n+1)}{1\cdot2\cdot3\ldots n}\right]x^n\\&=\frac{(\mu+\mu')(\mu+\mu'-1)(\mu+\mu'-2)\ldots(\mu+\mu'-n+1)}{1\cdot2\cdot3\ldots n}x^n.\end{aligned}$$

Given this, if we let $\varphi(\mu)$ denote the sum of the first of series (10) under the hypothesis that we have just made, that is if we suppose

(15) $$\varphi(\mu)=1+\frac{\mu}{1}x+\frac{\mu(\mu-1)}{1\cdot2}x^2+\ldots,$$

then under the same hypothesis the sums of series (10) and (12) are denoted $\varphi(\mu)$, $\varphi(\mu')$ and $\varphi(\mu+\mu')$, respectively, so that equation (13) becomes

(16) $$\varphi(\mu)\varphi(\mu')=\varphi(\mu+\mu').$$

Whenever we replace the series

$$b_0,\ b_1x,\ b_2x^2,\ \ldots$$

in equation (13) with a polynomial composed of a finite number of terms, we obtain a formula that never fails to be exact, as long as the series

$$a_0,\ a_1x,\ a_2x^2,\ \ldots$$

remains convergent. We will prove this directly by establishing the following theorem:

Theorem V. — *If series (1) is convergent and if we multiply the sum of this series by the polynomial*

(17) $$kx^m+lx^{m-1}+\ldots+px+q,$$

in which m denotes an integer number, the product we obtain is the [143] *sum of a new convergent series of the same form, the general term of which is*

$$(qa_n + pa_{n-1} + \ldots + la_{n-m+1} + ka_{n-m})x^m,$$

as long as, among the first terms, those quantities

$$a_{n-1},\ a_{n-2},\ \ldots,\ a_{n-m+1},\ a_{n-m}$$

that have negative indices are considered to be zero. In other words, we have[25]

$$(18)\qquad \left\{\begin{aligned} &\left(kx^m + lx^{m-1} + \ldots + px + q\right)\left(a_0 + a_1x + a_2x^2 + \ldots\right) \\ &\quad = qa_0 + (qa_1 + pa_0)x + \ldots \\ &\qquad + (qa_m + pa_{m-1} + \ldots + la_1 + ka_0)x^m \\ &\qquad + \ldots\ldots\ldots\ldots\ldots\ldots\ldots\ldots\ldots \\ &\qquad + (qa_n + pa_{n-1} + \ldots + la_{n-m+1} + ka_{n-m})x^n + \ldots. \end{aligned}\right.$$

Proof. — To multiply the sum of series (1) by the polynomial (17), it suffices to multiply it successively by the different terms of the polynomial. Thus, we have

$$\begin{aligned} &\left(kx^m + lx^{m-1} + \ldots + px + q\right)\left(a_0 + a_1x + a_2x^2 + \ldots\right) \\ &\quad = q\left(a_0 + a_1x + a_2x^2 + \ldots\right) + px\left(a_0 + a_1x + a_2x^2 + \ldots\right) \\ &\quad + \ldots\ldots\ldots\ldots\ldots\ldots\ldots\ldots\ldots\ldots\ldots\ldots \\ &\quad + lx^{m-1}\left(a_0 + a_1x + a_2x^2 + \ldots\right) + kx^m\left(a_0 + a_1x + a_2x^2 + \ldots\right). \end{aligned}$$

Because for any integer value of n we also have

$$\begin{aligned} &q\left(a_0 + a_1x + a_2x^2 + \ldots + a_{n-1}x^{n-1}\right) \\ &\quad = qa_0 + qa_1x + qa_2x^2 + \ldots + qa_{n-1}x^{n-1}, \end{aligned}$$

we conclude that, by making n increase indefinitely and passing to the limit,

$$q\left(a_0 + a_1x + a_2x^2 + \ldots\right) = qa_0 + qa_1x + qa_2x^2 + \ldots.$$

Similarly, we find

$$\begin{aligned} px\left(a_0 + a_1x + a_2x^2 + \ldots\right) &= pa_0x + pa_1x^2 + pa_2x^3 + \ldots, \\ \ldots\ldots\ldots\ldots\ldots\ldots\ldots\ldots&\ldots\ldots\ldots\ldots\ldots\ldots\ldots\ldots, \\ lx^{m-1}\left(a_0 + a_1x + a_2x^2 + \ldots\right) &= la_0x^{m-1} + la_1x^m + la_2x^{m+1} + \ldots, \\ kx^m\left(a_0 + a_1x + a_2x^2 + \ldots\right) &= ka_0x^m + ka_1x^{m+1} + ka_2x^{m+2} + \ldots. \end{aligned}$$

---

[25] The factor x^n on the last line of (18) is incorrectly written as x_n in [Cauchy 1897, p. 143]. It is correct in [Cauchy 1821, p. 160]. (tr.)

[144] If we add these last equations and form the sum of the right-hand sides, then by gathering together the coefficients of the same powers of x, we obtain precisely formula (18).

Imagine now that we vary the value of x in series (1) by insensible degrees. As long as the series remains convergent, that is as long as the value of x remains contained between the limits

$$-\frac{1}{A} \quad \text{and} \quad +\frac{1}{A},$$

the sum of the series is (by virtue of theorem I, § I) a continuous function of the variable x. Let $\varphi(x)$ be this continuous function. The equation

$$\varphi(x) = a_0 + a_1 x + a_2 x^2 + \dots$$

remains true for all values of x contained between the limits $-\frac{1}{A}$ and $+\frac{1}{A}$, which we indicate by writing these limits beside the series, as we see here:

$$(19) \qquad \varphi(x) = a_0 + a_1 x + a_2 x^2 + \dots \qquad \left(x = -\frac{1}{A},\ x = +\frac{1}{A}\right).$$

When the series is assumed to be known, we may sometimes deduce from it the value of the function $\varphi(x)$ in a finite form, and it is this that we call *summing* the series. However, more often the function $\varphi(x)$ is given, and we propose return from this function to the series, or in other words, to *expand*[26] the function into a convergent series ordered according to increasing integer powers of x. On this matter, it is easy to establish the proposition that I will state:

Theorem VI. — *A continuous function of the variable x can be expanded in only one way as a convergent series ordered according to the increasing integer powers of this variable.*

Proof. — Indeed, suppose that we have expanded the function $\varphi(x)$ by two [145] different methods, and let

$$a_0,\ a_1 x,\ a_2 x^2,\ \dots,\ a_n x^n,\ \dots,$$
$$b_0,\ b_1 x,\ b_2 x^2,\ \dots,\ b_n x^n,\ \dots$$

be the two expansions, that is two series, each convergent for values of x other than zero, and each having the function $\varphi(x)$ as its sum, as long as it remains convergent. Because these two series are constantly convergent for very small numerical values of x, for such values they have

$$a_0 + a_1 x + a_2 x^2 + \dots = b_0 + b_1 x + b_2 x^2 + \dots.$$

By making x vanish in the previous equation, we get

[26] Literally to "develop" (*développer*) (tr.).

$$a_0 = b_0.$$

In general, it follows that we may reduce that equation to

$$a_1 x + a_2 x^2 + \ldots = b_1 x + b_2 x^2 + \ldots,$$

or what amounts to the same thing, to

$$x(a_1 + a_2 x + \ldots) = x(b_1 + b_2 x + \ldots).$$

If we multiply both sides of this last equation by $\frac{1}{x}$, we obtain the following

$$a_1 + a_2 x + \ldots = b_1 + b_2 x + \ldots,$$

which must also remain true for very small numerical values of the variable x. By letting $x = 0$, we may conclude from this that

$$a_1 = b_1.$$

By continuing in the same way, we may show that the constants $a_0, a_1, a_2, \ldots$, are equal to the constants $b_0, b_1, b_2, \ldots$, respectively. From this it follows that the two expansions of the function $\varphi(x)$ are identical.

The differential Calculus gives very expeditious methods for expanding functions into series. We will describe these methods later on. [146] For now we will limit ourselves to expanding the function $(1+x)^\mu$, in which μ denotes any quantity, and using this to derive expansions of two other functions, which follow easily from the first, namely:

$$A^x \quad \text{and} \quad \log(1+x),$$

where A denotes a positive constant and log denotes the characteristic of the logarithm in a system chosen at will. As a consequence, we will solve the following three problems, one after another:

Problem I. — *When possible, to expand the function*

$$(1+x)^\mu$$

into a convergent series ordered according to increasing integer powers of the variable x.

Solution. — First of all, suppose that $\mu = m$, where m denotes any integer number. By the formula of Newton, we have

$$(1+x)^m = 1 + \frac{m}{1}x + \frac{m(m-1)}{1 \cdot 2}x^2 + \ldots.$$

The series whose sum constitutes the right-hand side of this formula is always composed of a finite number of terms. However, if we replace the integer number m by any quantity μ, the new series that we obtain, namely

(5) $$1, \frac{\mu}{1}x, \frac{\mu(\mu-1)}{1\cdot 2}x^2, \ldots,$$

is generally composed of an indefinite number of terms and is convergent only for numerical values of x less than 1. Under this hypothesis, let $\varphi(\mu)$ be the sum of the new series, so that we have

(15) $$\varphi(\mu) = 1 + \frac{\mu}{1}x + \frac{\mu(\mu-1)}{1\cdot 2}x^2 + \ldots \qquad (x=-1, \quad x=+1).$$

By virtue of theorem I (§ I), $\varphi(\mu)$ is a continuous function of the [147] variable μ, between arbitrary limits of this variable, and we have (see theorem III, Corollary IV)

(16) $$\varphi(\mu)\varphi(\mu') = \varphi(\mu+\mu').$$

Because this last equation is entirely similar to equation (2) of chapter V (§ I), it is solved in the same manner, and we conclude thereby that

$$\varphi(\mu) = [\varphi(1)]^{\mu} = (1+x)^{\mu}.$$

If we substitute the value of $\varphi(\mu)$ determined in this way into formula (15), we find that for all values of x contained between the limits $x=-1$ and $x=+1$,

(20) $$(1+x)^{\mu} = 1 + \frac{\mu}{1}x + \frac{\mu(\mu-1)}{1\cdot 2}x^2 + \ldots \qquad (x=-1, \quad x=+1).$$

Whenever the numerical value of x is greater than 1, series (5) is no longer convergent and ceases to have a sum, so that equation (20) no longer remains true. Under the same hypothesis, as we shall prove later with the aid of the infinitesimal Calculus, it is impossible to expand the function $(1+x)^{\mu}$ into a convergent series ordered according to the ascending powers of the variable x.

Corollary I. — If we replace μ by $\frac{1}{\alpha}$ and x by αx in equation (20), where α denotes an infinitely small quantity, then for all values of αx contained between the limits -1 and $+1$, or what amounts to the same thing, for all values of x contained between the limits $-\frac{1}{\alpha}$ and $+\frac{1}{\alpha}$, we have

$$(1+\alpha x)^{\frac{1}{\alpha}} = 1 + \tfrac{x}{1} + \tfrac{x^2}{1\cdot 2}(1-\alpha) + \tfrac{x^3}{1\cdot 2\cdot 3}(1-\alpha)(1-2\alpha) + \ldots$$

$$\left(x=-\tfrac{1}{\alpha}, \quad x=+\tfrac{1}{\alpha}\right).$$

This last equation ought to remain true, no matter how small the numerical value of α may be. If we denote as usual by the abbreviation lim placed in front of an expression that includes the variable α the [148] limit towards which this expression converges as the numerical value of α decreases indefinitely, then in passing to the limit, we find

(21) $$\lim(1+\alpha x)^{\frac{1}{\alpha}} = 1 + \frac{x}{1} + \frac{x^2}{1\cdot 2} + \frac{x^3}{1\cdot 2\cdot 3} + \ldots \quad (x=-\infty, x=+\infty).$$

It remains to seek the limit of $(1+\alpha x)^{\frac{1}{\alpha}}$. First, from the previous formula we get

$$\lim(1+\alpha)^{\frac{1}{\alpha}} = 1+\frac{1}{1}+\frac{1}{1\cdot 2}+\frac{1}{1\cdot 2\cdot 3}+\ldots,$$

or in other words,

(22) $$\lim(1+\alpha)^{\frac{1}{\alpha}} = e,$$

where e denotes the base of Napierian logarithms [see § I, eq. (6)]. We conclude immediately that

$$\lim(1+\alpha x)^{\frac{1}{\alpha x}} = e,$$

and as a consequence,

$$\lim(1+\alpha x)^{\frac{1}{\alpha}} = \lim\left[(1+\alpha x)^{\frac{1}{\alpha x}}\right]^{x} = e^{x}.$$

Now if we substitute the value of $\lim(1+\alpha)^{\frac{1}{\alpha}}$ into equation (21), then we obtain the following:

(23) $$e^{x} = 1+\frac{x}{1}+\frac{x^2}{1\cdot 2}+\frac{x^3}{1\cdot 2\cdot 3}+\ldots \quad (x=-\infty,\ x=+\infty).$$

We may derive equation (23) directly by observing that the series

(6) $$1,\ \frac{x}{1},\ \frac{x^2}{1\cdot 2},\ \frac{x^3}{1\cdot 2\cdot 3},\ \ldots$$

is convergent for all possible values of the variable x and then seeking that function of x which represents the sum of this same [149] series. Indeed, let $\varphi(x)$ be the sum of series (6), which has

$$\frac{x^n}{1\cdot 2\cdot 3\ldots n}$$

as its general term. Then $\varphi(y)$ is the sum of the series whose general term is

$$\frac{y^n}{1\cdot 2\cdot 3\ldots n}.$$

By virtue of theorem VI, § III, the product of these two sums is the sum of a new series that has

$$\frac{x^n}{1\cdot 2\cdot 3\ldots n}+\frac{x^{n-1}}{1\cdot 2\cdot 3\ldots(n-1)}\frac{y}{1}+\ldots$$
$$+\frac{x}{1}\frac{y^{n-1}}{1.2.3\ldots(n-1)}+\frac{x^n}{1\cdot 2\cdot 3\ldots n} = \frac{(x+y)^n}{1\cdot 2\cdot 3\ldots n}$$

as its general term. This product is therefore equal to $\varphi(x+y)$, and consequently, if we let

$$\varphi(x) = 1 + \frac{x}{1} + \frac{x^2}{1 \cdot 2} + \frac{x^3}{1 \cdot 2 \cdot 3} + \ldots,$$

then the function $\varphi(x)$ satisfies the equation

$$\varphi(x)\varphi(y) = \varphi(x+y).$$

Solving this equation, we find

$$\varphi(x) = [\varphi(1)]^x = \left(1 + \frac{1}{1} + \frac{1}{1 \cdot 2} + \frac{1}{1 \cdot 2 \cdot 3} + \ldots\right)^x.$$

That is

$$\varphi(x) = e^x.$$

Corollary II. — If we divide both sides of (20) by μ after subtracting 1 from both sides, the equation that we obtain may be written as follows:

$$\frac{(1+x)^\mu - 1}{\mu} = x - \frac{x^2}{2}(1-\mu) + \frac{x^3}{3}(1-\mu)\left(1 - \frac{1}{2}\mu\right) - \ldots$$
$$(x = -1, \quad x = +1).$$

[150] If we let μ converge towards the limit zero in this last equation we find, by passing to the limit, that[27]

$$\text{(24)} \qquad \lim \frac{(1+x)^\mu - 1}{\mu} = x - \frac{x}{2} + \frac{x^3}{3} + \ldots.$$

Furthermore, when ln denotes the characteristic of Napierian logarithm, taken in the system whose base is e, then we evidently have

$$1 + x = e^{\ln(1+x)}$$

and

$$(1+x)^\mu = e^{\mu \ln(a+x)} = 1 + \frac{\mu \ln(a+x)}{1} + \frac{\mu^2 [\ln(1+x)]^2}{1 \cdot 2} + \ldots.$$

We conclude that

$$\frac{(1+x)^\mu - 1}{\mu} = \ln(1+x) + \frac{\mu}{2}[\ln(1+x)]^2 + \ldots.$$

Consequently

$$\text{(25)} \qquad \lim \frac{(1+x)^\mu - 1}{\mu} = \ln(1+x).$$

Given this, formula (24) becomes

---

[27] Cauchy indeed uses a + sign following $\frac{x^3}{3}$ in [Cauchy 1821, p. 169, Cauchy 1897, p. 150]. Note the contrast with equation (26). (tr.)

$$(26)\qquad \ln(1+x) = x - \frac{x^2}{2} + \frac{x^3}{3} - \dots \qquad (x=-1, \quad x=+1).$$

The preceding equation remains true as long as the numerical value of x remains smaller than 1. In this case, the series

$$(27)\qquad x,\ -\frac{x^2}{2},\ +\frac{x^3}{3},\dots,\ \pm\frac{x^n}{n},\ \dots$$

is convergent, as is series (4), which differs from it only in the signs of the terms of odd order.[28] Because these same series are divergent when we suppose that the numerical value of x is greater than 1, equation (26) ceases to hold under this hypothesis.

In the particular case where we take $x = 1$, series (27) reduces to series (3) of the third section, which is convergent, [151] as we have shown. Thus, equation (26) ought to remain true, so that we have

$$(28)\qquad \ln(2) = 1 - \frac{1}{2} + \frac{1}{3} - \frac{1}{4} + \dots.$$

On the other hand, if we let $x = -1$, then series (27) is divergent and has no sum.

We may further note that, if after substituting $-x$ for x in formula (26), we change the signs on both sides of the equation, we obtain the following

$$(29)\qquad \ln\left(\frac{1}{1-x}\right) = x + \frac{x^2}{2} + \frac{x^3}{3} + \dots \qquad (x=-1, \quad x=+1).$$

Problem II. — *To expand the function*

$$A^x,$$

where A denotes an arbitrary number, into a convergent series ordered according to increasing integer powers of the variable x.

Solution. — We continue to let the characteristic ln denote the Napierian logarithm taken in the system whose base is e. From the the definition of this logarithm, we have

$$A = e^{\ln(A)},$$

and we thereby conclude that

$$(30)\qquad A^x = e^{x\ln(A)}.$$

Consequently, using equation (23), we have

[28] This use of "odd order" (*rang impair*) in both [Cauchy 1821, p. 170] and [Cauchy 1897, p. 150] may be an error. On the other hand, for Cauchy, the *rang* may mean the position of a term in a series, starting with order zero. So in series (27), terms of odd order may be the ones of even degree. However, this does not seem to have been the case in series (4) of this section [Cauchy 1821, p. 153, Cauchy 1897, p. 137].

(31) $$\begin{cases} A^x = 1 + \dfrac{x\ln(A)}{1} + \dfrac{x^2\,[\ln(A)]^2}{1\cdot 2} + \dfrac{x^3\,[\ln(A)]^3}{1\cdot 2\cdot 3} + \dots \\ (x = -\infty, \quad x = +\infty). \end{cases}$$

This last formula remains true for all possible real values of the variable x.

Problem III. — *Letting the characteristic* log *denote the logarithm taken* [152] *in the system whose base is A, to expand the function*

$$\log(1+x),$$

where possible, into a convergent series ordered according to increasing integer powers of the variable x.

Solution. — Still denoting the characteristic of the Napierian logarithm by ln, by virtue of well-known properties of the logarithm, we have

$$\log(1+x) = \frac{\log(1+x)}{\log(A)} = \frac{\ln(1+x)}{\ln(A)}.$$

Consequently, making use of equation (26), we find that for all values of x contained between the limits -1 and $+1$,

(32) $$\log(1+x) = \frac{1}{\ln(A)}\left(x - \frac{x^2}{2} + \frac{x^3}{3} - \dots\right) \quad (x = -1,\ x = +1).$$

This last formula remains true even in the case where we take $x = 1$, but it ceases to hold whenever we have $x = -1$ or $x^2 > 1$.

Chapter 7
On imaginary expressions and their moduli.

7.1 General considerations on imaginary expressions.

[153] In analysis, we call a *symbolic expression* or *symbol* any combination of algebraic signs that do not mean anything by themselves or to which we attribute a value different from that which they ought naturally to have. Likewise, we call *symbolic equations* all those that, taking the letters and the interpretations according to the generally established conventions, are inexact or do not make sense, but from which we can deduce exact results by modifying and altering either the equations themselves or the symbols which comprise them, according to fixed rules. The use of symbolic expressions or equations is often a means of simplifying calculations and of writing in a short form results that appear quite complicated. We have already seen this in the second section of the third chapter where formula (9) gives a very simple symbolic value to the unknown x satisfying equations (4).[1] Among those symbolic expressions or equations which are of some importance in analysis, we should distinguish above all those which we call *imaginary*. We are going to show how we can put them to good use.

We know that the sine and the cosine of the arc $a+b$ are given as functions of the sines and cosines of the arcs a and b by the formulas

$$(1) \qquad \begin{cases} \cos(a+b) = \cos a \cos b - \sin a \sin b, \\ \sin(a+b) = \sin a \cos b + \sin b \cos a. \end{cases}$$

[154] Now, without taking the trouble to remember these formulas, we have a very simple means of recovering them at will. Indeed, it suffices to consider the following remark.

Suppose that we multiply together the two symbolic expressions

[1] Formula (9) [Cauchy 1821, p. 80, Cauchy 1897, p. 79] is Cramer's rule. Equations (4) [Cauchy 1821, p. 77, Cauchy 1897, p. 76] comprise a system of n linear equations in n unknowns.

R.E. Bradley, C.E. Sandifer, *Cauchy's Cours d'analyse*, Sources and Studies in the History of Mathematics and Physical Sciences, DOI 10.1007/978-1-4419-0549-9_7,

$$\cos a + \sqrt{-1}\sin a \quad \text{and}$$
$$\cos b + \sqrt{-1}\sin b,$$

by applying the known rules of algebraic multiplication as if $\sqrt{-1}$ were a real quantity the square of which is equal to -1. The resulting product is composed of two parts, one entirely real and the other having a factor of $\sqrt{-1}$. The real part gives the value of $\cos(a+b)$ while the coefficient of $\sqrt{-1}$ gives the value of $\sin(a+b)$. To establish this remark, we write the formula

$$(2) \qquad \begin{cases} \cos(a+b) + \sqrt{-1}\sin(a+b) \\ \quad = \left(\cos a + \sqrt{-1}\sin a\right)\left(\cos b + \sqrt{-1}\sin b\right). \end{cases}$$

The three expressions that make up the preceding equation, namely

$$\cos a + \sqrt{-1}\sin a,$$
$$\cos b + \sqrt{-1}\sin b \quad \text{and}$$
$$\cos(a+b) + \sqrt{-1}\sin(a+b),$$

are three symbolic expressions that cannot be interpreted according to the generally established conventions, and they do not represent anything real. For this reason, they are called *imaginary expressions*. Equation (2) itself, taken literally, is inexact and it does not make sense. To get exact results, first we must expand its right-hand side by algebraic multiplication, and this reduces the expression to

$$(3) \qquad \begin{cases} \cos(a+b) + \sqrt{-1}\sin(a+b) \\ = \cos a\cos b - \sin a\sin b + \sqrt{-1}\,(\sin a\cos b + \sin b\cos a). \end{cases}$$

Secondly, we must equate the real part [155] of the left-hand side of equation (3) with the real part of the right-hand side, and then the coefficient of $\sqrt{-1}$ on the left-hand side with the coefficient of $\sqrt{-1}$ on the right. Thus we are brought back to equations (1), both of which we ought to consider as implicitly contained in formula (2).

In general, we call an *imaginary expression* any symbolic expression of the form

$$\alpha + \beta\sqrt{-1},$$

where α and β denote real quantities. We say that two imaginary expressions

$$\alpha + \beta\sqrt{-1} \quad \text{and} \quad \gamma + \delta\sqrt{-1}$$

are *equal* to each other when there is equality between corresponding parts: 1° between the real parts α and γ,[2] and 2° between the coefficients of $\sqrt{-1}$, namely β and δ. We indicate equality between two imaginary expressions in the same way

[2] This is incorrectly written as "real parts, α and β" in [Cauchy 1897, p. 155]. It was correct in [Cauchy 1821, p. 176]. (tr.)

that we indicate it between two real quantities, by the symbol =, and this results in what we call an *imaginary equation*. Given this, any imaginary equation is just the symbolic representation of two equations involving real quantities. For example, the symbolic equation

$$\alpha+\beta\sqrt{-1}=\gamma+\delta\sqrt{-1}$$

is just equivalent to the two real equations

$$\alpha=\gamma \quad \text{and} \quad \beta=\delta.$$

In the imaginary expression

$$\alpha+\beta\sqrt{-1},$$

when the coefficient β of $\sqrt{-1}$ vanishes, the term $\beta\sqrt{-1}$ is understood to be zero, and the expression itself reduces to the real quantity α. By virtue of this convention, imaginary expressions include the real quantities as special cases.

Imaginary expressions may be subjected to the same operations of Algebra as real quantities. In particular, if we perform addition, subtraction or multiplication [156] of two imaginary expressions, they operate according to the established rules for real quantities, and we obtain as a result a new imaginary expression that we call the *sum*, the *difference* or the *product* of the given expressions, and the ordinary notations are used to indicate that sum, difference or product. For example, if we are given two imaginary expressions,

$$\alpha+\beta\sqrt{-1} \quad \text{and} \quad \gamma+\delta\sqrt{-1},$$

we find

$$\left(\alpha+\beta\sqrt{-1}\right)+\left(\gamma+\delta\sqrt{-1}\right)=\alpha+\gamma+(\beta+\delta)\sqrt{-1}, \tag{4}$$

$$\left(\alpha+\beta\sqrt{-1}\right)-\left(\gamma+\delta\sqrt{-1}\right)=\alpha-\gamma+(\beta-\delta)\sqrt{-1}, \tag{5}$$

$$\left(\alpha+\beta\sqrt{-1}\right)\times\left(\gamma+\delta\sqrt{-1}\right)=\alpha\gamma-\beta\delta+(\alpha\delta+\beta\gamma)\sqrt{-1}. \tag{6}$$

It is worth remarking that the product of two or more imaginary expressions, like that of two or more real binomials, remains the same regardless of the order in which we multiply the different factors.[3]

To divide a first imaginary expression by a second is to find an imaginary expression which, when multiplied by the second, reproduces the first. The result of this operation is the *quotient* of the two given expressions. To indicate this, we use the ordinary symbol for division. So, for example,

$$\frac{\alpha+\beta\sqrt{-1}}{\gamma+\delta\sqrt{-1}}$$

---

[3] Although [Servois 1814] introduced the word "commutative" to describe this property, Cauchy has not adopted it here.

represents the quotient of the two imaginary expressions

$$\alpha+\beta\sqrt{-1} \quad \text{and} \quad \gamma+\delta\sqrt{-1}.$$

To raise an imaginary expression to the power m (where m denotes an integer number) is to form the product of m factors equal to that expression. We write the mth *power* of $\alpha+\beta\sqrt{-1}$ with the notation

$$\left(\alpha+\beta\sqrt{-1}\right)^m.$$

[157] To extract the nth root of the imaginary expression $\alpha+\beta\sqrt{-1}$, or in other words to raise this expression to the power $\frac{1}{n}$ (where n denotes any integer number), is to form a new imaginary expression whose nth power reproduces $\alpha+\beta\sqrt{-1}$. This problem has several solutions (see § IV), and as a result, the imaginary expression $\alpha+\beta\sqrt{-1}$ has several *roots* of degree n. When we do not wish to distinguish any one of these roots, we use the notation

$$\sqrt[n]{\!\!\sqrt{\alpha+\beta\sqrt{-1}}},$$

or the following,

$$\left(\left(\alpha+\beta\sqrt{-1}\right)\right)^{\frac{1}{n}}.$$

In the particular case where β vanishes, $\alpha+\beta\sqrt{-1}$ reduces to a real quantity α, and among the values of the expression

$$\sqrt[n]{\!\!\sqrt{\alpha}} = ((a))^{\frac{1}{n}}$$

we may find one or two real roots, as we will see below.

In addition to the integer powers and the corresponding roots of imaginary expressions, we must often consider what we call their fractional and negative powers. On this subject, we ought to make the following remarks.

To raise the imaginary expression $\alpha+\beta\sqrt{-1}$ to a fractional power $\frac{m}{n}$, supposing that the fraction $\frac{m}{n}$ is reduced to its lowest terms, we must: 1° extract the nth root of the given expression; and 2° raise this root to the integer power m. As this problem can be solved in several ways (see below, § IV), we denote indistinctly any one of the powers $\frac{m}{n}$ by the notation

$$\left(\left(\alpha+\beta\sqrt{-1}\right)\right)^{\frac{m}{n}}.$$

[158] In the particular case where β is zero, one or two of these powers can be real.

To raise the imaginary expression $\alpha+\beta\sqrt{-1}$ to a negative power, $-m$ or $-\frac{1}{n}$ or $-\frac{m}{n}$ is to divide 1 by the power m or $\frac{1}{n}$ or $\frac{m}{n}$ of the same expression. The problem has a unique solution in the first case, and several solutions in the two others. We denote the power $-m$ with the simple notation

$$\left(\alpha+\beta\sqrt{-1}\right)^{-m},$$

while the two notations

$$\left(\left(\alpha+\beta\sqrt{-1}\right)\right)^{-\frac{1}{n}}$$

and

$$\left(\left(\alpha+\beta\sqrt{-1}\right)\right)^{-\frac{m}{n}}$$

represent, in the first case, any of the powers $-\frac{1}{n}$, and in the second case, any of the powers $-\frac{m}{n}$.

We say that two imaginary expressions are *conjugate*[4] to each other when the two expressions differ only in the signs of the coefficient of $\sqrt{-1}$. The sum of two such expressions is always real, as is their product. Indeed, the two conjugate imaginary expressions

$$\alpha+\beta\sqrt{-1} \quad \text{and} \quad \alpha-\beta\sqrt{-1}$$

have as their sum 2α and as their product $\alpha^2+\beta^2$. The last part of this observation brings us to a theorem about numbers, which is stated here:

Theorem I.[5] — *If we multiply together two integer numbers that are each the sum of two squares, then the product is always a sum of two squares.*

Proof. — Let the integer numbers be

$$\alpha^2+\beta^2 \quad \text{and} \quad \alpha'^2+\beta'^2,$$

[159] where α^2, β^2, α'^2 and β'^2 denote perfect squares. We evidently have the two equations

$$(\alpha+\beta\sqrt{-1})(\alpha'+\beta'\sqrt{-1}) = \alpha\alpha'-\beta\beta'+(\alpha\beta'+\alpha'\beta)\sqrt{-1}$$

and

$$(\alpha-\beta\sqrt{-1})(\alpha'-\beta'\sqrt{-1}) = \alpha\alpha'-\beta\beta'-(\alpha\beta'+\alpha'\beta)\sqrt{-1}$$

and, by multiplying these term by term, we obtain the following

$$(7) \qquad \left(\alpha^2+\beta^2\right)\left(a'^2+\beta'^2\right) = \left(\alpha\alpha'-\beta\beta'\right)^2+\left(\alpha\beta'+\alpha'\beta\right)^2.$$

If we interchange the letters α' and β' in this last expression, we get

$$(8) \qquad \left(\alpha^2+\beta^2\right)\left(a'^2+\beta'^2\right) = \left(\alpha\beta'-\alpha'\beta\right)^2+\left(\alpha\alpha'+\beta\beta'\right)^2.$$

---

[4] According to [Smith 1958, vol. 2, p. 267], this is the first use of the term "conjugate" in this sense.

[5] This fact of number theory is sometimes called Lagrange's Theorem, though it is not originally due to Lagrange. See, for example, [Euler 1758].

Thus, in general we have two ways to decompose into two squares the product of two integer numbers each of which is the sum of two squares. Thus, for example, one draws from equations (7) and (8)

$$\left(2^2+1\right)\left(3^2+2^2\right)=4^2+7^2=1^2+8^2.$$

We see from these examples that the use of imaginary expressions can be of great use, not only in ordinary Algebra but also in the Theory of numbers.

Sometimes we represent an imaginary expression by a single letter. It is an artifice which augments the resources of Analysis and we will make use of it in what follows.

7.2 On the moduli of imaginary expressions and on reduced expressions.

A remarkable property of any imaginary expression

$$\alpha+\beta\sqrt{-1}$$

is that it can be put into the form

$$\rho\left(\cos\theta+\sqrt{-1}\sin\theta\right),$$

[160] where ρ denotes a positive quantity and θ a real arc. Indeed, if we write the symbolic equation

$$\alpha+\beta\sqrt{-1}=\rho\left(\cos\theta+\sqrt{-1}\sin\theta\right), \tag{1}$$

or what amounts to the same thing, the two real equations

$$\begin{cases}\alpha=\rho\cos\theta \quad \text{and}\\ \beta=\rho\sin\theta,\end{cases} \tag{2}$$

and we get

$$\alpha^2+\beta^2=\rho^2\left(\cos^2\theta+\sin^2\theta\right)=\rho^2 \quad \text{and}$$

$$\rho=\sqrt{\alpha^2+\beta^2}. \tag{3}$$

Having thus determined the value of the number ρ, all that remains to verify completely equations (2) is to find an arc θ such that its cosine and sine are, respectively,

$$\begin{cases}\cos\theta=\frac{\alpha}{\sqrt{\alpha^2+\beta^2}} \quad \text{and}\\ \sin\theta=\frac{\beta}{\sqrt{\alpha^2+\beta^2}}.\end{cases} \tag{4}$$

This last problem is always solvable because each of the quantities $\frac{\alpha}{\sqrt{\alpha^2+\beta^2}}$ and $\frac{\beta}{\sqrt{\alpha^2+\beta^2}}$ has a numerical value less than 1 and the sum of their squares is equal to 1. Moreover, it has infinitely many different solutions because, having calculated one suitable value of the arc θ, we can increase or decrease this arc by any number of circumferences without changing the value of the sine or the cosine.

When the imaginary expression $\alpha+\beta\sqrt{-1}$ is put into the form

$$\rho\left(\cos\theta+\sqrt{-1}\sin\theta\right),$$

the positive quantity ρ is called the *modulus*[6] of this imaginary expression. What remains after the suppression of the modulus, that is [161] to say the factor

$$\cos\theta+\sqrt{-1}\sin\theta,$$

is called the *reduced expression.* Because we take the quantities α and β to be known, we get only one unique value for the modulus ρ as determined by equation (3), and as a result, the modulus remains the same for any two imaginary expressions that are equal. Thus we can state the following theorem:

Theorem I. — *The equality of two imaginary expressions always entails the equality of their moduli, and as a consequence, the equality of their reduced expressions.*

If we compare two conjugate imaginary expressions to each other, we find that their moduli are equal. The square of their common modulus is simply their product.

When the second term β vanishes in the imaginary expression $\alpha+\beta\sqrt{-1}$, this expression reduces to a real quantity α. Under the same hypothesis, we get from equations (3) and (4): 1° when α is positive, that

$$\rho=\sqrt{\alpha^2},$$
$$\cos\theta=1 \quad \text{and} \quad \sin\theta=0,$$

and so

$$\theta=\pm 2k\pi,$$

where k denotes any integer number; and 2° when α is negative, that

$$\rho=\sqrt{\alpha^2},$$
$$\cos\theta=-1 \quad \text{and} \quad \sin\theta=0,$$

and so

$$\theta=\pm(2k+1)\pi.$$

[6] According to [Smith 1958, vol. 2, p. 267], this is the first use of the term "modulus" in this sense. However, [Grattan-Guinness 1990, vol. 2, p. 170] cites an earlier use in [Cauchy 1817].

Thus the modulus of a real quantity α is simply its numerical value $\sqrt{\alpha^2}$ and the reduced expression that corresponds to such a quantity is always $+1$ or -1, namely

$$+1 = \cos(\pm 2k\pi) + \sqrt{-1}\sin(\pm 2k\pi),$$

[162] whenever α is a positive quantity, and

$$-1 = \cos\left(\pm\overline{2k+1}\,\pi\right) + \sqrt{-1}\sin\left(\pm\overline{2k+1}\,\pi\right),$$

whenever α is a negative quantity.

Any imaginary expression that has modulus zero itself reduces to zero because its two terms vanish. Conversely, because the cosine and the sine of an arc are never zero at the same time, it follows that an imaginary expression cannot be reduced to zero unless its modulus vanishes.

Any imaginary expression which has 1 as its modulus is necessarily a reduced expression. Thus, for example,

$$\begin{array}{lcl} \cos a + \sqrt{-1}\sin a, & & \cos a - \sqrt{-1}\sin a, \\ -\cos a - \sqrt{-1}\sin a & \text{and} & -\cos a + \sqrt{-1}\sin a \end{array}$$

are four reduced expressions forming two conjugate pairs. In fact, to get these four expressions from the formula

$$\cos\theta + \sqrt{-1}\sin\theta,$$

it is enough to take successively

$$\begin{array}{lcl} \theta = \pm 2k\pi + a, & & \theta = \pm 2k\pi - a, \\ \theta = \pm(2k+1)\pi + a & \text{and} & \theta = \pm(2k+1)\pi - a, \end{array}$$

where k denotes any integer number.

Calculations involving imaginary expressions can be simplified by considering reduced expressions. It is important to take note of their properties. These properties are contained in the theorems that I am about to state.

Theorem II. — *To multiply together two reduced expressions*

$$\cos\theta + \sqrt{-1}\sin\theta \quad \textit{and} \quad \cos\theta' + \sqrt{-1}\sin\theta',$$

it suffices to add the corresponding arcs θ and θ'.

[163] *Proof.* — Indeed, we have

$$(5) \qquad \begin{cases} \left(\cos\theta + \sqrt{-1}\sin\theta\right)\left(\cos\theta' + \sqrt{-1}\sin\theta'\right) \\ \quad = \cos(\theta+\theta') + \sqrt{-1}\sin(\theta+\theta'). \end{cases}$$

Corollary. — If we make $\theta' = -\theta$ in the previous theorem, we find, as we might expect,

$$(6) \qquad \left(\cos\theta + \sqrt{-1}\sin\theta\right)\left(\cos\theta - \sqrt{-1}\sin\theta\right) = 1.$$

Theorem III. — *To multiply together several reduced expressions,*

$$\cos\theta + \sqrt{-1}\sin\theta,\ \cos\theta' + \sqrt{-1}\sin\theta',\ \cos\theta'' + \sqrt{-1}\sin\theta'',\ \ldots,$$

it suffices to add the corresponding arcs, θ, θ', θ'', ...

Proof. — Indeed, we have successively,[7]

$$\begin{aligned}&(\cos\theta + \sqrt{-1}\sin\theta)(\cos\theta' + \sqrt{-1}\sin\theta')\\ &\quad = \cos(\theta+\theta') + \sqrt{-1}\sin(\theta+\theta'),\end{aligned}$$

$$\begin{aligned}&(\cos\theta + \sqrt{-1}\sin\theta)(\cos\theta' + \sqrt{-1}\sin\theta')(\cos\theta'' + \sqrt{-1}\sin\theta'')\\ &\quad = \left[\cos(\theta+\theta') + \sqrt{-1}\sin(\theta+\theta')\right](\cos\theta'' + \sqrt{-1}\sin\theta'')\\ &\quad = \cos(\theta+\theta'+\theta'') + \sqrt{-1}\sin(\theta+\theta'+\theta''),\\ &\ldots\ldots\ldots\ldots\ldots\ldots\ldots\ldots\ldots\ldots\ldots\ldots,\end{aligned}$$

and, continuing in the same way, we find in general that whatever the number of arcs, θ, θ', θ'', ... may be,

$$(7) \qquad \begin{aligned}&\left(\cos\theta + \sqrt{-1}\sin\theta\right)\left(\cos\theta' + \sqrt{-1}\sin\theta'\right)\left(\cos\theta'' + \sqrt{-1}\sin\theta''\right)\ldots\\ &\quad = \cos\left(\theta+\theta'+\theta''+\ldots\right) + \sqrt{-1}\sin\left(\theta+\theta'+\theta''+\ldots\right).\end{aligned}$$

Corollary. — If we expand the left-hand side of equation (7) by ordinary multiplication,[8] the expansion will consist of two parts, one real and the other having a factor $\sqrt{-1}$. Given this, the real part will take on the value

$$\cos\left(\theta+\theta'+\theta''+\ldots\right),$$

[164] and the coefficient of $\sqrt{-1}$ in the second part will have the value

$$\sin\left(\theta+\theta'+\theta''+\ldots\right).$$

For example, suppose that we are considering only three arcs, θ, θ' and θ''. Then equation (7) becomes

$$\begin{aligned}&\left(\cos\theta + \sqrt{-1}\sin\theta\right)\left(\cos\theta' + \sqrt{-1}\sin\theta'\right)\left(\cos\theta'' + \sqrt{-1}\sin\theta''\right)\\ &\quad = \cos\left(\theta+\theta'+\theta''\right) + \sqrt{-1}\sin\left(\theta+\theta'+\theta''\right)\end{aligned}$$

---

[7] In the fourth line of this calculation, then term $\cos(\theta+\theta')$ was missing the right parenthesis in [Cauchy 1897, p. 163]. [Cauchy 1821, p. 187] was parenthesized properly. (tr.)

[8] Cauchy calls this *multiplication immédiate*. It is not clear how this is different from what he calls "algebraic multiplication."

and, after expanding the left-hand side of this last equation by algebraic multiplication, we conclude that

$$\cos\left(\theta+\theta'+\theta''\right) = \begin{array}{l} \cos\theta\cos\theta'\cos\theta''-\cos\theta\sin\theta'\sin\theta'' \\ -\sin\theta\cos\theta'\sin\theta''-\sin\theta\sin\theta'\cos\theta'' \end{array}$$

and[9]

$$\sin\left(\theta+\theta'+\theta''\right) = \begin{array}{l} \sin\theta\cos\theta'\cos\theta''+\cos\theta\sin\theta'\cos\theta'' \\ +\cos\theta\cos\theta'\sin\theta''-\sin\theta\sin\theta'\sin\theta''. \end{array}$$

Theorem IV. — *To divide the reduced expression*

$$\cos\theta+\sqrt{-1}\sin\theta$$

by the following

$$\cos\theta'+\sqrt{-1}\sin\theta',$$

it suffices to subtract the arc θ' that corresponds to the second expression from the arc θ corresponding to the first.

Proof. — Let x be the quotient we are seeking, so that

$$x=\frac{\cos\theta+\sqrt{-1}\sin\theta}{\cos\theta'+\sqrt{-1}\sin\theta'}.$$

This quotient ought to be a new imaginary expression chosen so that when it is multiplied by $\cos\theta'+\sqrt{-1}\sin\theta'$ it reproduces $\cos\theta+\sqrt{-1}\sin\theta$. In other words, x ought to satisfy the equation

$$\left(\cos\theta'+\sqrt{-1}\sin\theta'\right)x=\cos\theta+\sqrt{-1}\sin\theta.$$

To solve this equation for x, it suffices to multiply both sides by

$$\cos\theta'-\sqrt{-1}\sin\theta'.$$

[165] In this way we reduce the coefficient of x to 1 (see theorem II, corollary I), and we find that

$$\begin{aligned} x &= \left(\cos\theta+\sqrt{-1}\sin\theta\right)\left(\cos\theta'-\sqrt{-1}\sin\theta'\right) \\ &= \left(\cos\theta+\sqrt{-1}\sin\theta\right)\left[\cos\left(-\theta'\right)+\sqrt{-1}\sin\left(-\theta'\right)\right] \\ &= \cos\left(\theta-\theta'\right)+\sqrt{-1}\sin\left(\theta-\theta'\right). \end{aligned}$$

---

[9] The minus sign preceding the final term in the next equation is a plus sign in [Cauchy 1897, p. 164]. This error was not present in [Cauchy 1821, p. 188]. (tr.)

Thus, we definitely have

$$\text{(8)} \qquad \frac{\cos\theta+\sqrt{-1}\sin\theta}{\cos\theta'+\sqrt{-1}\sin\theta'}=\cos\left(\theta-\theta'\right)+\sqrt{-1}\sin\left(\theta-\theta'\right).$$

Corollary. — If we take $\theta=0$ in equation (8), we have

$$\text{(9)} \qquad \frac{1}{\cos\theta'+\sqrt{-1}\sin\theta'}=\cos\theta'-\sqrt{-1}\sin\theta'.$$

Theorem V.[10] — *To raise the imaginary expression*

$$\cos\theta+\sqrt{-1}\sin\theta$$

to the power m (where m denotes any integer number), it suffices to multiply the arc θ in this expression by the number m.

Proof. — Indeed, because the arcs θ, θ', θ'', ... could be arbitrary in formula (7), if we suppose that they are all equal to θ, and that there are m of them, we find

$$\text{(10)} \qquad \left(\cos\theta+\sqrt{-1}\sin\theta\right)^m=\cos m\theta+\sqrt{-1}\sin m\theta.$$

Corollary. — If in equation (10) we take successively $\theta=z$ and then $\theta=-z$, we get the following two equations:

$$\text{(11)} \qquad \begin{cases} \left(\cos z+\sqrt{-1}\sin z\right)^m=\cos mz+\sqrt{-1}\sin mz \quad \text{and} \\ \left(\cos z-\sqrt{-1}\sin z\right)^m=\cos mz-\sqrt{-1}\sin mz. \end{cases}$$

Because they are always the product of m equal factors, the left-hand sides of each of these last equations can be expanded by ordinary multiplication of these factor, or what amounts to the same thing, by the [166] formula of Newton.[11] After expanding the equation, if we equate corresponding parts in each equation: 1° the real parts and 2° the coefficients of $\sqrt{-1}$, we conclude

$$\text{(12)} \qquad \begin{cases} \cos mz=\cos^m z-\frac{m(m-1)}{1\cdot 2}\cos^{m-2}z\sin^2 z \\ \qquad\qquad +\frac{m(m-1)(m-2)(m-3)}{1\cdot 2\cdot 3\cdot 4}\cos^{m-4}z\sin^4 z-\ldots, \\ \sin mz=\frac{m}{1}\cos z^{m-1}\sin z \\ \qquad\qquad -\frac{m(m-1)(m-2)}{1\cdot 2\cdot 3}\cos^{m-3}z\sin^3 z+\ldots. \end{cases}$$

[10] This theorem is known as de Moivre's Theorem, although it seems de Moivre did not give the result in this form; see [Kline 1990, vol. 2, p. 409]. Euler treated this material in more detail in [Euler 1748, vol. 1, ch. VIII, § 132 ff].

[11] That is, Newton's binomial formula.

Supposing $m = 2$, for example, we find

$$\cos 2z = \cos^2 z - \sin^2 z \quad \text{and}$$
$$\sin 2z = 2\sin z \cos z.$$

Supposing $m = 3$,

$$\cos 3z = \cos^3 z - 3\cos z \sin^2 z \quad \text{and}$$
$$\sin 3z = 3\cos^2 z \sin z - \sin^3 z,$$

and so on.

Theorem VI. — *To raise the imaginary expression*

$$\cos\theta + \sqrt{-1}\sin\theta$$

to the power $-m$, (where m denotes any integer number), it suffices to multiply the arc θ in this expression by the degree $-m$.

Proof. — Indeed, from the definition we have given of negative powers (see § I), we get

$$\left(\cos\theta + \sqrt{-1}\sin\theta\right)^{-m} = \frac{1}{\left(\cos\theta + \sqrt{-1}\sin\theta\right)^m} = \frac{1}{\cos m\theta + \sqrt{-1}\sin m\theta}.$$

Consequently, using formula (9) we get

(13) $$\left(\cos\theta + \sqrt{-1}\sin\theta\right)^{-m} = \cos m\theta - \sqrt{-1}\sin m\theta,$$

[167] or what amounts to the same thing,

(14) $$\left(\cos\theta + \sqrt{-1}\sin\theta\right)^{-m} = \cos(-m\theta) + \sqrt{-1}\sin(-m\theta).$$

After establishing the principal properties of reduced expressions, as we have just done, it becomes easy to multiply or divide two or more imaginary expressions if we know their moduli, as well as to raise any imaginary expression to a power m or $-m$, (where m denotes an integer number). Indeed, we can easily perform these different operations with the aid of the following theorems:

Theorem VII. — *To obtain the product of two or more imaginary expressions, it suffices to multiply the product of the reduced expressions to which they correspond by the product of the moduli.*

Proof. — The stated theorem follows immediately from the principle that the product of several factors, real or imaginary, remains the same regardless of the order in which one multiplies them. Indeed, let

$$\rho\left(\cos\theta+\sqrt{-1}\sin\theta\right),\ \rho'\left(\cos\theta'+\sqrt{-1}\sin\theta'\right),$$
$$\rho''\left(\cos\theta''+\sqrt{-1}\sin\theta''\right),\ \ldots$$

be several imaginary expressions, where ρ, ρ', ρ'', ... denote their moduli. When we want to multiply these expressions together, where each expression is the product of a modulus and a reduced expression, we can, by virtue of the principle just mentioned, form one part as the product of the moduli, and the other as the product of all the reduced expressions, then multiply together these two products. In this way, we find that the final result is

$$\rho\rho'\rho''\ldots\left[\cos\left(\theta+\theta'+\theta''+\ldots\right)+\sqrt{-1}\sin\left(\theta+\theta'+\theta''+\ldots\right)\right]. \tag{15}$$

Corollary I. — The product of several imaginary expressions is a new imaginary expression which has as its modulus the product of the moduli of all the others.

Corollary II. — Because an imaginary expression can never vanish [168] unless its modulus vanishes, and because in order to make the product of several moduli vanish, it is necessary that one of them reduces to zero, it is clear that one may draw from theorem VII the following conclusion:

The product of two or more imaginary expressions cannot vanish except when one of the factors reduces to zero.

Theorem VIII. — *To obtain the quotient of two imaginary expressions, it suffices to multiply the quotient of their corresponding reduced expressions by the quotient of their moduli.*

Proof. — Suppose that it is a matter of dividing the imaginary expression

$$\rho\left(\cos\theta+\sqrt{-1}\sin\theta\right),$$

where the modulus is ρ, by the following

$$\rho'\left(\cos\theta'+\sqrt{-1}\sin\theta'\right),$$

where the modulus is ρ'. If we denote by x the desired quotient, then x must be a new imaginary expression satisfying the equation

$$\rho'\left(\cos\theta'+\sqrt{-1}\sin\theta'\right)x=\rho\left(\cos\theta+\sqrt{-1}\sin\theta\right).$$

To solve this equation for the value of x, we multiply both sides by the product of the two factors

$$\frac{1}{\rho'} \quad \text{and} \quad \cos\theta' + \sqrt{-1}\sin\theta'.$$

In this way we find, writing $\frac{\rho}{\rho'}$ in place of $\rho\frac{1}{\rho'}$, that

$$x = \frac{\rho}{\rho'}\left[\cos\left(\theta-\theta'\right) + \sqrt{-1}\sin\left(\theta-\theta'\right)\right].$$

Thus, in the final analysis we have

$$(16) \qquad \frac{\rho\left(\cos\theta + \sqrt{-1}\sin\theta\right)}{\rho'\left(\cos\theta' + \sqrt{-1}\sin\theta'\right)} = \frac{\rho}{\rho'}\left[\cos\left(\theta-\theta'\right) + \sqrt{-1}\sin\left(\theta-\theta'\right)\right].$$

[169] Because, by virtue of theorem IV,

$$\cos\left(\theta-\theta'\right) + \sqrt{-1}\sin\left(\theta-\theta'\right)$$

is precisely the quotient of the two reduced expressions

$$\cos\theta + \sqrt{-1}\sin\theta \quad \text{and} \quad \cos\theta' + \sqrt{-1}\sin\theta',$$

it is clear that, having established formula (16), we ought to consider theorem VIII as being proved.

Corollary. — If we take $\theta = 0$ in equation (16),[12] we have

$$(17) \qquad \frac{1}{\rho'\left(\cos\theta' + \sqrt{-1}\sin\theta'\right)} = \frac{1}{\rho'}\left(\cos\theta' - \sqrt{-1}\sin\theta'\right).$$

Theorem IX. — *To obtain the mth power of an imaginary expression (where m denotes any integer number), it suffices to multiply the mth power of the corresponding reduced expression by the mth power of the modulus.*

Proof. — Indeed, if in theorem VII we take the imaginary expressions

$$\begin{array}{l} \rho\ \left(\cos\theta + \sqrt{-1}\sin\theta\ \right), \\ \rho'\ \left(\cos\theta' + \sqrt{-1}\sin\theta'\right), \\ \rho''\left(\cos\theta'' + \sqrt{-1}\sin\theta''\right), \\ \cdots\cdots\cdots\cdots\cdots\cdots \end{array}$$

all to be equal to each other and to be m in number, their product will be equivalent to the mth power of the first one, that is to say, equal to

$$\left[\rho\left(\cos\theta + \sqrt{-1}\sin\theta\right)\right]^m.$$

Under this hypothesis expression (15) becomes

[12] Apparently Cauchy means to take $\rho = 1$ as well as $\theta = 0$.

$$\rho^m \left(\cos m\theta + \sqrt{-1}\sin m\theta\right).$$

Ultimately, we have

$$\left[\rho \left(\cos\theta + \sqrt{-1}\sin\theta\right)\right]^m = \rho^m \left(\cos m\theta + \sqrt{-1}\sin m\theta\right). \tag{18}$$

[170] The reduced expression

$$\cos m\theta + \sqrt{-1}\sin m\theta$$

is equal (by virtue of theorem V) to

$$\left(\cos\theta + \sqrt{-1}\sin\theta\right)^m.$$

Thus, having established formula (18), it follows that we ought to consider theorem IX to be proved.

Theorem X. — *To raise an imaginary expression to the power* $-m$ (where m denotes an integer number), it suffices to form the same powers of the modulus and of the reduced expression, then to multiply the two parts together.

Proof. — Suppose that it is a matter of raising the following imaginary expression to the power $-m$

$$\rho \left(\cos\theta + \sqrt{-1}\sin\theta\right),$$

where the modulus is ρ. By virtue of the definition of negative powers, we have

$$\begin{aligned}\left[\rho \left(\cos\theta + \sqrt{-1}\sin\theta\right)\right]^{-m} &= \frac{1}{\left[\rho \left(\cos\theta + \sqrt{-1}\sin\theta\right)\right]^m} \\ &= \frac{1}{\rho \left(\cos m\theta + \sqrt{-1}\sin m\theta\right)}.\end{aligned}$$

Consequently, making use of formula (17), we find

$$\left[\rho \left(\cos\theta + \sqrt{-1}\sin\theta\right)\right]^{-m} = \frac{1}{\rho^m} \left(\cos m\theta - \sqrt{-1}\sin m\theta\right)$$

or what amounts to the same thing,

$$\left[\rho \left(\cos\theta + \sqrt{-1}\sin\theta\right)\right]^{-m} = \rho^{-m} \left(\cos m\theta - \sqrt{-1}\sin m\theta\right). \tag{19}$$

This last formula, together with equation (13), gives the complete proof of theorem X.

7.3 On the real and imaginary roots of the two quantities $+1$ and -1 and on their fractional powers.

[171] Suppose that m and n denote two relatively prime integer numbers. If we use the notations adopted in § I, the nth roots of unity, or what amounts to the same thing, the powers of degree $\frac{1}{n}$, are the various values of the expression

$$\sqrt[n]{1} = ((1))^{\frac{1}{n}},$$

and likewise, the fractional positive or negative powers of unity of degree $\frac{m}{n}$ or $-\frac{m}{n}$ are the various values of

$$((1))^{\frac{m}{n}} \quad \text{or} \quad ((1))^{-\frac{m}{n}}.$$

Thus we conclude that to determine these roots and powers it suffices to solve the following three problems, one after another.

Problem I. — *To find the various real and imaginary values of the expression*

$$((1))^{\frac{1}{n}}.$$

Solution. — Let x be one of these values, and in order to present it in a general form that includes the real quantities and the imaginary quantities at the same time, suppose that

$$x = r\left(\cos t + \sqrt{-1}\sin t\right),$$

where r denotes a positive quantity and t denotes a real arc. Because of the definition of the expression $((1))^{\frac{1}{n}}$, we have that

$$x^n = 1, \tag{1}$$

or what amounts to the same thing,

$$r^n\left(\cos nt + \sqrt{-1}\sin nt\right) = 1.$$

[172] We can draw from this last equation (with the aid of theorem I, § II)

$$r^n = 1 \quad \text{and}$$
$$\cos nt + \sqrt{-1}\sin nt = 1,$$

and so,

$$r = 1,$$
$$\cos nt = 1, \quad \sin nt = 0, \quad nt = \pm 2k\pi \quad \text{and}$$
$$t = \pm\frac{2k\pi}{n},$$

where k represents any integer number. The quantities r and t being thereby determined, the various values that satisfy equation (1) are evidently contained in the formula

(2) $$x = \cos\frac{2k\pi}{n} \pm \sqrt{-1}\sin\frac{2k\pi}{n}.$$

In other words, the various values of $((1))^{\frac{1}{n}}$ are given by the equation

(3) $$((1))^{\frac{1}{n}} = \cos\frac{2k\pi}{n} \pm \sqrt{-1}\sin\frac{2k\pi}{n}.$$

Now let h be the integer number closest to the ratio $\frac{k}{n}$. The difference between the numbers h and $\frac{k}{n}$ will be at most equal to $\frac{1}{2}$, so that we have

$$\frac{k}{n} = h \pm \frac{k'}{n},$$

where $\frac{k'}{n}$ denotes a fraction equal to or less than $\frac{1}{2}$, and as a consequence k' is a integer number less than, or at most equal to $\frac{n}{2}$. From this we conclude

$$\tfrac{2k\pi}{n} = 2h\pi \pm \tfrac{2k'\pi}{n} \quad \text{and}$$
$$\cos\tfrac{2k\pi}{n} \pm \sqrt{-1}\sin\tfrac{2k\pi}{n} = \cos\tfrac{2k'\pi}{n} \pm \sqrt{-1}\sin\tfrac{2k'\pi}{n}.$$

Consequently, all the values of $((1))^{\frac{1}{n}}$ are contained in the formula

$$\cos\frac{2k'\pi}{n} \pm \sqrt{-1}\sin\frac{2k'\pi}{n},$$

[173] if we suppose that k' is contained between the limits 0 and $\frac{n}{2}$, or what amounts to the same thing, if we suppose that k is contained between the same limits in formula (3).

Corollary I. — When n is even, the various values that the integer number k can assume without going outside the limits 0 and $\frac{n}{2}$ are, respectively,

$$0, \quad 1, \quad 2, \quad \ldots, \quad \frac{n-2}{2} \quad \text{and} \quad \frac{n}{2}.$$

In general, for each of these values of k, formula (3) gives two conjugate imaginary values of the expression $((1))^{\frac{1}{n}}$, that is to say, two conjugate imaginary roots of unity of degree n. However, for $k = 0$ we find but a single real root, $+1$, and for $k = \frac{n}{2}$ another real root, -1. In summary, when n is even, the expression

$$((1))^{\frac{1}{n}}$$

admits two real values, namely

$$+1 \quad \text{and} \quad -1,$$

along with $n-2$ imaginary values, conjugate in pairs, namely

$$(4)\qquad \begin{cases} \cos\frac{2\pi}{n}+\sqrt{-1}\sin\frac{2\pi}{n}, & \cos\frac{2\pi}{n}-\sqrt{-1}\sin\frac{2\pi}{n}, \\ \cos\frac{4\pi}{n}+\sqrt{-1}\sin\frac{4\pi}{n}, & \cos\frac{4\pi}{n}-\sqrt{-1}\sin\frac{4\pi}{n}, \\ \dots\dots\dots\dots, & \dots\dots\dots\dots, \\ \cos\frac{(n-2)\pi}{n}+\sqrt{-1}\sin\frac{(n-2)\pi}{n}, & \cos\frac{(n-2)\pi}{n}-\sqrt{-1}\sin\frac{(n-2)\pi}{n}. \end{cases}$$

The total number of these values, real and imaginary, is equal to n.

Suppose, for example, that $n=2$. We find that there exist two values of the expression

$$((1))^{\frac{1}{2}},$$

[174] or what amounts to the same thing, two values of x that satisfy the equation

$$x^2=1,$$

and that these values, both real, are, respectively,

$$+1 \quad \text{and} \quad -1.$$

Now suppose that $n=4$. We find that there are four values of the expression

$$((1))^{\frac{1}{4}},$$

or what amounts to the same thing, four values of x that satisfy the equation

$$x^4=1.$$

Among these four values, two of them are real, namely

$$+1 \quad \text{and} \quad -1.$$

The other two are imaginary and are, respectively, equal to

$$\cos\frac{\pi}{2}+\sqrt{-1}\sin\frac{\pi}{2}=+\sqrt{-1},$$

and to

$$\cos\frac{\pi}{2}-\sqrt{-1}\sin\frac{\pi}{2}=-\sqrt{-1}.$$

Corollary II. — When n is odd, the various values that the integer number k can assume without going outside the limits 0 and $\frac{n}{2}$ are, respectively,

$$0, \quad 1, \quad 2, \quad \dots, \quad \frac{n-1}{2}.$$

In general, for each of these values of k, formula (3) gives two conjugate imaginary values of the expression $((1))^{\frac{1}{n}}$, that is to say two conjugate imaginary roots of unity of degree n. However, for $k = 0$ we find but a single real root, namely $+1$. In summary, when n is odd, the expression

$$((1))^{\frac{1}{n}}$$

[175] admits the single real value

$$+1,$$

along with $n-1$ imaginary values, conjugate in pairs, namely

$$(5) \qquad \begin{cases} \cos\frac{2\pi}{n} + \sqrt{-1}\sin\frac{2\pi}{n}, & \cos\frac{2\pi}{n} - \sqrt{-1}\sin\frac{2\pi}{n}, \\ \cos\frac{4\pi}{n} + \sqrt{-1}\sin\frac{4\pi}{n}, & \cos\frac{4\pi}{n} - \sqrt{-1}\sin\frac{4\pi}{n}, \\ \dots\dots\dots\dots, & \dots\dots\dots\dots, \\ \cos\frac{(n-1)\pi}{n} + \sqrt{-1}\sin\frac{(n-1)\pi}{n}, & \cos\frac{(n-1)\pi}{n} - \sqrt{-1}\sin\frac{(n-1)\pi}{n}. \end{cases}$$

The total number of these values, real and imaginary, is equal to n.

Suppose, for example, that $n = 3$. We find that there exist three values of the expression

$$((1))^{\frac{1}{3}},$$

or what amounts to the same thing, three values of x that satisfy the equation

$$x^3 = 1,$$

and these values, of which one is real, are, respectively,

$$+1,$$

$$\cos\tfrac{2\pi}{3} + \sqrt{-1}\sin\tfrac{2\pi}{3} \quad \text{and} \quad \cos\tfrac{2\pi}{3} - \sqrt{-1}\sin\tfrac{2\pi}{3}.$$

Moreover, as we know, the side of the hexagon is equal to its radius and the supplement of the arc subtended by this side has as its measure $\frac{2\pi}{3}$, so we can easily obtain the equations

$$\cos\frac{2\pi}{3} = -\frac{1}{2} \quad \text{and} \quad \sin\frac{2\pi}{3} = +\frac{3^{\frac{1}{2}}}{2}.$$

By virtue of these equations, the imaginary values of the expression $((1))^{\frac{1}{3}}$ reduce to

$$-\frac{1}{2} + \frac{3^{\frac{1}{2}}}{2}\sqrt{-1} \quad \text{and} \quad -\frac{1}{2} - \frac{3^{\frac{1}{2}}}{2}\sqrt{-1}.$$

[176] *Corollary III.* — If n is any integer number, the number of values, real and imaginary, of the expression $((1))^{\frac{1}{n}}$, or what amounts to the same thing, the number of values of x that satisfy the equation $x^n = 1$, is always equal to n.

Problem II. — *To find the various values, real and imaginary, of the expression*

$$((1))^{\frac{m}{n}}.$$

Solution. — The numbers m and n are assumed to be relatively prime. Because of the definition of the expression $((1))^{\frac{m}{n}}$, we have that

$$((1))^{\frac{m}{n}} = \left[((1))^{\frac{1}{n}}\right]^m.$$

Substituting the general value for $((1))^{\frac{1}{n}}$ found in equation (3), we get

$$((1))^{\frac{m}{n}} = \left[\cos\frac{2k\pi}{n} \pm \sqrt{-1}\sin\frac{2k\pi}{n}\right]^m$$

and so

(6) $$((1))^{\frac{m}{n}} = \cos\frac{m\cdot 2k\pi}{n} \pm \sqrt{-1}\sin\frac{m\cdot 2k\pi}{n}.$$

To deduce all of the values of $((1))^{\frac{m}{n}}$ from this last formula, one needs only to give k the integer values between 0 and $\frac{n}{2}$ successively. Let k' and k'' be two such values, assumed to be unequal. I say that the cosines

$$\cos\frac{m\cdot 2k'\pi}{n} \quad \text{and} \quad \cos\frac{m\cdot 2k''\pi}{n}$$

are necessarily different from each other. Indeed, these cosines cannot be equal except in the case where the arcs to which they correspond are related to each other by an equation of the form

$$\frac{m\cdot 2k'\pi}{n} = \pm 2h\pi \pm \frac{m\cdot 2k''\pi}{n},$$

[177]where h denotes an integer number. Now from this equation we get

$$h = \frac{m\,(\pm k' \pm k'')}{n}.$$

Thus, because m is relatively prime to n, it is necessary that $\pm k' \pm k''$ be divisible by n, which cannot happen because the numbers k' and k'' are unequal and each of them cannot exceed $\frac{1}{2}n$, so their sum and their differences are necessarily less than n. Thus, two different values of k contained between the limits 0 and $\frac{1}{2}n$ give two different values of

$$\cos\frac{m\cdot 2k\pi}{n}.$$

From this remark, we easily conclude that the values, real or imaginary, of the expression $((1))^{\frac{m}{n}}$ given by equation (6) are the same in number as the real and imaginary roots of $((1))^{\frac{1}{n}}$ determined by equation (3). Moreover, because we evidently have

$$\left(\cos\tfrac{m\cdot 2k\pi}{n} \pm \sqrt{-1}\sin\tfrac{m\cdot 2k\pi}{n}\right)^n = \cos(m\cdot 2k\pi) \pm \sqrt{-1}\sin(m\cdot 2k\pi) = 1,$$

it follows that every value of $((1))^{\frac{m}{n}}$ is a real or imaginary expression, the nth power of which equals 1, and thus is a value of $((1))^{\frac{1}{n}}$. These observations lead to the formula

(7) $$((1))^{\frac{m}{n}} = ((1))^{\frac{1}{n}}$$

in which the sign = indicates only that each of the values on the left-hand side is always equal to one of the values on the right-hand side.[13]

Problem III. — *To find the various values, real and imaginary, of the expression*

$$((1))^{-\frac{m}{n}}.$$

Solution. — From the definition of negative powers, we have that

$$((1))^{-\frac{m}{n}} = \frac{1}{((1))^{\frac{m}{n}}}.$$

[178] Substituting the general value for $((1))^{\frac{m}{n}}$ found in equation (6), and considering formula (9) of the preceding section, we get

(8) $$((1))^{-\frac{m}{n}} = \cos\frac{m\cdot 2k\pi}{n} \mp \sqrt{-1}\sin\frac{m\cdot 2k\pi}{n}.$$

It follows from this last equation that the various values of $((1))^{-\frac{m}{n}}$ are the same as those of $((1))^{\frac{m}{n}}$ and consequently are equal to those of $((1))^{\frac{1}{n}}$. Thus we have

(9) $$((1))^{-\frac{m}{n}} = ((1))^{\frac{1}{n}},$$

where the sign = ought to be interpreted as in equation (7).

Corollary. — If we make $m = 1$ in formula (9), it gives

(10) $$((1))^{-\frac{1}{n}} = ((1))^{\frac{1}{n}}.$$

Now suppose that we seek roots and fractional powers, not of unity, but of the quantity −1. The nth roots of this quantity, or what amounts to the same thing, its powers of degree $\frac{1}{n}$, are the various values of the expression

[13] This is a remarkable proof. The modern reader might have trouble even stating the conclusion without recourse to such notions as "set," "subset," "one-to-one" and "cardinality," none of which were available to Cauchy.

$$\sqrt[n]{-1} = ((-1))^{\frac{1}{n}},$$

and likewise the fractional powers of -1, positive or negative, of degree $\frac{m}{n}$ or $-\frac{m}{n}$ are the various values of

$$((1))^{\frac{m}{n}} \quad \text{or} \quad ((1))^{-\frac{m}{n}}.$$

As a consequence, to determine these roots and powers, it suffices to solve, one after another, the three problems that I will pose.

Problem IV. — *To find the various real and imaginary powers of the expression*

$$((-1))^{\frac{1}{n}}.$$

[179] *Solution.* — Let

$$x = r\left(\cos t + \sqrt{-1}\sin t\right)$$

be one of these values, where r denotes a positive quantity and t denotes a real arc. From the definition itself of the expression $((-1))^{\frac{1}{n}}$ we have

$$x^n = -1, \tag{11}$$

or what amounts to the same thing,

$$r^n\left(\cos nt + \sqrt{-1}\sin nt\right) = -1.$$

We conclude from this last equation (with the aid of theorem I, § II),

$$r^n = 1 \quad \text{and}$$
$$\cos nt + \sqrt{-1}\sin nt = -1.$$

It follows that

$$r = 1,$$
$$\cos nt = -1, \quad \sin nt = 0, \quad nt = \pm(2k+1)\pi \quad \text{and}$$
$$t = \pm\frac{(2k+1)\pi}{n},$$

where k represents any integer number. The quantities r and t being thereby determined, the various values of x that satisfy equation (11) are evidently contained in the formula

$$x = \cos\frac{(2k+1)\pi}{n} \pm \sqrt{-1}\sin\frac{(2k+1)\pi}{n}. \tag{12}$$

In other words, the various values of $((-1))^{\frac{1}{n}}$ are given by the equation

$$(13) \qquad ((-1))^{\frac{1}{n}} = \cos\frac{(2k+1)\pi}{n} \pm \sqrt{-1}\sin\frac{(2k+1)\pi}{n}.$$

Now let h be the integer number closest to the ratio $\frac{2k+1}{2n}$. The difference between the two numbers h and $\frac{2k+1}{n}$ is obviously a fraction with an odd numerator, less than or at most [180] equal to $\frac{1}{2}$. Thus it follows that we have

$$\frac{2k+1}{2n} = h \pm \frac{2k'+1}{2n},$$

where $2k'+1$ denotes an odd number less than or equal to n. We then conclude that

$$\frac{(2k+1)\pi}{n} = 2h\pi \pm \frac{(2k'+1)\pi}{n} \quad \text{and}$$

$$\cos\tfrac{(2k+1)\pi}{n} \pm \sqrt{-1}\sin\tfrac{(2k+1)\pi}{n} = \cos\tfrac{(2k'+1)\pi}{n} \pm \sqrt{-1}\sin\tfrac{(2k'+1)\pi}{n}.$$

Consequently, if we suppose that $2k'+1$ is contained between the limits 0 and n, then all the values of $((-1))^{\frac{1}{n}}$ are contained in the formula

$$\cos\frac{(2k'+1)\pi}{n} \pm \sqrt{-1}\sin\frac{(2k'+1)\pi}{n},$$

or what amounts to the same thing, the values are contained in formula (13), if we suppose that $2k+1$ is contained between the same limits.

Corollary I. — When n is even, the various values that $2k+1$ can assume without going outside the limits 0 and n are, respectively,

$$1, \quad 3, \quad 5, \quad \ldots, \quad \overline{n-1}.$$

For each of these values of $2k+1$, formula (13) always gives two conjugate imaginary values of the expression $((-1))^{\frac{1}{n}}$. Consequently, in the case we are considering here, this expression does not admit any real values, but only n imaginary values, conjugate in pairs, namely

$$(14) \qquad \begin{cases} \cos\frac{\pi}{n} + \sqrt{-1}\sin\frac{\pi}{n}, & \cos\frac{\pi}{n} - \sqrt{-1}\sin\frac{\pi}{n}, \\ \cos\frac{3\pi}{n} + \sqrt{-1}\sin\frac{3\pi}{n}, & \cos\frac{3\pi}{n} - \sqrt{-1}\sin\frac{3\pi}{n}, \\ \dots\dots\dots\dots\dots\dots, & \dots\dots\dots\dots\dots\dots, \\ \cos\frac{(n-1)\pi}{n} + \sqrt{-1}\sin\frac{(n-1)\pi}{n}, & \cos\frac{(n-1)\pi}{n} - \sqrt{-1}\sin\frac{(n-1)\pi}{n}. \end{cases}$$

[181] Suppose for example that $n = 2$. We find that there are two values of the expression $((-1))^{\frac{1}{2}}$, or what amounts to the same thing, two values of x that satisfy the equation

$$x^2 = -1,$$

and that these values, both imaginary, are, respectively,

$$\cos\tfrac{\pi}{2}+\sqrt{-1}\sin\tfrac{\pi}{2}=+\sqrt{-1}\quad\text{and}$$

$$\cos\tfrac{\pi}{2}-\sqrt{-1}\sin\tfrac{\pi}{2}=-\sqrt{-1}.$$

Now suppose that $n=4$. We see that there are four values of the expression $((-1))^{\frac{1}{4}}$, or in other words, four values of x that satisfy the equation

$$x^4=-1,$$

and that these four values are contained in the two formulas

$$\cos\tfrac{\pi}{4}\pm\sqrt{-1}\sin\tfrac{\pi}{4}\quad\text{and}$$

$$\cos\tfrac{3\pi}{4}\pm\sqrt{-1}\sin\tfrac{3\pi}{4},$$

or what amounts to the same thing, in the single formula

$$\pm\cos\frac{\pi}{4}\pm\sqrt{-1}\sin\frac{\pi}{4}.$$

Moreover, because we have

$$\cos\frac{\pi}{4}=\sin\frac{\pi}{4}=\frac{1}{\sqrt{2}},$$

we finally find that

$$((-1))^{\frac{1}{4}}=\pm\frac{1}{2^{\frac{1}{2}}}\pm\frac{1}{2^{\frac{1}{2}}}\sqrt{-1}.$$

Corollary II. — When n is odd, the various values that $2k+1$ can assume without going outside the limits 0 and n are, respectively,

$$1,\quad 3,\quad 5,\quad \ldots,\quad n-2\quad\text{and}\quad n.$$

[182] In general, for each of these values of $2k+1$, formula (13) gives two conjugate imaginary values of the expression $((-1))^{\frac{1}{n}}$, that is to say two conjugate imaginary roots of -1 of degree n. However, for $2k+1=n$, we find but a single real root, namely -1. In summary, when n is odd, the expression $((-1))^{\frac{1}{n}}$ admits only the one real value,

$$-1,$$

along with $n-1$ imaginary roots, conjugate in pairs, namely

$$(15)\qquad\begin{cases}\cos\frac{\pi}{n}+\sqrt{-1}\sin\frac{\pi}{n}, & \cos\frac{\pi}{n}-\sqrt{-1}\sin\frac{\pi}{n},\\ \cos\frac{3\pi}{n}+\sqrt{-1}\sin\frac{3\pi}{n}, & \cos\frac{3\pi}{n}-\sqrt{-1}\sin\frac{3\pi}{n},\\ \ldots\ldots\ldots\ldots\ldots\ldots, & \ldots\ldots\ldots\ldots\ldots\ldots,\\ \cos\frac{(n-2)\pi}{n}+\sqrt{-1}\sin\frac{(n-2)\pi}{n}, & \cos\frac{(n-2)\pi}{n}-\sqrt{-1}\sin\frac{(n-2)\pi}{n}.\end{cases}$$

The total number of these values, real and imaginary, is equal to n.

Suppose, for example, that $n = 3$. We find that there exist three values of the expression $((-1))^{\frac{1}{3}}$, or what amounts to the same thing, values of x that satisfy the equation

$$x^3 = -1,$$

and these values, of which one is real, are, respectively,

$$-1,$$

$$\cos\tfrac{\pi}{3} + \sqrt{-1}\sin\tfrac{\pi}{3} = \tfrac{1}{2} + \tfrac{3^{\frac{1}{2}}}{2}\sqrt{-1} \quad \text{and}$$

$$\cos\tfrac{\pi}{3} + \sqrt{-1}\sin\tfrac{\pi}{3} = \tfrac{1}{2} - \tfrac{3^{\frac{1}{2}}}{2}\sqrt{-1}.$$

Corollary III. — If n is any integer number, the number of values, real and imaginary, of the expression $((-1))^{\frac{1}{n}}$, or what amounts to the same thing, the number of values of x that satisfy the equation $x^n = -1$, is always equal to n.

[183] **Problem V.** — To find the various values, real and imaginary, of the expression

$$((-1))^{\frac{m}{n}}.$$

Solution. — The numbers m and n are assumed to be relatively prime. Because of the definition of the expression $((-1))^{\frac{m}{n}}$, we have that

$$((-1))^{\frac{m}{n}} = \left[((-1))^{\frac{1}{n}}\right]^m.$$

Substituting the general value for $((-1))^{\frac{1}{n}}$ found in equation (13), we get

$$(16) \qquad ((-1))^{\frac{m}{n}} = \cos\frac{m\,(2k+1)\,\pi}{n} \pm \sqrt{-1}\sin\frac{m\,(2k+1)\,\pi}{n}.$$

To deduce all of the values of $((-1))^{\frac{m}{n}}$ from this last formula, one needs only successively to give to $2k+1$ all the odd, integer values between 0 and n. Let $2k'+1$ and $2k''+1$ be two such values, assumed to be unequal. I say that the cosines

$$\cos\frac{m\,(2k'+1)\,\pi}{n} \quad \text{and} \quad \cos\frac{m\,(2k''+1)\,\pi}{n}$$

are necessarily different from each other. Indeed, these cosines cannot be equal except in the case where the arcs to which they correspond are related to each other by an equation of the form

$$\frac{m\,(2k'+1)\,\pi}{n} = \pm 2h\pi \pm \frac{m\,(2k''+1)\,\pi}{n},$$

where h denotes an integer number. Now from this equation we get

$$h = \frac{m\left[\frac{\pm(2k'+1)\pm(2k''+1)}{2}\right]}{n}.$$

Thus, because m is relatively prime to n, it is necessary that the integer number

$$\frac{\pm(2k'+1)\pm(2k''+1)}{2}$$

[184] be divisible by n, which cannot happen because the numbers $2k'+1$ and $2k''+1$ are unequal and each of them cannot exceed n, so their half-sum, and *a fortiori* their half-difference, is necessarily less than n. Thus, two different values of $2k+1$ between 0 and n give two different values of

$$\cos\frac{m(2k+1)\pi}{n}.$$

From this remark we easily conclude that the values, real or imaginary, of the expression $((-1))^{\frac{m}{n}}$ given by equation (16) are n in number, like those of $((1))^{\frac{1}{n}}$ and of $((-1))^{\frac{1}{n}}$. Moreover, because we evidently have

$$\left[\cos\frac{m(2k+1)\pi}{n} \pm \sqrt{-1}\sin\frac{m(2k+1)\pi}{n}\right]^n$$
$$= \cos m(2k+1)\pi \pm \sqrt{-1}\sin m(2k+1)\pi = (-1)^m = \pm 1,$$

it follows that every value of $((-1))^{\frac{m}{n}}$ is a real or imaginary expression, the nth power of which equals ± 1, and thus is a value of $((1))^{\frac{1}{n}}$ or of $((-1))^{\frac{1}{n}}$. This remark leads to the equation

(17) $$((-1))^{\frac{m}{n}} = ((1))^{\frac{1}{n}}$$

every time that $(-1)^m = 1$, that is to say, whenever m is an even number, and it leads to

(18) $$((-1))^{\frac{m}{n}} = ((-1))^{\frac{1}{n}}$$

whenever $(-1)^m = -1$, that is to say, whenever m is an odd number. Let us add that we could combine equations (17) and (18) into a single formula by writing

(19) $$((-1))^{\frac{m}{n}} = (((-1)^m))^{\frac{1}{n}}.$$

[185] **Problem VI.** — *To find the various values, real and imaginary, of the expression*

$$((-1))^{-\frac{m}{n}}.$$

Solution. — From the definition of negative powers, we have that

$$((-1))^{-\frac{m}{n}} = \frac{1}{((-1))^{\frac{m}{n}}}.$$

Substituting the general value for $((-1))^{-\frac{m}{n}}$ found in equation (16), and considering formula (9) of the preceding section, we get

$$(20)\qquad ((-1))^{-\frac{m}{n}} = \cos\frac{m(2k+1)\pi}{n} \mp \sqrt{-1}\sin\frac{m(2k+1)\pi}{n}.$$

It follows from this last equation that the various values of $((-1))^{-\frac{m}{n}}$ are the same as those of $((1))^{\frac{m}{n}}$. As a consequence we get

$$(21)\qquad ((-1))^{-\frac{m}{n}} = ((1))^{\frac{1}{n}} \quad \text{if } m \text{ is even}$$

and

$$(22)\qquad ((-1))^{-\frac{m}{n}} = ((-1))^{\frac{1}{n}} \quad \text{if } m \text{ is odd.}$$

In place of the two preceding formulas, we could content ourselves by writing instead

$$(23)\qquad ((-1))^{-\frac{m}{n}} = (((-1)^m))^{\frac{1}{n}}.$$

Corollary. — If we make $m = 1$ in formula (23), it gives

$$(24)\qquad ((-1))^{-\frac{1}{n}} = ((-1))^{\frac{1}{n}}.$$

To complete this section, we remark that equations (3), (6), (8), (13), (16) and (20), with the aid of which we have determined the values of the expressions

$$((1))^{\frac{1}{n}},\quad ((1))^{\frac{m}{n}},\quad ((1))^{-\frac{m}{n}},$$
$$((-1))^{\frac{1}{n}},\ ((-1))^{\frac{m}{n}},\ ((-1))^{-\frac{m}{n}},$$

[186] can be replaced by two formulas. Indeed, if we denote by a a quantity, positive or negative, but with a fractional numerical value, the value of $((1))^a$ determined by equation (3), (6) or (8) is evidently

$$(25)\qquad ((1))^a = \cos 2ka\pi \pm \sqrt{-1}\sin 2ka\pi,$$

while the value of $((-1))^a$ determined by equation (13), (16) or (20) is

$$(26)\qquad ((-1))^a = \cos(2k+1)a\pi \pm \sqrt{-1}\sin(2k+1)a\pi.$$

In the two preceding formulas, we may take any integer number whatsoever for k.

7.4 On the roots of imaginary expressions, and on their fractional and irrational powers.

Let

$$\alpha+\beta\sqrt{-1}$$

be any imaginary expression. We can always find (see § II) a positive value of ρ and infinitely many real values of θ that satisfy the equation

$$\alpha+\beta\sqrt{-1}=\rho\left(\cos\theta+\sqrt{-1}\sin\theta\right). \tag{1}$$

Given this, imagine that we denote two relatively prime integer numbers by m and n. If we use the notations adopted in § I, the nth roots of the expression $\alpha+\beta\sqrt{-1}$, or what amounts to the same thing, its powers of degree $\frac{1}{n}$, are the various values of[14]

$$\sqrt[n]{\alpha+\beta\sqrt{-1}}=\left(\left(\alpha+\beta\sqrt{-1}\right)\right)^{\frac{1}{n}}$$

and likewise, the fractional powers of $\alpha+\beta\sqrt{-1}$, positive or negative, of degree $\frac{m}{n}$ or $-\frac{m}{n}$, are the various values of

$$\left(\left(\alpha+\beta\sqrt{-1}\right)\right)^{\frac{m}{n}} \quad \text{or} \quad \left(\left(\alpha+\beta\sqrt{-1}\right)\right)^{-\frac{m}{n}}.$$

[187] As a consequence, to determine these roots and these powers, it suffices to solve, one after another, the following three problems:

Problem I. — *To find the various values of the expression*

$$\left(\left(\alpha+\beta\sqrt{-1}\right)\right)^{\frac{1}{n}}.$$

Solution. — Let

$$x=r\left(\cos t+\sqrt{-1}\sin t\right)$$

be one of these values, where r denotes a positive quantity and t a real arc. From the definition itself of the expression $((\alpha+\beta\sqrt{-1}))^{\frac{1}{n}}$, we have

$$x^n=\alpha+\beta\sqrt{-1}=\rho\left(\cos\theta+\sqrt{-1}\sin\theta\right), \tag{2}$$

or what amounts to the same thing,

$$r^n\left(\cos nt+\sqrt{-1}\sin nt\right)=\rho\left(\cos\theta+\sqrt{-1}\sin\theta\right).$$

With the aid of theorem I, § II, we conclude from this last equation

$$r^n=\rho \quad \text{and}$$
$$\cos nt+\sqrt{-1}\sin nt=\cos\theta+\sqrt{-1}\sin\theta$$

[14] The n was omitted from the radical sign in [Cauchy 1897, p. 186]. This error was not present in [Cauchy 1821, p. 218]. (tr.)

and it follows that

$$r = \rho^{\frac{1}{n}},$$

$$\cos nt = \cos\theta, \quad \sin nt = \sin\theta, \quad nt = \theta \pm 2k\pi \quad \text{and}$$

$$t = \frac{\theta \pm 2k\pi}{n},$$

where k represents any integer number. The quantities r and t are thus determined, and so the various values of x that satisfy equation (1) are evidently given by the formula

$$\begin{aligned} x &= \rho^{\frac{1}{n}}\left(\cos\frac{\theta \pm 2k\pi}{n} + \sqrt{-1}\sin\frac{\theta \pm 2k\pi}{n}\right) \\ &= \rho^{\frac{1}{n}}\left(\cos\frac{\theta}{n} + \sqrt{-1}\sin\frac{\theta}{n}\right)\left(\cos\frac{2k\pi}{n} \pm \sqrt{-1}\sin\frac{2k\pi}{n}\right), \end{aligned}$$

or what amounts to the same thing, by the following:

(3) $$x = \rho^{\frac{1}{n}}\left(\cos\frac{\theta}{n} \pm \sqrt{-1}\sin\frac{\theta}{n}\right)((1))^{\frac{1}{n}}.$$

[188] In other words, the expression $((\alpha + \beta\sqrt{-1}))^{\frac{1}{n}}$, as well as $((1))^{\frac{1}{n}}$, gives n different values, determined by the equation

(4) $$\left(\left(\alpha + \beta\sqrt{-1}\right)\right)^{\frac{1}{n}} = \rho^{\frac{1}{n}}\left(\cos\frac{\theta}{n} \pm \sqrt{-1}\sin\frac{\theta}{n}\right)((1))^{\frac{1}{n}}.$$

Corollary I. — Suppose that $n = 2$. We find that there exist two values of the expression

$$\left(\left(\alpha + \beta\sqrt{-1}\right)\right)^{\frac{1}{2}},$$

or what amounts to the same thing, two values of x that satisfy the equation

$$x^2 = \alpha + \beta\sqrt{-1} = \rho\left(\cos\theta + \sqrt{-1}\sin\theta\right),$$

and that these two values are contained in the formula

$$\pm\rho^{\frac{1}{2}}\left(\cos\frac{\theta}{2} + \sqrt{-1}\sin\frac{\theta}{2}\right).$$

Corollary II. — Suppose now that $n = 3$. We find that there exist three values of the expression

$$\left(\left(\alpha + \beta\sqrt{-1}\right)\right)^{\frac{1}{3}},$$

or what amounts to the same thing, three values of x that satisfy the equation

$$x^3 = \alpha + \beta\sqrt{-1} = \rho\left(\cos\theta + \sqrt{-1}\sin\theta\right),$$

and that these three[15] values are, respectively,

$$\begin{aligned}
&\rho^{\frac{1}{3}}\left(\cos\tfrac{\theta}{3} + \sqrt{-1}\sin\tfrac{\theta}{3}\right), \\
&\rho^{\frac{1}{3}}\left(\cos\tfrac{\theta}{3} + \sqrt{-1}\sin\tfrac{\theta}{3}\right)\left(\cos\tfrac{2\pi}{3} + \sqrt{-1}\sin\tfrac{2\pi}{3}\right) \\
&\quad = \rho^{\frac{1}{2}}\left(\cos\tfrac{\theta+2\pi}{3} + \sqrt{-1}\sin\tfrac{\theta+2\pi}{3}\right) \quad \text{and} \\
&\rho^{\frac{1}{3}}\left(\cos\tfrac{\theta}{3} + \sqrt{-1}\sin\tfrac{\theta}{3}\right)\left(\cos\tfrac{2\pi}{3} - \sqrt{-1}\sin\tfrac{2\pi}{3}\right) \\
&\quad = \rho^{\frac{1}{3}}\left(\cos\tfrac{\theta-2\pi}{3} + \sqrt{-1}\sin\tfrac{\theta-2\pi}{3}\right).
\end{aligned}$$

[189] *Corollary III.* — Suppose finally that $n = 4$. We find that there are four values of the expression

$$\left(\left(\alpha + \beta\sqrt{-1}\right)\right)^{\frac{1}{4}},$$

or what amounts to the same thing, four values of x that satisfy the equation

$$x^4 = \alpha + \beta\sqrt{-1} = \rho\left(\cos\theta + \sqrt{-1}\sin\theta\right),$$

and that these four values are contained in the two formulas

$$\begin{aligned}
&\pm\rho^{\frac{1}{4}}\left(\cos\tfrac{\theta}{4} + \sin\tfrac{\theta}{4}\right) \quad \text{and} \\
&\pm\rho^{\frac{1}{4}}\left(\sin\tfrac{\theta}{4} - \cos\tfrac{\theta}{4}\right).
\end{aligned}$$

Problem II. — *To find the various values of the expression*

$$\left(\left(\alpha \mid \beta\sqrt{1}\right)\right)^{\frac{m}{n}}.$$

Solution. — The numbers m and n are assumed to be relatively prime. Because of the definition itself of the expression $((\alpha + \beta\sqrt{-1}))^{\frac{m}{n}}$, we have

$$\left(\left(\alpha + \beta\sqrt{-1}\right)\right)^{\frac{m}{n}} = \left[\left(\left(\alpha + \beta\sqrt{-1}\right)\right)^{\frac{1}{n}}\right]^m.$$

Substituting the general value for $((\alpha + \beta\sqrt{-1}))^{\frac{1}{n}}$ found in equation (4), we get

$$\left(\left(\alpha + \beta\sqrt{-1}\right)\right)^{\frac{m}{n}} = \rho^{\frac{m}{n}}\left(\cos\frac{m\theta}{n} + \sqrt{-1}\sin\frac{m\theta}{n}\right)((1))^{\frac{m}{n}}. \tag{5}$$

---

[15] [Cauchy 1897, p. 188] read "two values" instead of "three values" here. Also in [Cauchy 1897, p. 188], the first of these three values, displayed on the following line, had $\sqrt{-1}\sin\frac{3}{\theta}$ instead of $\sqrt{-1}\sin\frac{\theta}{3}$. Neither error was present in [Cauchy 1821, p. 220]. (tr.)

Corollary I. — If in equation (5) we substitute for $((1))^{\frac{m}{n}}$ its value given by formula (6) (§ III), we obtain the following:

$$(6)\qquad \left(\left(\alpha+\beta\sqrt{-1}\right)\right)^{\frac{m}{n}} = \rho^{\frac{m}{n}}\left[\cos\frac{m(\theta\pm 2k\pi)}{n}+\sqrt{-1}\sin\frac{m(\theta\pm 2k\pi)}{n}\right].$$

Problem III. — *To find the various values of the expression*

$$\left(\left(\alpha+\beta\sqrt{-1}\right)\right)^{-\frac{m}{n}}.$$

[190] *Solution.* — From the definition itself of negative powers, we have

$$\left(\left(\alpha+\beta\sqrt{-1}\right)\right)^{-\frac{m}{n}} = \frac{1}{((\alpha+\beta\sqrt{-1}))^{\frac{m}{n}}}.$$

Substituting the value for $((\alpha+\beta\sqrt{-1}))^{\frac{m}{n}}$ found in equation (6), and considering formula (17) of § II, we get

$$\begin{aligned} &((\alpha+\beta\sqrt{-1}))^{-\frac{m}{n}}\\ &= \rho^{-\frac{m}{n}}\left[\cos\frac{m(\theta\pm 2k\pi)}{n}-\sqrt{-1}\sin\frac{m(\theta\pm 2k\pi)}{n}\right]\\ &= \rho^{-\frac{m}{n}}\left(\cos\frac{m\theta}{n}-\sqrt{-1}\sin\frac{m\theta}{n}\right)\left(\cos\frac{m\cdot 2k\pi}{n}\mp\sqrt{-1}\sin\frac{m\cdot 2k\pi}{n}\right), \end{aligned}$$

or in other words,

$$(7)\qquad \left(\left(\alpha+\beta\sqrt{-1}\right)\right)^{-\frac{m}{n}} = \rho^{-\frac{m}{n}}\left(\cos\frac{m\theta}{n}-\sqrt{-1}\sin\frac{m\theta}{n}\right)((1))^{-\frac{m}{n}}.$$

Corollary I. — If we make $m=1$, then equation (7) gives

$$(8)\qquad \left(\left(\alpha+\beta\sqrt{-1}\right)\right)^{-\frac{1}{n}} = \rho^{-\frac{1}{n}}\left(\cos\frac{\theta}{n}-\sqrt{-1}\sin\frac{\theta}{n}\right)((1))^{-\frac{1}{n}}.$$

Having determined, as we have just done, the various values of the four expressions

$$\begin{aligned} &((\alpha+\beta\sqrt{-1}))^{\frac{1}{n}}, &&((\alpha+\beta\sqrt{-1}))^{\frac{m}{n}},\\ &((\alpha+\beta\sqrt{-1}))^{-\frac{1}{n}} &&\text{and } ((\alpha+\beta\sqrt{-1}))^{-\frac{m}{n}}, \end{aligned}$$

we see without trouble that equations (4), (5), (8) and (7), with the aid of which these values are determined, can be replaced by a single formula. If we let a be a quantity, positive or negative, with a numerical value that is fractional,[16] the formula in question is

[16] Here, as in Chapter V, Cauchy carefully treats the case where a is rational before extending his results to irrational values of a.

$$\left(\left(\alpha+\beta\sqrt{-1}\right)\right)^{a}=\rho^{a}\left(\cos a\theta+\sqrt{-1}\sin a\theta\right)((1))^{a}. \tag{9}$$

[191] In the above calculations, ρ always denotes the modulus of the imaginary expression $\alpha+\beta\sqrt{-1}$, that is to say the positive quantity $\sqrt{\alpha^2+\beta^2}$, and θ denotes one of the arcs that satisfy equation (1), or what amounts to the same thing, equations (4) of § II, namely

$$\begin{cases} \cos\theta=\dfrac{\alpha}{\sqrt{\alpha^2+\beta^2}} \quad \text{and} \\ \sin\theta=\dfrac{\beta}{\sqrt{\alpha^2+\beta^2}}. \end{cases} \tag{10}$$

By dividing these two formulas, we conclude

$$\tan\theta=\frac{\beta}{\alpha}. \tag{11}$$

Consequently, if we let ζ be the smallest arc, ignoring the sign, which has $\frac{\beta}{\alpha}$ as its tangent, in other words, if we make

$$\zeta=\arctan\frac{\beta}{\alpha}, \tag{12}$$

then we find

$$\tan\theta=\tan\zeta. \tag{13}$$

Given this, it becomes easy to introduce the arc ζ, whose value is completely determined, in place of the arc θ in the various formulas given above. Indeed, we arrive at this through the following considerations.

Because the arcs θ and ζ have the same tangent, they also have the same sine and the same cosine, ignoring the sign. Furthermore, because equation (13) can be put into the form

$$\frac{\sin\theta}{\cos\theta}=\frac{\sin\zeta}{\cos\zeta},$$

it is clear that, in order to satisfy that equation, we must either have both

$$\cos\theta=\cos\zeta \quad \text{and} \quad \sin\theta=\sin\zeta \tag{14}$$

or else both

$$\cos\theta=-\cos\zeta \quad \text{and} \quad \sin\theta=-\sin\zeta. \tag{15}$$

[192] Moreover, because the value of $\cos\theta$ determined by the first of equations (10) is evidently of the same sign as α, whereas the arc ζ, being contained between the limits $-\frac{\pi}{2}$ and $+\frac{\pi}{2}$, always has a positive cosine, it follows that equations (14) hold if α is positive, and equations (15) hold if α is negative. Now let us see how formulas (1) and (9) reduce under these two hypotheses.

First, if we suppose that α is positive, equations (10) can be replaced by equations (14), and we derive infinitely many values of θ, among which we ought to distinguish the following one:

$$\theta = \zeta. \tag{16}$$

When we use this value, formulas (1) and (9) become, respectively,

$$\alpha+\beta\sqrt{-1} = \rho\left(\cos\zeta+\sqrt{-1}\sin\zeta\right) \quad \text{and} \tag{17}$$

$$\left(\left(\alpha+\beta\sqrt{-1}\right)\right)^a = \rho^a\left(\cos a\zeta+\sqrt{-1}\sin a\zeta\right)((1))^a. \tag{18}$$

Second, if we suppose that α is negative, then equations (10) can be replaced by equations (15), from which we derive, among other values of θ,

$$\theta = \zeta+\pi. \tag{19}$$

Consequently, under this hypothesis, we can substitute the following for formulas (1) and (9):

$$\alpha+\beta\sqrt{-1} = -\rho\left(\cos\zeta+\sqrt{-1}\sin\zeta\right) \quad \text{and} \tag{20}$$

$$\begin{cases} ((\alpha+\beta\sqrt{-1}))^a \\ = \rho^a\left[\cos(a\zeta+a\pi)+\sqrt{-1}\sin(a\zeta+a\pi)\right]((1))^a \\ = \rho^a\left(\cos a\zeta+\sqrt{-1}\sin a\zeta\right)\left(\cos a\pi+\sqrt{-1}\sin a\pi\right)((1))^a. \end{cases} \tag{21}$$

In particular, if we make $\alpha+\beta\sqrt{-1}=-1$, that is to say $\alpha=-1$ and [193] $\beta=0$, then we find that

$$\zeta = \arctan\frac{0}{-1} = 0,$$

and formula (21) becomes

$$((-1))^a = \left(\cos a\pi+\sqrt{-1}\sin a\pi\right)((1))^a. \tag{22}$$

As a result, under the given hypothesis, we have in general

$$\left(\left(\alpha+\beta\sqrt{-1}\right)\right)^a = \rho^a\left(\cos a\zeta+\sqrt{-1}\sin a\zeta\right)((-1))^a. \tag{23}$$

By combining formulas (17), (18), (20) and (23) with equations (25) and (26) of § III, we finally obtain the following conclusions.

Let $\alpha+\beta\sqrt{-1}$ be any imaginary expression, let a be a quantity, positive or negative, with a numerical value that is fractional, and let k be an integer number chosen arbitrarily. Moreover, if we make

$$\rho = \sqrt{\alpha^2+\beta^2} \quad \text{and} \quad \zeta = \arctan\frac{\beta}{\alpha}, \tag{24}$$

then for positive values of α we have

$$(25)\qquad \begin{cases} \alpha+\beta\sqrt{-1}=\rho\left(\cos\zeta+\sqrt{-1}\sin\zeta\right), \\ ((\alpha+\beta\sqrt{-1}))^a=\rho^a\left(\cos a\zeta+\sqrt{-1}\sin a\zeta\right)((1))^a, \\ ((1))^a=\cos 2ka\pi\pm\sqrt{-1}\sin 2ka\pi, \end{cases}$$

and for negative values of α,

$$(26)\qquad \begin{cases} \alpha+\beta\sqrt{-1}=-\rho\left(\cos\zeta+\sqrt{-1}\sin\zeta\right), \\ ((\alpha+\beta\sqrt{-1}))^a=\rho^a\left(\cos a\zeta+\sqrt{-1}\sin a\zeta\right)((-1))^a, \\ ((-1))^a=\cos\left(\overline{2k+1}a\pi\right)\pm\sqrt{-1}\sin\left(\overline{2k+1}a\pi\right). \end{cases}$$

We ought to add that if we denote the denominator of the simplest fraction that represents the numerical value of a by n, then n is precisely the number of distinct values of each of the expressions

$$((1))^a, \quad ((-1))^a \quad \text{and} \quad \left(\left(\alpha+\beta\sqrt{-1}\right)\right)^a,$$

and that to deduce these same values from formulas (25) and (26), it [194] suffices to substitute successively in place of $2k$ and $2k+1$ all the integer numbers within the limits 0 and n.

If the numerical value of a becomes irrational, then each of the reduced expressions

$$\cos 2ka\pi\pm\sqrt{-1}\sin 2ka\pi \quad \text{and}$$
$$\cos\left(\overline{2k+1}a\pi\right)\pm\sqrt{-1}\sin\left(\overline{2k+1}a\pi\right)$$

has an indefinite number of values corresponding to the various integer values of k. Consequently, in the calculations we could no longer use the notations

$$((1))^a, \quad ((-1))^a \quad \text{and} \quad \left(\left(\alpha+\beta\sqrt{-1}\right)\right)^a$$

unless we consider each of them as representing an infinity of imaginary expressions, each different from the others. To avoid this inconvenience, we will never use these notations except in the case where the numerical value of a is fractional.

Among the various values of $((1))^a$, there is always one that is real and positive, namely $+1$, which we indicate by $(1)^a$, if we are using the single parentheses, or by 1^a if we leave them out entirely. If we substitute this particular value of $((1))^a$ into the second of equations (25), we get one corresponding value of

$$\left(\left(\alpha+\beta\sqrt{-1}\right)\right)^a,$$

which analogy leads us to indicate, with the aid of simple parentheses, by the notation

$$\left(\alpha+\beta\sqrt{-1}\right)^a.$$

This is what we will do from now on. As a consequence, by supposing that α is positive and that the quantities ρ and ζ are determined by equations (24), we have

(27) $$\left(\alpha+\beta\sqrt{-1}\right)^a = \rho^a\left(\cos a\zeta+\sqrt{-1}\sin a\zeta\right).$$

Because this last equation holds whenever the numerical value of a is integer or fractional, analogy again leads us to consider it to be true in the case where this numerical value [195] becomes irrational. Consequently, we agree to denote the product $\rho^a\left(\cos a\zeta+\sqrt{-1}\sin a\zeta\right)$ by

$$\left(\alpha+\beta\sqrt{-1}\right)^a$$

in the case where α is positive, whatever real value is given to the quantity a. In other words, if we denote by ζ an arc contained between the limits $-\frac{\pi}{2}$ and $+\frac{\pi}{2}$, whatever a is, we have,

$$\left[\rho\left(\cos\zeta+\sqrt{-1}\sin\zeta\right)\right]^a = \rho^a\left(\cos a\zeta+\sqrt{-1}\sin a\zeta\right).$$

If we take $\rho = 1$ in the preceding equation, it becomes[17]

(28) $$\left(\cos\zeta+\sqrt{-1}\sin\zeta\right)^a = \cos a\zeta+\sqrt{-1}\sin a\zeta.$$

This last formula is entirely similar to equations (10) and (14) of § II, with the only difference being that it applies only for values of ζ between the limits $-\frac{\pi}{2}$ and $+\frac{\pi}{2}$, while the other equations apply for any values of θ.

When the quantity α becomes negative, even if we suppose that suppose that the numerical value of a is fractional, then it is no longer clear that it is the value of the expression $((\alpha+\beta\sqrt{-1}))^a$ that we may distinguish from the others by using the notation

$$\left(\alpha+\beta\sqrt{-1}\right)^a.$$

However, because $-\alpha$ is a positive quantity, it is easy to establish the formula

(29) $$\left(-\alpha-\beta\sqrt{-1}\right)^a = \rho^a\left(\cos a\zeta+\sqrt{-1}\sin a\zeta\right)$$

for any value of a.

We finish this section by making the observation that, in the case where the numerical value of a is fractional, formulas (27) and (29) reduce equations (18) and (23) to those that follow:

(30) $$\left(\left(\alpha+\beta\sqrt{-1}\right)\right)^a = \left(\alpha+\beta\sqrt{-1}\right)((1))^a \quad \text{and}$$

[17] In his first published proof that $e^{i\theta} = \cos\theta + i\sin\theta$ [Euler 1748, §132–134 and §138], Euler made this leap from the rational to the irrational without a second thought. Compare this to Cauchy's remarks on page iii of his Introduction.

$$(31) \qquad \left(\left(\alpha+\beta\sqrt{-1}\right)\right)^{a}=\left(-\alpha-\beta\sqrt{-1}\right)((1))^{a},$$

[196] where equation (30) holds only for positive values of the quantity α, and equation (31) holds for negative values of the same quantity.

7.5 Applications of the principles established in the preceding sections.

We are going to apply the principles established in the preceding sections to the solution of three problems about sines and cosines.

Problem I. — *To transform* $\sin mz$ *and* $\cos mz$ (*where* m *denotes any integer number*) *into a polynomial ordered according to the ascending integer powers of* $\sin z$, *or at least into a product formed by the multiplication of such a polynomial and* $\cos z$.

Solution. — When we substitute the even powers of $\cos z$ for the integer powers of $1-\sin^2 z$ in equations (12) of § II, these equations become, for even values of m,

$$\cos mz = \left(1-\sin^2 z\right)^{\frac{m}{2}} - \frac{m(m-1)}{1\cdot 2}\left(1-\sin^2 z\right)^{\frac{m-2}{2}}\sin^2 z + \frac{m(m-1)(m-2)(m-3)}{1\cdot 2\cdot 3\cdot 4}\left(1-\sin^2 z\right)^{\frac{m-4}{2}}\sin^4 z - \dots \quad \text{and}$$

$$\sin mz = \cos z\left[\frac{m}{1}\left(1-\sin^2 z\right)^{\frac{m-2}{2}}\sin z - \frac{m(m-1)(m-2)}{1\cdot 2\cdot 3}\left(1-\sin^2 z\right)^{\frac{m-4}{2}}\sin^3 z + \dots\right],$$

and, for odd values of m,

$$\cos mz = \cos z\left[\left(1-\sin^2 z\right)^{\frac{m-1}{2}} - \frac{m(m-1)}{1\cdot 2}\left(1-\sin^2 z\right)^{\frac{m-3}{2}}\sin^2 z + \frac{m(m-1)(m-2)(m-3)}{1\cdot 2\cdot 3\cdot 4}\left(1-\sin^2 z\right)^{\frac{m-5}{2}}\sin^4 z - \dots\right] \quad \text{and}$$

$$\sin mz = \frac{m}{1}\left(1-\sin^2 z\right)^{\frac{m-1}{2}}\sin z - \frac{m(m-1)(m-2)}{1\cdot 2\cdot 3}\left(1-\sin^2 z\right)^{\frac{m-3}{2}}\sin^3 z + \dots .$$

[197] If we expand the right-hand sides of the four preceding formulas, or at least the coefficients of $\cos z$ on the right-hand sides, into polynomials ordered according to the ascending integer powers of $\sin z$, we find that for even values of m,

$$(1)\quad \begin{cases} \cos mz = 1 - \frac{m}{1}\left(\frac{m-1}{2}+\frac{1}{2}\right)\sin^2 z \\ \qquad + \frac{m(m-2)}{1\cdot 3}\left[\frac{(m-1)(m-3)}{2\cdot 4}+\frac{m-1}{2}\frac{3}{2}+\frac{3\cdot 1}{2\cdot 4}\right]\sin^4 z - \dots \quad \text{and} \\ \sin mz = \cos z\left\{\frac{m}{1}\sin z - \frac{m(m-2)}{1\cdot 3}\left(\frac{m-1}{2}+\frac{3}{2}\right)\sin^3 z \right. \\ \qquad \left. + \frac{m(m-2)(m-4)}{1\cdot 3\cdot 5}\left[\frac{(m-1)(m-3)}{2\cdot 4}+\frac{m-1}{2}\frac{5}{2}+\frac{5\cdot 3}{2\cdot 4}\right]\sin^5 z - \dots\right\}, \end{cases}$$

and for odd values of m,

$$(2)\quad \begin{cases} \cos mz = \cos z\left\{1 - \frac{m-1}{1}\left(\frac{m}{2}+\frac{1}{2}\right)\sin^2 z \right. \\ \qquad \left. + \frac{(m-1)(m-3)}{1\cdot 3}\left[\frac{m(m-2)}{2\cdot 4}+\frac{m}{2}\frac{3}{2}+\frac{3\cdot 1}{2\cdot 4}\right]\sin^4 z \dots\right\} \quad \text{and} \\ \sin mz = \frac{m}{1}\sin z - \frac{m(m-1)}{1\cdot 3}\left(\frac{m-2}{2}+\frac{3}{2}\right)\sin^3 z \\ \qquad + \frac{m(m-1)(m-3)}{1\cdot 3\cdot 5}\left[\frac{(m-2)(m-4)}{2\cdot 4}+\frac{m-2}{2}\frac{5}{2}+\frac{5\cdot 3}{2\cdot 4}\right]\sin^5 z - \dots. \end{cases}$$

Equations (1) and (2) evidently contain the solution to the given question. It only remains to present them in a simpler form. To do that, it suffices to observe that the coefficient of each integer power of $\sin z$ generally contains a sum of fractions into which equation (5) of Chapter IV (§ III) permits us to substitute a unique fraction. As a consequence of this reduction, the expansions of $\cos mz$ and $\sin mz$ become, for even values of m,

$$(3)\quad \begin{cases} \cos mz = 1 - \frac{m\cdot m}{1\cdot 2}\sin^2 z + \frac{(m+2)m\cdot m(m-2)}{1\cdot 2\cdot 3\cdot 4}\sin^4 z \\ \qquad - \frac{(m+4)(m+2)m\cdot m(m-2)(m-4)}{1\cdot 2\cdot 3\cdot 4\cdot 5\cdot 6}\sin^6 z + \dots \end{cases}$$

[198] and

$$(4)\quad \begin{cases} \sin mz = \cos z\left[\frac{m}{1}\sin z - \frac{(m+2)m(m-2)}{1\cdot 2\cdot 3}\sin^3 z \right. \\ \qquad \left. + \frac{(m+4)(m+2)m(m-2)(m-4)}{1\cdot 2\cdot 3\cdot 4\cdot 5}\sin^5 z - \dots\right], \end{cases}$$

and for odd values of m,

$$(5)\quad \begin{cases} \cos mz = \cos z\left[1 - \frac{(m+1)(m-1)}{1\cdot 2}\sin^2 z \right. \\ \qquad \left. + \frac{(m+3)(m+1)(m-1)(m-3)}{1\cdot 2\cdot 3\cdot 4}\sin^4 z - \dots\right] \end{cases}$$

and

$$(6)\quad \begin{cases} \sin mz = \frac{m}{1}\sin z - \frac{(m+1)m(m-1)}{1\cdot 2\cdot 3}\sin^3 z \\ \qquad + \frac{(m+3)(m+1)m(m-1)(m-3)}{1\cdot 2\cdot 3\cdot 4\cdot 5}\sin^5 z - \dots. \end{cases}$$

Corollary I. — If in equation (3) we successively make

$$m = 2, \quad m = 4, \quad m = 6, \quad \dots,$$

we obtain the following:

$$
(7)\qquad \begin{cases} \cos 2z = 1 - \ \ 2\sin^2 z, \\ \cos 4z = 1 - \ \ 8\sin^2 z + \ \ 8\sin^4 z, \\ \cos 6z = 1 - 18\sin^2 z + 48\sin^4 z - 32\sin^6 z, \\ \dots\dots\dots\dots\dots\dots\dots\dots\dots \end{cases}
$$

Corollary II. — If in equation (6) we successively make

$$
m = 1, \quad m = 3, \quad m = 5, \quad \dots,
$$

we get:

$$
(8)\qquad \begin{cases} \sin\ z = \ \ \sin z, \\ \sin 3z = 3\sin z - \ \ 4\sin^3 z, \\ \sin 5z = 5\sin z - 20\sin^3 z + 16\sin^5 z, \\ \dots\dots\dots\dots\dots\dots\dots\dots \end{cases}
$$

Problem II. — *To transform* $\sin mz$ *and* $\cos mz$ *(where* m *denotes any integer number) into a polynomial ordered according to the ascending integer powers* [199] *of* $\cos z$*, or at least into a product formed by the multiplication of such a polynomial and* $\sin z$*.*

Solution. — To obtain the formulas that solve the given problem, it suffices to replace z with $\frac{\pi}{2} - z$ in equations (3), (4), (5) and (6), and to observe moreover that for even values of m, we have

$$
\cos\left(\frac{m\pi}{2} - mz\right) = (-1)^{\frac{m}{2}} \cos mz \quad \text{and}
$$
$$
\sin\left(\frac{m\pi}{2} - mz\right) = (-1)^{\frac{m}{2}+1} \sin mz,
$$

and for odd values of m,

$$
\cos\left(\frac{m\pi}{2} - mz\right) = (-1)^{\frac{m-1}{2}} \sin mz \quad \text{and}
$$
$$
\sin\left(\frac{m\pi}{2} - mz\right) = (-1)^{\frac{m-1}{2}} \sin mz.
$$

In this way we find that if m is an even number

$$
(9)\qquad \begin{cases} (-1)^{\frac{m}{2}} \cos mz = 1 - \frac{m\cdot m}{2}\cos^2 z + \frac{(m+2)m\cdot m(m-2)}{1\cdot2\cdot3\cdot4}\cos^4 z \\ \qquad\qquad - \frac{(m+4)(m+2)m\cdot m(m-2)(m-4)}{1\cdot2\cdot3\cdot4\cdot5\cdot6}\cos^6 z + \dots, \end{cases}
$$

and

$$
(10)\qquad \begin{cases} (-1)^{\frac{m}{2}+1} \sin mz = \sin z \left[\frac{m}{1}\cos z - \frac{(m+2)m(m-2)}{1\cdot2\cdot3}\cos^3 z \right. \\ \qquad\qquad \left. + \frac{(m+4)(m+2)m(m-2)(m-4)}{1\cdot2\cdot3\cdot4\cdot5}\cos^5 z - \dots\right], \end{cases}
$$

and if m is an odd number,

$$(11)\qquad \begin{cases} (-1)^{\frac{m-1}{2}}\sin mz = \sin z\Big[1-\frac{(m+1)(m-1)}{1\cdot 2}\cos^2 z \\ \qquad\qquad +\frac{(m+3)(m+1)(m-1)(m-3)}{1\cdot 2\cdot 3\cdot 4}\cos^4 z-\dots\Big], \end{cases}$$

and

$$(12)\qquad \begin{cases} (-1)^{\frac{m-1}{2}}\cos mz = \frac{m}{1}\cos z-\frac{(m+1)m(m-1)}{1\cdot 2\cdot 3}\cos^3 z \\ \qquad\qquad +\frac{(m+3)(m+1)m(m-1)(m-3)}{1\cdot 2\cdot 3\cdot 4\cdot 5}\cos^5 z-\dots. \end{cases}$$

[200] *Corollary I.* — If in formula (9) we successively make

$$m=2,\quad m=4,\quad m=6,\quad \dots,$$

we obtain the following:

$$(13)\qquad \begin{cases} -\cos 2z = 1-2\cos^2 z, \\ \cos 4z = 1-8\cos^2 z+8\cos^4 z, \\ -\cos 6z = 1-18\cos^2 z+48\cos^4 z-32\cos^6 z, \\ \dots\dots\dots\dots\dots\dots\dots\dots \end{cases}$$

Corollary II. — If in equation (12) we successively make

$$m=1,\quad m=3,\quad m=5,\quad \dots,$$

we conclude that

$$(14)\qquad \begin{cases} \cos\ z = \cos z, \\ -\cos 3z = 3\cos z-4\cos^3 z, \\ \cos 5z = 5\cos z-20\cos^3 z+16\cos^5 z, \\ \dots\dots\dots\dots\dots\dots\dots\dots \end{cases}$$

Problem III. — *To express the integer powers of* $\sin z$ *and of* $\cos z$ *as a linear function of the sines and cosines of the arcs* z, $2z$, $3z$, ...

Solution. — We solve this problem easily by considering the properties of two conjugate imaginary expressions,

$$\cos z+\sqrt{-1}\sin z \quad\text{and}\quad \cos z-\sqrt{-1}\sin z.$$

If we denote the first of these by u and the second by v, we have

$$2\cos z = u+v \quad\text{and}\quad 2\sqrt{-1}\sin z = u-v.$$

By raising both sides of each of the two preceding equations to the integer power m, then dividing each by 2 or by $2\sqrt{-1}$, then making the reductions indicated by the formulas

$$uv = 1,$$

$$\frac{u^n + v^n}{2} = \cos nz \quad \text{and} \quad \frac{u^n - v^n}{2\sqrt{-1}} = \sin nz,$$

where the last two equations apply for any integer value [201] of n, we find that if m represents an even number, then

$$(15)\quad \begin{cases} 2^{m+1}\cos^m z = \cos mz + \frac{m}{1}\cos\left(\overline{m-2}\cdot z\right) \\ \qquad + \frac{m(m-1)}{1\cdot 2}\cos\left(\overline{m-4}\cdot z\right) + \dots \\ \qquad + \frac{1}{2}\frac{m(m-1)\dots\left(\frac{m}{2}+1\right)}{1\cdot 2\cdot 3\dots\frac{m}{2}}, \end{cases}$$

and

$$(16)\quad \begin{cases} (-1)^{\frac{m}{2}}\,2^{m-1}\sin^m z = \cos mz - \frac{m}{1}\cos\left(\overline{m-2}\cdot z\right) \\ \qquad + \frac{m(m-1)}{1\cdot 2}\cos\left(\overline{m-4}\cdot z\right) - \dots \\ \qquad \pm \frac{1}{2}\frac{m(m-1)\dots\left(\frac{m}{2}+1\right)}{1\cdot 2\cdot 3\dots\frac{m}{2}}, \end{cases}$$

and if m represents an odd number, then

$$(17)\quad \begin{cases} 2^{m-1}\cos^m z = \cos mz - \frac{m}{1}\cos\left(\overline{m-2}\cdot z\right) \\ \qquad + \frac{m(m-1)}{1\cdot 2}\cos\left(\overline{m-4}\cdot z\right) + \dots \\ \qquad + \frac{m(m-1)\dots\frac{m+3}{2}}{1\cdot 2\cdot 3\dots\frac{m-1}{2}}\cos z, \end{cases}$$

and

$$(18)\quad \begin{cases} (-1)^{\frac{m-1}{2}}\,2^{m-1}\sin^m z = \sin mz - \frac{m}{1}\sin\left(\overline{m-2}\cdot z\right) \\ \qquad + \frac{m(m-1)}{1\cdot 2}\sin\left(\overline{m-4}\cdot z\right) - \dots \\ \qquad \pm \frac{m(m-1)\dots\frac{m+3}{2}}{1\cdot 2\cdot 3\dots\frac{m-1}{2}}\sin z. \end{cases}$$

Corollary I. — If in formula (15) we successively make

$$m = 2, \quad m = 4, \quad m = 6, \quad \dots,$$

[202] we conclude

$$(19)\quad \begin{cases} 2\cos^2 z = \cos 2z + 1, \\ 8\cos^4 z = \cos 4z + 4\cos 2z + 3, \\ 32\cos^6 z = \cos 6z + 6\cos 4z + 15\cos 2z + 10, \\ \dots\dots\dots\dots\dots\dots\dots\dots\dots\dots \end{cases}$$

We would arrive at the same equations if we sought to deduce the successive values of

$$\cos^2 z, \quad \cos^4 z, \quad \cos^6 z, \quad \ldots$$

from formulas (13) as lincar functions of

$$\cos 2z, \quad \cos 4z, \quad \cos 6z, \quad \ldots.$$

Corollary II. — If in formula (16) we successively make

$$m = 2, \quad m = 4, \quad m = 6, \quad \ldots,$$

we obtain the equations

$$(20) \qquad \begin{cases} -2\sin^2 z = \cos 2z - 1, \\ 8\sin^4 z = \cos 4z - 4\cos 2z + 3, \\ -32\sin^6 z = \cos 6z - 6\cos 4z + 15\cos 2z - 10, \\ \ldots\ldots\ldots\ldots\ldots\ldots\ldots\ldots\ldots\ldots, \end{cases}$$

which we could equally well deduce from formulas (7) by eliminating the quantities

$$\sin^2 z, \quad \sin^4 z, \quad \sin^6 z, \quad \ldots.$$

Corollary III. — If in formula (17) we successively make

$$m = 1, \quad m = 3, \quad m = 5, \quad \ldots,$$

we conclude

$$(21) \qquad \begin{cases} \cos z = \cos z, \\ 4\cos^3 z = \cos 3z + 3\cos z, \\ 16\cos^5 z = \cos 5z + 5\cos 3z + 10\cos z, \\ \ldots\ldots\ldots\ldots\ldots\ldots\ldots\ldots\ldots\ldots \end{cases}$$

[203] We would arrive at the same equations if we sought to deduce the successive values of

$$\cos z, \quad \cos^3 z, \quad \cos^5 z, \quad \ldots$$

from formulas (14) as linear functions of

$$\cos z, \quad \cos 3z, \quad \cos 5z, \quad \ldots.$$

Corollary IV. — If in formula (18) we successively take

$$m = 1, \quad m = 3, \quad m = 5, \quad \ldots,$$

we obtain the equations

$$
(22) \qquad \begin{cases} \sin z \;\; = \sin z, \\ -4\sin^3 z = \sin 3z - 3\sin z, \\ 16\sin^5 z = \sin 5z - 5\sin 3z + 10\sin z, \\ \dots\dots\dots\dots\dots\dots\dots\dots\dots\dots\dots\dots, \end{cases}
$$

which we could equally well deduce from formulas (8) by eliminating the quantities

$$
\sin z, \quad \sin^3 z, \quad \sin^5 z, \quad \dots.
$$

Chapter 8
On imaginary functions and variables.

8.1 General considerations on imaginary functions and variables.

[204] When we suppose that one or both of the two real quantities u and v are variables, then the expression

$$u+v\sqrt{-1}$$

is called an *imaginary variable*. If also the variable u converges towards the limit U and the variable v towards the limit V, then

$$U+V\sqrt{-1}$$

is the *limit* towards which the imaginary expression

$$u+v\sqrt{-1}$$

converges.

When the constants or variables contained in a given function, having been considered real are later supposed to be imaginary, the notation that was used to express the function cannot be retained in the calculation except by virtue of new conventions able to determine the sense of this notation under the new hypotheses. Thus, for example, by virtue of the conventions established in the preceding Chapter, the values of the notations

$$a+x, \quad a-x, \quad ax \quad \text{and} \quad \frac{a}{x}$$

are completely determined in the case where the constant a and [205] the variable x become imaginary. Suppose, in order to clarify these ideas, that the constant a remains real, and the variable x has the imaginary value

$$\alpha+\beta\sqrt{-1}=\rho\left(\cos\theta+\sqrt{-1}\sin\theta\right),$$

R.E. Bradley, C.E. Sandifer, *Cauchy's Cours d'analyse*, Sources and Studies in the History of Mathematics and Physical Sciences, DOI 10.1007/978-1-4419-0549-9_8,

where α and β denote two real quantities which can be replaced by the modulus ρ and the real arc θ. We conclude from Chapter VII (§§ I and II) that the four notations

$$a+x, \quad a-x, \quad ax \quad \text{and} \quad \frac{a}{x}$$

denote, respectively, the four imaginary expressions

$$\begin{aligned}
&a+\rho\cos\theta+\rho\sin\theta\sqrt{-1},\\
&a-\rho\cos\theta-\rho\sin\theta\sqrt{-1},\\
&a\rho\cos\theta+a\rho\sin\theta\sqrt{-1} \quad \text{and}\\
&\frac{a}{\rho}\cos\theta-\frac{a}{\rho}\sin\theta\sqrt{-1},
\end{aligned}$$

or in other words, the following quantities:

$$a+\alpha+\beta\sqrt{-1}, \quad a-\alpha-\beta\sqrt{-1}, \quad a\alpha+a\beta\sqrt{-1}$$
$$\text{and} \quad \frac{a\alpha}{\alpha^2+\beta^2}-\frac{a\beta}{\alpha^2+\beta^2}\sqrt{-1}.$$

In general, by means of the principles established in Chapter VII, we can clarify without difficulty the values of algebraic expressions in which several imaginary variables or constants are related to each other by the signs of addition, subtraction, multiplication or division. We see without trouble that these expressions retain all the properties as imaginary variables and constants that they have when they are real. For example, if we denote by

$$x,\ y,\ z,\ \ldots,\ u,\ v,\ w,\ \ldots$$

several variables, either real or imaginary, we have, in every [206] possible case

$$(1) \qquad \begin{cases}
x+y+z+\ldots-(u+v+w+\ldots)\\
\qquad =x+y+z+\ldots-u-v-w-\ldots,\\
xy=yx,\\
u\,(x+y+z+\ldots)=ux+uy+uz+\ldots,\\
\dfrac{x+y+z+\ldots}{u}=\dfrac{x}{u}+\dfrac{y}{u}+\dfrac{z}{u}+\ldots,\\
\dfrac{x}{u}\times\dfrac{y}{v}\times\dfrac{z}{w}\times\ldots=\dfrac{xyz\ldots}{uvw\ldots},\\
\dfrac{x}{\left(\frac{u}{v}\right)}=\dfrac{vx}{u}=\dfrac{v}{u}\times x,\\
\ldots\ldots\ldots\ldots\ldots\ldots\ldots
\end{cases}$$

Now consider the notation

$$x^a,$$

in the case where the constant a remains real and the variable x takes the imaginary value

$$\alpha+\beta\sqrt{-1}=\rho\left(\cos\theta+\sqrt{-1}\sin\theta\right).$$

If we take for the value of a a quantity for which the numerical value is an integer number m, in this same notation, namely

$$x^a=x^{\pm m},$$

we have, for any real values of α and β, a precise meaning. It is given by the imaginary expression

$$\rho^m\cos m\theta+\rho^m\sin m\theta\sqrt{-1},$$

if $a=+m$, and the following

$$\rho^{-m}\cos m\theta-\rho^{-m}\sin m\theta\sqrt{-1},$$

if $a=-m$ (see Chapter VII, § II, equations (18) and (19)). But any time that the constant a takes a fractional [207] or irrational numerical value, the notation

$$x^a$$

no longer has a precise and determined value, at least when the real part α of the imaginary expression x is not positive. If in this particular case we make

$$\zeta=\arctan\frac{\beta}{\alpha},$$

then the arc ζ is contained between the limits $-\frac{\pi}{2}$ and $+\frac{\pi}{2}$, and writing x in place of $\alpha+\beta\sqrt{-1}$ in § IV of Chapter VII (equations (17) and (27)), we find

$$\begin{aligned} x &=\rho\left(\cos\zeta+\sqrt{-1}\sin\zeta\right) \quad \text{and}\\ x^a &=\rho^a\left(\cos a\zeta+\sqrt{-1}\sin a\zeta\right), \end{aligned}$$

so that the notation x^a denotes the imaginary expression

$$\rho^a\cos a\zeta+\rho^a\sin a\zeta\sqrt{-1}.$$

It follows also from the conventions and the principles established above (Chap. VII, §§ III and IV) that, for a fractional value of the constant a, the notation

$$((x))^a$$

represents all at once many imaginary expressions, the values of which are given by the two formulas

$$((x))^a=x^a\,((1))^a \quad \text{and} \quad ((1))^a=\cos 2ka\pi\pm\sqrt{-1}\sin 2ka\pi,$$

when the real part α of the imaginary expression x is positive, and by the two formulas

$$((x))^a = (-x)^a\,((-1))^a \quad \text{and}$$
$$((-1))^a = \cos(2k+1)\,a\pi \pm \sqrt{-1}\sin(2k+1)\,a\pi,$$

when the quantity α becomes negative (on this subject, see § IV of Chapter VII, equations (25) and (26)). The same notation can no longer be employed in the case where the numerical value of a becomes irrational.

[208] Expressions of the form

$$x^a$$

retain the same properties for real and for imaginary values of the variable as long as the numerical value of the exponent is an integer number. But otherwise these properties do not hold except under certain conditions. For example, let

$$x = \alpha + \beta\sqrt{-1}, \quad y = \alpha' + \beta'\sqrt{-1}, \quad z = \alpha'' + \beta''\sqrt{-1}, \quad \ldots$$

be several imaginary expressions, which would be reduced to real quantities if β, β', β'' vanish. Moreover, denote by a, b, c, ... any real quantities with numerical values that are fractional or irrational, and by m, m', m'', ... several integer numbers. By virtue of the principles established in Chapter VII, we always have

$$(2) \qquad \begin{cases} x^m x^{m'} x^{m''} \ldots = x^{m+m'+m''+\ldots}, \\ x^{-m} x^{-m'} x^{-m''} \ldots = x^{-m-m'-m''-\ldots}, \\ x^{\pm m} x^{\pm m'} x^{\pm m''} \ldots = x^{\pm m \pm m' \pm m'' \pm \ldots}, \end{cases}$$

where each of the numbers m, m', m'', ... is given the same sign on both sides of the equation. Also

$$(3) \qquad \begin{cases} x^m\, y^m\, z^m \ldots = (xyz\ldots)^m, \\ x^{-m} y^{-m} z^{-m} \ldots = (xyz\ldots)^{-m} \end{cases}$$

and

$$(4) \qquad \begin{cases} (x^m)^{m'} = (x^{-m})^{-m'} = x^{mm'}, \\ (x^m)^{-m'} = (x^{-m})^{m'} = x^{-mm'}. \end{cases}$$

On the other hand, we find that of the three formulas

$$(5) \qquad x^a x^b x^c \ldots = x^{a+b+c+\ldots},$$
$$(6) \qquad x^a y^a z^a \ldots = (xyz\ldots)^a \quad \text{and}$$
$$(7) \qquad (x^a)^b = x^{ab},$$

the first remains always true only if the real part α [209] of the imaginary expression x is positive. The second remains true when α, α', α'', ... are positive and the sum

$$\arctan\frac{\beta}{\alpha} + \arctan\frac{\beta'}{\alpha'} + \arctan\frac{\beta''}{\alpha''} + \ldots$$

remains contained between the limits $-\frac{\pi}{2}$ and $+\frac{\pi}{2}$. The last remains true when α is positive and the product

$$a \arctan \frac{\beta}{\alpha}$$

is contained between the same limits.

The conventions adopted in Chapter VII do not yet suffice to determine precisely the meanings of the notations

$$A^x, \ \log x, \ \sin x, \ \cos x, \ \arcsin x \ \text{and} \ \arccos x,$$

in the case where the variable x becomes imaginary. The simplest means of arriving at such precise meanings is by considering imaginary series. We will revisit this subject in Chapter IX.

From what has been said above, any algebraic notation that includes imaginary constants along with the variables $x, y, z, \ldots$, assumed to be real, cannot be used in calculation except in the case where, by virtue of established conventions, it has a determined imaginary expression as its value. Such an expression, in which the real part and the coefficient of $\sqrt{-1}$ are necessarily real functions of the variables x, y, z, ..., is called an *imaginary function* of these same variables. Thus, for example, if we denote by $\varphi(x)$ and $\chi(x)$ two real functions of x, an imaginary function of this variable is

$$\varphi(x) + \chi(x)\sqrt{-1}.$$

Sometimes we indicate such a function with the aid of a single symbol ϖ, and we write

$$\varpi(x) = \varphi(x) + \chi(x)\sqrt{-1}.$$

[210] Similarly, if we denote two real functions of the variables x, y, z, ... by $\varphi(x,y,z,\ldots)$ and $\chi(x,y,z,\ldots)$, then

$$\varpi(x,y,z,\ldots) = \varphi(x,y,z,\ldots) + \chi(x,y,z,\ldots)\sqrt{-1}$$

is an imaginary function of these several variables.

The imaginary function

$$\varphi(x,y,z,\ldots) + \chi(x,y,z,\ldots)\sqrt{-1}$$

is called *algebraic* or *exponential* or *logarithmic* or *circular*, etc., and, in the first case, is called *rational* or *irrational*, *integer* or *fractional*, etc., whenever both of the real functions $\varphi(x,y,z,\ldots)$ and $\chi(x,y,z,\ldots)$ enjoy the properties associated with the name in question. Thus, in particular, the general form of a linear imaginary function of the variables x, y, z, ... is

$$(a + bx + cy + dz + \ldots) + (a' + b'x + c'y + d'z + \ldots)\sqrt{-1}$$

or what amounts to the same thing,

$$\left(a+a'\sqrt{-1}\right)+\left(b+b'\sqrt{-1}\right)x+\left(c+c'\sqrt{-1}\right)y+\left(d+d'\sqrt{-1}\right)z+\ldots,$$

where $a, b, c, d, \ldots, a', b', c', d', \ldots$ denote real constants.

Again, we ought to distinguish among imaginary functions, as we do among real functions, the ones we call *explicit*, and which are immediately expressed by means of the variables, as opposed to those which we call *implicit*, for which the values are determined by certain equations but which cannot be known explicitly until the equations have been solved. Let

$$\varpi(x) \quad \text{or} \quad \varpi(x,y,z,\ldots)$$

be an implicit imaginary function determined by a single equation. We can represent this function by $u+v\sqrt{-1}$, where u and v denote two real quantities. If in the imaginary equation which must [211] be satisfied, we write

$$u+v\sqrt{-1}$$

instead of $\varpi(x)$ or $\varpi(x,y,z,\ldots)$, then after expanding both sides of the equation and equating both the real parts and the coefficients of $\sqrt{-1}$, we get two real equations between the unknown functions u and v. When we can solve these last equations, the solutions determine the explicit values of u and v, and consequently we get the explicit value of the imaginary expression

$$u+v\sqrt{-1}.$$

For an imaginary function of a single variable to be completely determined, it is necessary and it suffices that for each particular value attributed to the variable, we can deduce the corresponding value of the function.[1] Sometimes, for each value of the variable, the given function obtains several values, different from one another. Conforming to the conventions that we have already established, we ordinarily denote these multiple values of an imaginary function with the notation of doubled signs or doubled parentheses. Thus, for example,

$$\sqrt[n]{\!\sqrt{\cos z+\sqrt{-1}\sin z}}$$

or

$$\left(\left(\cos z+\sqrt{-1}\sin z\right)\right)^{\frac{1}{n}}$$

indicates any one of the roots of degree n of the imaginary expression

$$\cos z+\sqrt{-1}\sin z.$$

[1] This is tantalizingly close to the modern definition of function, but deceptively so. Cauchy is still thinking of functions given by a formula, either implicitly or explicitly, and here he is merely distinguishing explicit functions and those implicit functions that are single-valued for a particular value of x.

8.2 On infinitely small imaginary expressions and on the continuity of imaginary functions.

An imaginary expression is called *infinitely small* when it converges to the limit zero, which implies that in the given expression, the real part and the coefficient of $\sqrt{-1}$ converge at the same [212] time towards this limit. Given this, represent a variable imaginary expression by

$$\alpha+\beta\sqrt{-1}=\rho\left(\cos\theta+\sqrt{-1}\sin\theta\right),$$

where α and β denote two real quantities for which we can substitute the modulus ρ and the real arc θ. Because this expression is infinitely small, it is evidently necessary and sufficient[2] that its modulus

$$\rho=\sqrt{\alpha^2+\beta^2}$$

itself be infinitely small.

An imaginary function of the real variable x is called *continuous* between two given limits of this variable when between these limits, an infinitely small increase in the variable always produces an infinitely small increase in the function itself. As a result, the imaginary function

$$\varphi(x)+\chi(x)\sqrt{-1}$$

is continuous between two limits of x if the real functions $\varphi(x)$ and $\chi(x)$ are continuous between these limits.

We say that an imaginary function of the variable x is a *continuous* function of that variable in the neighborhood of a particular value of x whenever it remains continuous between two limits which contain that value, even if they are very close to each other.

Finally, when an imaginary function of the variable x ceases to be continuous in the neighborhood of a particular value of this variable, we say that it then becomes *discontinuous*, and that there is a *solution of continuity* for this particular value.

On the basis of the concepts which we have just established relative to the continuity of imaginary functions, we easily recognize that theorems I, II and III of Chapter II (§ II) remain true even in the case where we replace the real functions

$$f(x)\quad\text{and}\quad f(x,y,z,\ldots)$$

[213] with the imaginary functions

$$\varphi(x)+\chi(x)\sqrt{-1}\quad\text{and}\quad\varphi(x,y,z,\ldots)+\chi(x,y,z,\ldots)\sqrt{-1}.$$

---

[2] Up to here, Cauchy has used the expression "it is necessary and it suffices." Here for the first time he writes the more familiar "necessary and sufficient" (*nécessaire et suffisant*).

As a consequence, we can state the following propositions.

Theorem I. — *If the real variables x, y, z, ... have for limits the fixed and determined quantities X, Y, Z, ..., and if the imaginary function*

$$\varphi(x,y,z,\ldots)+\chi(x,y,z,\ldots)\sqrt{-1}$$

is continuous with respect to each of the variables x, y, z, ... in the neighborhood of the system of particular values

$$x=X, \quad y=Y, \quad z=Z, \quad \ldots,$$

then $\varphi(x,y,z,\ldots)+\chi(x,y,z,\ldots)\sqrt{-1}$ has as its limit

$$\varphi(X,Y,Z,\ldots)+\chi(X,Y,Z,\ldots)\sqrt{-1},$$

or more briefly, if we write

$$\varphi(x,y,z,\ldots)+\chi(x,y,z,\ldots)\sqrt{-1}=\varpi(x,y,z,\ldots),$$

then $\varpi(x,y,z,\ldots)$ has as its limit

$$\varpi(X,Y,Z,\ldots).$$

Theorem II. — *Let x, y, z, ... denote several real functions of the variable t which are continuous with respect to this variable in the neighborhood of the real value $t=T$. Furthermore, let X, Y, Z, ... be the particular values of x, y, z, ... corresponding to $t=T$. Suppose that in the neighborhood of these particular values, the imaginary function*

$$\varpi(x,y,z,\ldots)=\varphi(x,y,z,\ldots)+\chi(x,y,z,\ldots)\sqrt{-1}$$

is simultaneously continuous with respect to x, with respect to y, with respect to z, etc. Then $\varpi(x,y,z,\ldots)$, considered as an imaginary function of t, is also continuous with respect to t in the neighborhood of the particular value $t=T$.

In the preceding theorem, if we reduce the variables x, y, z, ... to a single variable, we get the following statement:

[214] **Theorem III.** — *Suppose that in the expression*

$$\varpi(x)=\varphi(x)+\chi(x)\sqrt{-1}$$

the variable x is a real function of another variable t. Imagine further that the variable x is a continuous function of t in the neighborhood of the particular value $t=T$ and that $\varpi(x)$ is a continuous function of x in the neighborhood of the particular value $x=X$ corresponding to $t=T$. The imaginary expression $\varpi(x)$, considered as *a function of t, is also continuous with respect to this variable in the neighborhood of the particular value t = T.*

8.3 On imaginary functions that are symmetric, alternating or homogeneous.

In extending the definitions that we gave (Chapter III) of symmetric, alternating or homogeneous functions of several variables x, y, z, ... to imaginary functions, we recognize immediately that

$$\varphi(x,y,z,\ldots)+\chi(x,y,z,\ldots)\sqrt{-1}$$

is symmetric, alternating or homogeneous of degree a with respect to the variables x, y, z, ... when the real functions

$$\varphi(x,y,z,\ldots) \quad \text{and} \quad \chi(x,y,z,\ldots)$$

are both symmetric, homogeneous or alternating of degree a with respect to these same variables.

8.4 On imaginary integer functions of one or several variables.

By virtue of what has been said above (§ I),

$$\varphi(x)+\chi(x)\sqrt{-1}$$

and

$$\varphi(x,y,z,\ldots)+\chi(x,y,z,\ldots)\sqrt{-1}$$

[215] are two imaginary integer functions, the first of the variable x and the second of the variables x, y, z, ..., when

$$\varphi(x) \quad \text{and} \quad \chi(x), \qquad \varphi(x,y,z,\ldots) \quad \text{and} \quad \chi(x,y,z,\ldots)$$

are real integer functions of the same variables. Consequently, if $\varpi(x)$ represents an imaginary integer function of the variable x, then the value of $\varpi(x)$ is determined by an equation of the form

$$\begin{aligned}\varpi(x) &= \varphi(x)+\chi(x)\sqrt{-1}\\ &= a_0+a_1x+a_2x^2+\ldots+\left(b_0+b_1x+b_2x^2+\ldots\right)\sqrt{-1},\end{aligned}$$

where a_0, a_1, a_2, ..., b_0, b_1, b_2, ... denote real constants. We conclude from this equation, by combining the coefficients of similar powers of x, that

(1)
$$\varpi(x) = \left(a_0 + b_0\sqrt{-1}\right) + \left(a_1 + b_1\sqrt{-1}\right)x + \left(a_2 + b_2\sqrt{-1}\right)x^2 + \ldots.$$

For the function $\varpi(x)$ determined by the previous formula to vanish with x, it is necessary that we have

$$a_0 + b_0\sqrt{-1} = 0,$$

that is to say $a_0 = 0$ and $b_0 = 0$, in which case the value of $\varpi(x)$ reduces to

$$\begin{aligned}\varpi(x) &= \left(a_1 + b_1\sqrt{-1}\right)x + \left(a_2 + b_2\sqrt{-1}\right)x^2 + \ldots \\ &= x\left[a_1 + b_1\sqrt{-1} + \left(a_2 + b_2\sqrt{-1}\right)x + \ldots\right].\end{aligned}$$

Thus, any imaginary integer function of the variable x that vanishes with that variable is the product of the factor x by a second function of the same kind, or in other words, is divisible by x. On the basis of this remark, we easily extend theorems I and II of Chapter IV (§ I) to the case where the integer functions which are mentioned there are also imaginary. I will add that these two theorems remain true even if we replace the particular real values given to the variable x, such as

$$x_0,\ x_1,\ x_2,\ \ldots$$

[216] with the imaginary values[3]

$$a\alpha_0 + \beta_0\sqrt{-1}, \quad \alpha_1 + \beta_1\sqrt{-1}, \quad \alpha_2 + \beta_2\sqrt{-1}, \quad \ldots.$$

To prove this assertion, it suffices to establish the two following propositions:

Theorem I. — *If an imaginary integer function of the variable x vanishes for a particular value of that variable, for example, for*

$$x = \alpha_0 + \beta_0\sqrt{-1},$$

then this function is algebraically divisible by

$$x - \alpha_0 - \beta_0\sqrt{-1}.$$

Proof. — Indeed, let

$$\varpi(x) = \varphi(x) + \chi(x)\sqrt{-1}$$

be the imaginary function under consideration. If we let

$$x = \alpha_0 + \beta_0\sqrt{-1} + z,$$

[3] This word was changed to *variables* in [Cauchy 1897, p. 216] from the correct word *valeurs* in [Cauchy 1821, p. 255]. (tr.)

where z denotes a new variable, then by substituting this, we evidently obtain as a result an imaginary integer function of z, namely

$$\varpi\left(\alpha_0+\beta_0\sqrt{-1}+z\right).$$

Because this function of z ought to vanish for $z=0$, we conclude that

$$\varpi(x)=\varpi\left(\alpha_0+\beta_0\sqrt{-1}+z\right)$$

is divisible by

$$z=x-\alpha_0-\beta_0\sqrt{-1}.$$

Corollary I. — The preceding proposition remains true even in the case where the function $\chi(x)$ vanishes, that is to say in the case where $\varpi(x)$ reduces to a real function $\varphi(x)$.

Corollary II. — The preceding theorem also remains true when we [217] suppose that $\beta=0$, and consequently when the particular value assigned to the variable x is real.

Theorem II. — *If an imaginary integer function of the variable x vanishes for each of the particular values of x contained in the sequence*

$$\alpha_0+\beta_0\sqrt{-1},\ \alpha_1+\beta_1\sqrt{-1},\ \alpha_2+\beta_2\sqrt{-1},\ \ldots,\ \alpha_{n-1}+\beta_{n-1}\sqrt{-1},$$

where n denotes any integer number, this function is equivalent to the product of the factors

$$x-\alpha_0-\beta_0\sqrt{-1},\quad x-\alpha_1-\beta_1\sqrt{-1},\quad x-\alpha_2-\beta_2\sqrt{-1},\quad \ldots,$$
$$\ldots,\quad x-\alpha_{n-1}-\beta_{n-1}\sqrt{-1}$$

by a new imaginary integer function of the variable x.

Proof. — Let

$$\varpi(x)=\varphi(x)+\chi(x)\sqrt{-1}$$

be the given function. Because it should vanish for

$$x=\alpha_0+\beta_0\sqrt{-1},$$

by virtue of theorem I it is algebraically divisible by

$$x-\alpha_0-\beta_0\sqrt{-1}.$$

As a consequence, we have

(2) $$\varpi(x)=\left(x-\alpha_0-\beta_0\sqrt{-1}\right)Q_0,$$

where Q_0 denotes a new imaginary integer function of the variable x. The function $\varpi(x)$ also ought to vanish when we suppose that

$$x=\alpha_1+\beta_1\sqrt{-1},$$

so this assumption necessarily reduces the right-hand side of equation (2) to zero, and consequently it reduces to zero one of the two factors which compose it (see Chapter VII, § II, theorem VII, corollary II). [218] Furthermore, because the first factor

$$x-\alpha_0-\beta_0\sqrt{-1}$$

cannot become zero for

$$x=\alpha_1+\beta_1\sqrt{-1}$$

as long as the particular values

$$\alpha_0+\beta_0\sqrt{-1} \quad \text{and} \quad \alpha_1+\beta_1\sqrt{-1}$$

are distinct from each other, it is clear that by assigning the second of these two values of x, we ought to reduce to zero the integer function Q_0, and consequently this integer function is algebraically divisible by

$$x-\alpha_1-\beta_1\sqrt{-1}.$$

Thus we have

$$Q_0=\left(x-\alpha_1-\beta_1\sqrt{-1}\right)Q_1,$$

where Q_1 denotes a new imaginary integer function of the variable x. Consequently equation (2) can be put into the form

$$\varpi(x)=\left(x-\alpha_0-\beta_0\sqrt{-1}\right)\left(x-\alpha_1-\beta_1\sqrt{-1}\right)Q_1. \tag{3}$$

By reasoning again in this way we find: 1° that the function $\varpi(x)$ ought to vanish by virtue of the assumption that

$$x=\alpha_2+\beta_2\sqrt{-1},$$

so this assumption necessarily reduces the right-hand side of equation (3) to zero, and consequently it reduces to zero one of its three factors; 2° that the factor which reduces to zero cannot be any other than the integer function Q_1, as long as the three particular values of x,

$$\alpha_0+\beta_0\sqrt{-1}, \quad \alpha_1+\beta_1\sqrt{-1}, \quad \alpha_2+\beta_2\sqrt{-1},$$

are distinct from one another; and 3° that because the integer function Q_1 ought to vanish for

$$x=\alpha_2+\beta_2\sqrt{-1},$$

[219] it is algebraically divisible by

$$x-\alpha_2-\beta_2\sqrt{-1}.$$

Consequently, we have

$$Q_1=(x-\alpha_2-\beta_2)\,Q_2$$

and it follows that

$$\varpi(x)=(x-\alpha_0-\beta_0\sqrt{-1})(x-\alpha_1-\beta_1\sqrt{-1})(x-\alpha_2-\beta_2\sqrt{-1})Q_2, \tag{4}$$

where Q_2 again denotes an integer imaginary function of the variable x. Continuing in the same way, we eventually recognize that, in the case where the integer function $\varpi(x)$ vanishes for n different values of x, respectively denoted by

$$\alpha_0+\beta_0\sqrt{-1},\ \alpha_1+\beta_1\sqrt{-1},\ \alpha_2+\beta_2\sqrt{-1},\ \ldots,\ \alpha_{n-1}+\beta_{n-1}\sqrt{-1},$$

then we necessarily have

$$\left\{\begin{aligned}\varpi(x)=&\left(x-\alpha_0-\beta_0\sqrt{-1}\right)\left(x-\alpha_1-\beta_1\sqrt{-1}\right)\\&\times\left(x-\alpha_2-\beta_2\sqrt{-1}\right)\ldots\left(x-\alpha_{n-1}-\beta_{n-1}\sqrt{-1}\right)Q,\end{aligned}\right. \tag{5}$$

where Q denotes a new integer function of the variable x.

It is almost unnecessary to observe that the preceding theorem remains true when we suppose that

$$\chi(x)=0,$$

or else

$$\beta_0=0,\quad \beta_1=0,\quad \beta_2=0,\quad \ldots,\quad \beta_{n-1}=0,$$

that is to say when the function $\varpi(x)$ or the particular values assigned to the variable x become real.

With the aid of the principles established in this section, we can prove without difficulty that in Chapter IV (§ I), theorems III and IV, along with formula (1), can be extended to the case where the functions and the variables, at the same time as the particular values attributed to those functions and variables, become imaginary. We can prove as well that propositions I, II and III, along with formulas (1) and (2) in § II of Chapter IV, and formulas (2), (3), (4), (5) and (6) of § III in the same Chapter, remain true whatever the real or imaginary values [220] of the variables, functions and constants may be. Thus, for example, we see, in particular, that equation (6) of § III, namely

$$\left\{\begin{aligned}\frac{(x+y)^n}{1\cdot2\cdot3\ldots n}=&\frac{x^n}{1\cdot2\cdot3\ldots n}+\frac{x^{n-1}}{1\cdot2\cdot3\ldots(n-1)}\frac{y}{1}+\ldots\\&+\frac{x}{1}\frac{y^{n-1}}{1\cdot2\cdot3\ldots(n-1)}+\frac{y^n}{1\cdot2\cdot3\ldots n},\end{aligned}\right. \tag{6}$$

holds for any imaginary values of the variables x and y.

8.5 Determination of continuous imaginary functions of a single variable that satisfy certain conditions.

Let

$$\varpi(x) = \varphi(x) + \sqrt{-1}\,\chi(x)$$

be a continuous imaginary function of the variable x, where $\varphi(x)$ and $\chi(x)$ are two real continuous functions. The imaginary function $\varpi(x)$ is completely determined if for all the possible real values of the variables x and y, it is required to satisfy one of the equations

(1) $$\varpi(x+y) = \varpi(x) + \varpi(y) \quad \text{or}$$
(2) $$\varpi(x+y) = \varpi(x) \times \varpi(y),$$

or else, for all real positive values of the same variables, one of the following equations:

(3) $$\varpi(xy) = \varpi(x) + \varpi(y) \quad \text{or}$$
(4) $$\varpi(xy) = \varpi(x) \times \varpi(y).$$

We will solve these four equations successively, which will provide us with four problems analogous to those we have already treated in § I of Chapter V.

Problem I. — *To determine the imaginary function* $\varpi(x)$ *in such a manner that it remains continuous between any two real limits of the variable* [221] *x and so that for all real values of the variables x and y, we have*

(1) $$\varpi(x+y) = \varpi(x) + \varpi(y).$$

Solution. — If, with the aid of the formula

$$\varpi(x) = \varphi(x) + \chi(x)\sqrt{-1},$$

we replace the imaginary function ϖ in equation (1) with the real functions φ and χ, this equation becomes

$$\varphi(x+y) + \chi(x+y)\sqrt{-1} = \varphi(x) + \chi(x)\sqrt{-1} + \varphi(y) + \chi(y)\sqrt{-1},$$

then by equating the real parts and the coefficients of $\sqrt{-1}$ on both sides, we conclude

$$\varphi(x+y) = \varphi(x) + \varphi(y) \quad \text{and}$$
$$\chi(x+y) = \chi(x) + \chi(y).$$

From these last formulas (see Chapter V, § I, problem I), we get

$$\varphi(x) = x\varphi(1) \quad \text{and}$$
$$\chi(x) = x\chi(1).$$

Consequently

(5) $$\varpi(x) = x\left[\varphi(1) + \chi(1)\sqrt{-1}\right],$$

or what amounts to the same thing,

(6) $$\varpi(x) = x\varpi(1).$$

It follows from equation (5) that any value of $\varpi(x)$ that satisfies the given question is necessarily of the form

(7) $$\varpi(x) = \left(a + b\sqrt{-1}\right)x,$$

where a and b denote two constant quantities. Moreover, it is easy to assure ourselves that any such value of $\varpi(x)$ satisfies equation (1), whatever the values of a and b. These quantities are thus two arbitrary constants.

[222] We could remark that to obtain the preceding value of $\varpi(x)$, it suffices to replace the arbitrary real constant a in the value of $\varphi(x)$ given by equation (7) of Chapter V (§ I) by the arbitrary but imaginary constant

$$a + b\sqrt{-1}.$$

Problem II. — *To determine the imaginary function $\varpi(x)$ in such a manner that it remains continuous between any two real limits of the variable x and so that for all real values of the variables x and y, we have*

(2) $$\varpi(x+y) = \varpi(x)\varpi(y).$$

Solution.[4] — If we make $x = 0$ in equation (2), we get

$$\varpi(0) = 1,$$

or, because of the formula

$$\varpi(x) = \varphi(x) + \chi(x)\sqrt{-1},$$

we get what amounts to the same thing,

$$\varphi(0) + \chi(0)\sqrt{-1} = 1.$$

Consequently,

[4] Note that this solution is very different from Cauchy's solution to the corresponding problem II in Chapter V, § I. By contrast, problem I in this section followed as an easy corollary of problem I of Chapter V, § I.

$$\varphi(0) = 1 \quad \text{and} \quad \chi(0) = 0.$$

The function $\varphi(x)$ reduces to 1 for the particular value 0 assigned to the variable x, and because it is assumed to be continuous between any limits, it is clear that in the neighborhood of this particular value, it is only very slightly different from 1, and consequently it is positive. Thus, if α denotes a very small number, we can choose this number in such a way that the function $\varphi(x)$ remains constantly positive between the limits

$$x = 0 \quad \text{and} \quad x = \alpha.$$

With this condition satisfied, because the quantity $\varphi(\alpha)$ is itself positive, if we take

$$\rho = \sqrt{\varphi(\alpha)^2 + \chi(\alpha)^2} \quad \text{and} \quad \zeta = \arctan\frac{\chi(\alpha)}{\varphi(\alpha)},$$

we conclude that

$$\varpi(\alpha) = \varphi(\alpha) + \chi(\alpha)\sqrt{-1} = \rho\left(\cos\zeta + \sqrt{-1}\sin\zeta\right).$$

[223] Now imagine that in equation (2) we successively replace y by $y+z$, then z by $z+u$, ... We conclude that

$$\varpi(x+y+z+\ldots) = \varpi(x)\varpi(y)\varpi(z)\ldots,$$

however many variables, $x, y, z, \ldots$, there may be. If we also denote by m the number of variables, and if we make

$$x = y = z = \ldots = \alpha,$$

then the equation we have just found becomes

$$\varpi(m\alpha) = [\varpi(\alpha)]^m = \rho^m\left(\cos m\zeta + \sqrt{-1}\sin m\zeta\right).$$

I add that the formula

$$\varpi(m\alpha) = \rho^m\left(\cos m\zeta + \sqrt{-1}\sin m\zeta\right)$$

remains true if we replace the integer number m by a fraction, or even by an arbitrary number μ. We will prove this easily in what follows.

If in equation (2) we make

$$x = \frac{1}{2}\alpha \quad \text{and} \quad y = \frac{1}{2}\alpha,$$

then we conclude

$$\left[\varpi\left(\frac{1}{2}\alpha\right)\right]^2 = \varpi(\alpha) = \rho\left[\cos\zeta + \sqrt{-1}\sin\zeta\right].$$

Then, by taking square roots of both sides in such a way that the real parts are positive, and by observing that the two functions $\varphi(x)$ and $\cos x$ remain positive, the first between the limits $x=0$ and $x=\alpha$, and the second between the limits $x=0$ and $x=\zeta$, we find that

$$\varpi\left(\frac{1}{2}\alpha\right)=\varphi\left(\frac{1}{2}\alpha\right)+\chi\left(\frac{1}{2}\alpha\right)\sqrt{-1}=\rho^{\frac{1}{2}}\left(\cos\frac{\zeta}{2}+\sqrt{-1}\sin\frac{\zeta}{2}\right).$$

Likewise, if in equation (2) we make

$$x=\frac{1}{4}\alpha \quad \text{and} \quad y=\frac{1}{4}\alpha,$$

[224] then we conclude

$$\left[\varpi\left(\frac{1}{4}\alpha\right)\right]^2=\varpi\left(\frac{1}{2}\alpha\right)=\rho^{\frac{1}{2}}\left(\cos\frac{\zeta}{2}+\sqrt{-1}\sin\frac{\zeta}{2}\right).$$

Then, by taking square roots of both sides so as to obtain positive real parts, we find

$$\varpi\left(\frac{1}{4}\alpha\right)=\rho^{\frac{1}{4}}\left(\cos\frac{\zeta}{4}+\sqrt{-1}\sin\frac{\zeta}{4}\right).$$

By similar reasoning, we can establish successively the formulas

$$\varpi\left(\frac{1}{8}\alpha\right)=\rho^{\frac{1}{8}}\left(\cos\frac{\zeta}{8}+\sqrt{-1}\sin\frac{\zeta}{8}\right),$$
$$\varpi\left(\frac{1}{16}\alpha\right)=\rho^{\frac{1}{16}}\left(\cos\frac{\zeta}{16}+\sqrt{-1}\sin\frac{\zeta}{16}\right),$$
$$\cdots\cdots\cdots\cdots\cdots\cdots\cdots\cdots,$$

and in general, where n denotes any integer number,

$$\varpi\left(\frac{1}{2^n}\alpha\right)=\rho^{\frac{1}{2^n}}\left[\cos\left(\frac{1}{2^n}\zeta\right)+\sqrt{-1}\sin\left(\frac{1}{2^n}\zeta\right)\right].$$

If we operate on the preceding value of $\varpi\left(\frac{1}{2^n}\alpha\right)$ to derive the value of $\varpi\left(\frac{m}{2^n}\alpha\right)$ the same way we operate on the value of $\varpi(\alpha)$ to derive that of $\varpi(m\alpha)$, we find that

$$\varpi\left(\frac{m}{2^n}\alpha\right)=\rho^{\frac{m}{2^n}}\left[\cos\left(\frac{m}{2^n}\zeta\right)+\sqrt{-1}\sin\left(\frac{m}{2^n}\zeta\right)\right],$$

or what amounts to the same thing,

$$\varphi\left(\frac{m}{2^n\alpha}\right)+\chi\left(\frac{m}{2^n\alpha}\right)\sqrt{-1}=\rho^{\frac{m}{2^n}}\left[\cos\left(\frac{m}{2^n}\zeta\right)+\sqrt{-1}\sin\left(\frac{m}{2^n}\zeta\right)\right].$$

Consequently,

$$\varphi\left(\frac{m}{2^n}\alpha\right)=\rho^{\frac{m}{2^n}}\cos\left(\frac{m}{2^n}\zeta\right) \quad \text{and}$$
$$\chi\left(\frac{m}{2^n}\alpha\right)=\rho^{\frac{m}{2^n}}\sin\left(\frac{m}{2^n}\zeta\right).$$

[225] Then, by supposing that the fraction $\frac{m}{2^n}$ varies in such a way as to approach indefinitely the number μ and passing to the limit, we get the equations

$$\varphi(\mu\alpha)=\rho^{\mu}\cos\mu\zeta \quad \text{and} \quad \chi(\mu\alpha)=\rho^{\mu}\sin\mu\zeta,$$

from which we conclude that

(8) $$\varpi(\mu\alpha)=\rho^{\mu}\left(\cos\mu\zeta+\sqrt{-1}\sin\mu\zeta\right).$$

Moreover, if in equation (2) we set

$$x=\mu\alpha \quad \text{and} \quad y=-\mu\alpha,$$

we get

$$\varpi(-\mu\alpha)=\frac{\varpi(0)}{\varpi(\mu\alpha)}=\rho^{-\mu}\left[\cos(-\mu\zeta)+\sqrt{-1}\sin(-\mu\zeta)\right].$$

Thus, formula (8) remains true when we replace μ by $-\mu$. In other words, for all real values of the variable x, both positive and negative, we have

(9) $$\varpi(\alpha x)=\rho^{x}\left[\cos\zeta x+\sqrt{-1}\sin\zeta x\right]=[\varpi(\alpha)]^{x}.$$

In this last formula, if we write $\frac{x}{\alpha}$ instead of x, it becomes

(10) $$\varpi(x)=\rho^{\frac{x}{\alpha}}\left[\cos\left(\frac{\zeta}{\alpha}x\right)+\sqrt{-1}\sin\left(\frac{\zeta}{\alpha}x\right)\right]=[\varpi(\alpha)]^{\frac{x}{\alpha}}.$$

If, for brevity, we make

(11) $$\rho^{\frac{1}{\alpha}}=A \quad \text{and} \quad \frac{\zeta}{\alpha}=b,$$

we find

(12) $$\varpi(x)=A^{x}\left(\cos bx+\sqrt{-1}\sin bx\right).$$

Thus any value of $\varpi(x)$ that satisfies the given question is necessarily of the form

$$A^{x}\left(\cos bx+\sqrt{-1}\sin bx\right),$$

where A and b denote two real quantities, of which the first must be [226] positive. Moreover, it is easy to assure ourselves that such a value of $\varpi(x)$ satisfies equation (2), whatever the values of the number A and the quantity b may be. This number and this quantity are thus arbitrary constants.

Corollary. — In the particular case where the function $\varphi(x)$ remains positive between the limits $x = 0$ and $x = 1$, we can, instead of supposing that α is very small, set $\alpha = 1$. Then we conclude immediately from equations (9) and (10) that

$$\varpi(x) = [\varpi(1)]^x. \tag{13}$$

Problem III. — *To determine the imaginary function $\varpi(x)$ in such a manner that it remains continuous between any two positive limits of the variable x and so that for all positive values of the variables x and y,*

$$\varpi(xy) = \varpi(x) + \varpi(y). \tag{3}$$

Solution. — If, with the aid of the formula

$$\varpi(x) = \varphi(x) + \chi(x)\sqrt{-1},$$

we replace the imaginary function ϖ in equation (3) by the real functions φ and χ, then we equate the real parts and the coefficients of $\sqrt{-1}$ on both sides, we find

$$\begin{aligned} \varphi(xy) &= \varphi(x) + \varphi(y) \quad \text{and} \\ \chi(xy) &= \chi(x) + \chi(y). \end{aligned}$$

Moreover, if A denotes any number and log denotes the characteristic of logarithms in the system for which the base is A, we get from the preceding equations (see Chapter V, § I, problem III)

$$\begin{aligned} \varphi(x) &= \varphi(A)\log(x) \quad \text{and} \\ \chi(x) &= \chi(A)\log(x). \end{aligned}$$

We conclude that

$$\varpi(x) = \left[\varphi(A) + \chi(A)\sqrt{-1}\right]\log(x), \tag{14}$$

[227] or what amounts to the same thing,

$$\varpi(x) = \varpi(A)\log(x). \tag{15}$$

It follows from formula (14) that any value of $\varpi(x)$ that satisfies the given question is necessarily of the form

$$\varpi(x) = \left(a + b\sqrt{-1}\right)\log(x), \tag{16}$$

where a and b denote constant quantities. Moreover, it is easy to assure ourselves that such a value of $\varpi(x)$ satisfies equation (3), whatever the quantities a and b may be. Thus these quantities are two arbitrary constants.

We could remark that to obtain the preceding value of $\varpi(x)$, it suffices to replace the arbitrary real constant a in the value of $\varphi(x)$ given by equation (12) of Chapter V (§ I) by the arbitrary but imaginary constant

$$a+b\sqrt{-1}.$$

Note. — We could arrive very simply at equation (15) in the following manner. By virtue of the identities

$$x=A^{\log x} \quad \text{and} \quad y=A^{\log y},$$

equation (3) becomes

$$\varpi\left(A^{\log x+\log y}\right)=\varpi\left(A^{\log x}\right)+\varpi\left(A^{\log y}\right).$$

Because in this last formula the variable quantities $\log x$ and $\log y$ take on all real values, both positive and negative, as a result we have, for all possible real values of the variables x and y,

$$\varpi\left(A^{x+y}\right)=\varpi\left(A^{x}\right)+\varpi\left(A^{y}\right).$$

We conclude [see problem I, equation (6)] that

$$\varpi\left(A^{x}\right)=x\varpi\left(A^{1}\right)=x\varpi\left(A\right),$$

and consequently

$$\varpi\left(A^{\log x}\right)=\varpi\left(A\right)\log x,$$

[228] or what amounts to the same thing,

$$\varpi(x)=\varpi\left(A\right)\log x.$$

Problem IV. — *To determine the imaginary function $\varpi(x)$ in such a manner that it remains continuous between any two positive limits of the variable x and so that for all positive values of the variables x and y, we have*

(4) $$\varpi\left(xy\right)=\varpi(x)\varpi\left(y\right).$$

Solution. — It would be easy to apply a method similar to that which we used to solve the second problem to the solution of this problem. However, we will arrive more promptly at the solution we seek if we observe that, by denoting by log the characteristic of logarithms in the system for which the base is A, we can put equation (4) into the form

$$\varpi\left(A^{\log x+\log y}\right)=\varpi\left(A^{\log x}\right)\varpi\left(A^{\log y}\right).$$

Because in this last equation the variable quantities $\log x$ and $\log y$ admit any real values, positive and negative, it follows that we have, for all possible real values of the variables x and y,

$$\varpi\left(A^{x+y}\right)=\varpi\left(A^{x}\right)\varpi\left(A^{y}\right).$$

If α represents a very small number and if we replace $\varpi(x)$ with $\varpi(A^x)$ in equation (10) of the second problem, we conclude that

$$\varpi(A^x) = [\varpi(A^\alpha)]^{\frac{x}{\alpha}}.$$

Consequently, we find that

$$\varpi\left(A^{\log x}\right) = [\varpi(A^\alpha)]^{\frac{\log x}{\alpha}},$$

or what amounts to the same thing,

$$\varpi(x) = [\varpi(A^\alpha)]^{\frac{\log x}{\alpha}}. \tag{17}$$

It is essential to observe that the imaginary function $\varpi(A^x)$, and consequently its real part $\varphi(A^x)$, reduce to 1 for $x = 0$, [229] or in other words, that the imaginary function $\varpi(x)$ and its real part $\varphi(x)$ reduce to 1 for $x = 1$. We can prove this directly by taking

$$x = A^0 = 1$$

in equation (4). As for the number α, it need only be small enough that the real part of the imaginary function $\varpi(A^x)$ remain constantly positive between the limits $x = 0$ and $x = \alpha$. When this condition is satisfied, the real part of the imaginary expression

$$\varpi(A^\alpha) = \varphi(A^\alpha) + \chi(A^\alpha)\sqrt{-1}$$

is itself positive. Consequently, if we make

$$\rho = \sqrt{[\varphi(A^\alpha)]^2 + [\chi(A^\alpha)]^2} \quad \text{and} \quad \zeta = \arctan\frac{\chi(A^\alpha)}{\varphi(A^\alpha)},$$

we have

$$\varpi(A^\alpha) = \rho\left(\cos\zeta + \sqrt{-1}\sin\zeta\right).$$

Given this, equation (17) becomes

$$\begin{cases} \varpi(x) = \rho^{\frac{\log x}{\alpha}}\left[\cos\left(\frac{\zeta}{\alpha}\log x\right) + \sqrt{-1}\sin\left(\frac{\zeta}{\alpha}\log x\right)\right] \\ \qquad = x^{\frac{\log\rho}{\alpha}}\left[\cos\left(\frac{\zeta}{\alpha}\log x\right) + \sqrt{-1}\sin\left(\frac{\zeta}{\alpha}\log x\right)\right]. \end{cases} \tag{18}$$

By virtue of this last equation, any value of $\varpi(x)$ that satisfies the given question is necessarily of the form

$$\varpi(x) = x^a\left[\cos(b\log x) + \sqrt{-1}\sin(b\log x)\right], \tag{19}$$

where a and b denote two constant quantities. Moreover, it is easy to assure ourselves that these two constant quantities ought to remain entirely arbitrary.

Chapter 9
On convergent and divergent imaginary series. Summation of some convergent imaginary series. Notations used to represent imaginary functions that we find by evaluating the sum of such series.

9.1 General considerations on imaginary series.

[230] Let

$$(1) \qquad p_0, \quad p_1, \quad p_2, \ldots, p_n, \quad \ldots \quad \text{and}$$
$$(2) \qquad q_0, \quad q_1, \quad q_2, \ldots, q_n, \quad \ldots$$

be two real series. The sequence of imaginary expressions

$$(3) \qquad p_0+q_0\sqrt{-1},\ p_1+q_1\sqrt{-1},\ p_2+q_2\sqrt{-1},\ \ldots,\ p_n+q_n\sqrt{-1},\ \ldots$$

forms what we call an *imaginary series*. Moreover, let

$$(4) \qquad \begin{cases} s_n = \left(p_0+q_0\sqrt{-1}\right)+\left(p_1+q_1\sqrt{-1}\right)+\ldots \\ \qquad +\left(p_{n-1}+q_{n-1}\sqrt{-1}\right) \\ \quad = (p_0+p_1+\ldots+p_{n-1})+(q_0+q_1+\ldots+q_{n-1})\sqrt{-1} \end{cases}$$

be the sum of the first n terms of this series. Depending on whether or not s_n converges towards a fixed limit for increasing values of n, we say that series (3) is *convergent* and that it has this limit as its sum, or else that it is *divergent* and it does not have a sum. The first case evidently occurs if the two sums

$$\begin{gathered} p_0+p_1+\ldots+p_{n-1} \quad \text{and} \\ q_0+q_1+\ldots+q_{n-1} \end{gathered}$$

themselves converge towards [231] fixed limits, for increasing values of n, and the second in the opposite case. In other words, series (3) is always convergent at the same time as the real series (1) and (2) are convergent. If even one of these series is divergent, then series (3) is divergent as well.

In every possible case, the term of series (3) that corresponds to the index n, namely

R.E. Bradley, C.E. Sandifer, *Cauchy's Cours d'analyse*, Sources and Studies in the History of Mathematics and Physical Sciences, DOI 10.1007/978-1-4419-0549-9_9,

$$p_n + q_n\sqrt{-1},$$

is called its *general term.*

The simplest of these imaginary series is the one we get by attributing an imaginary value to the variable x in the geometric progression

$$1, \quad x, \quad x^2, \quad \dots, \quad x^n, \quad \dots.$$

Imagine, to clarify these ideas, that we make

$$x = z\left(\cos\theta + \sqrt{-1}\sin\theta\right),$$

where z denotes a new real variable and θ a real arc. The geometric progression in question becomes

$$(5) \qquad \begin{cases} 1, \ z\left(\cos\theta + \sqrt{-1}\sin\theta\right), \ z^2\left(\cos 2\theta + \sqrt{-1}\sin 2\theta\right), \ \dots \\ \quad \dots, \ z^n\left(\cos n\theta + \sqrt{-1}\sin n\theta\right), \ \dots. \end{cases}$$

To obtain the equation that determines the sum of the first n terms of the preceding series, it suffices to replace x by $z\left(\cos\theta + \sqrt{-1}\sin\theta\right)$ in the formula

$$1 + x + x^2 + \dots + x^{n-1} = \frac{1}{1-x} - \frac{x^n}{1-x}.$$

In this way, we find that

$$(6) \qquad \begin{cases} 1 + z\left(\cos\theta + \sqrt{-1}\sin\theta\right) + z^2\left(\cos 2\theta + \sqrt{-1}\sin 2\theta\right) + \dots \\ \qquad + z^{n-1}\left[\cos(n-1)\theta + \sqrt{-1}\sin(n-1)\theta\right] \\ = \dfrac{1}{1 - z\left(\cos\theta + \sqrt{-1}\sin\theta\right)} - \dfrac{z^{n-1}}{1 - z\left(\cos\theta + \sqrt{-1}\sin\theta\right)}. \end{cases}$$

For increasing values of n, the modulus of the imaginary [232] expression

$$\frac{z^n\left(\cos n\theta + \sqrt{-1}\sin n\theta\right)}{1 - z\cos\theta - z\sin\theta\sqrt{-1}},$$

namely

$$\frac{\pm z^n}{(1 - 2z\cos\theta + z^2)^{\frac{1}{2}}},$$

converges towards the limit zero or grows beyond all limits, depending on whether we suppose that the numerical value of z is less than or greater than 1. Thus we ought to conclude from equation (6) that under the first hypothesis, series (5) is a convergent series that has

$$\frac{1}{1 - z\cos\theta - z\sin\theta\sqrt{-1}}$$

for its sum, and under the second hypothesis, it is a divergent series that does not have a sum.

We indicate the sum of an imaginary series the same way we do for a real series, by the sum of its first terms followed by an ellipsis ....

Given this, if we denote the sum of series (3) by s, assuming it is convergent, and if we make n grow indefinitely in formula (4), we find by passing to the limit that

$$(7)\qquad \begin{cases} s = \left(p_0+q_0\sqrt{-1}\right)+\left(p_1+q_1\sqrt{-1}\right)+\left(p_2+q_2\sqrt{-1}\right)+\ldots \\ = (p_0+p_1+p_2+\ldots)+(q_0+q_1+q_2+\ldots)\sqrt{-1}. \end{cases}$$

In the same way, when we suppose that the numerical value of z is less than 1 and we make n grow beyond any assignable limit, we get from equation (6),

$$(8)\qquad \begin{cases} 1+z\left(\cos\theta+\sqrt{-1}\sin\theta\right)+z^2\left(\cos 2\theta+\sqrt{-1}\sin 2\theta\right)+\ldots \\ \quad = \dfrac{1}{1-z\cos\theta-z\sin\theta\sqrt{-1}} = \dfrac{1-z\cos\theta+z\sin\theta\sqrt{-1}}{1-2z\cos\theta+z^2}. \end{cases}$$

By virtue of formula (7), the first part of equation (8) can [233] be written in the following form:

$$\left(1+z\cos\theta+z^2\cos 2\theta+\ldots\right)+\left(z\sin\theta+z^2\sin 2\theta+\ldots\right)\sqrt{-1}.$$

Thus, for numerical values of z less than 1, we have

$$(9)\qquad \begin{cases} \left(1+z\cos\theta+z^2\cos 2\theta+\ldots\right)+\left(z\sin\theta+z^2\sin 2\theta+\ldots\right)\sqrt{-1} \\ \quad = \dfrac{1-z\cos\theta}{1-2z\cos\theta+z^2}+\dfrac{z\sin\theta\sqrt{-1}}{1-2z\cos\theta+z^2}\sqrt{-1}. \end{cases}$$

Thus we conclude that[1]

$$(10)\qquad \begin{cases} 1+z\cos\theta+z^2\cos 2\theta+z^3\cos 3\theta+\ldots = \dfrac{1-z\cos\theta}{1-2z\cos\theta+z^2}, \\ z\sin\theta+z^2\sin 2\theta+z^3\sin 3\theta+\ldots = \dfrac{z\sin\theta}{1-2z\cos\theta+z^2} \end{cases}$$

$$(z=-1,\ z=+1).$$

Thus the substitution of an imaginary value for x in the geometric progression

$$1,\quad x,\quad x^2,\quad \ldots,\quad x^n,\quad \ldots$$

is enough to lead to the summation of the two series

$$(11)\qquad \begin{cases} 1,\ z\cos\theta,\ z^2\cos 2\theta,\ \ldots,\ z^n\cos n\theta,\ \ldots \\ z\sin\theta,\ z^2\sin 2\theta,\ \ldots,\ z^n\sin n\theta,\ \ldots \end{cases}$$

[1] Euler summed a similar series by similar means in [Euler 1774].

whenever the variable z remains contained between the limits

$$z = -1 \quad \text{and} \quad z = +1,$$

that is to say, whenever the two series are convergent.

The left-hand sides of equations (10) are (by virtue of theorem I, Chapter VI, § I) continuous functions of the variable z in the neighborhood of any particular value contained between the limits

$$z = -1 \quad \text{and} \quad z = +1,$$

and so the left-hand side of equation (9) is itself a continuous function of z in the neighborhood of the same value. Now, this left-hand side is nothing but the sum of series (5), of which the different [234] terms remain continuous functions of z between any limits whatsoever. By generalizing the remark that we have just made, we obtain the following proposition:

Theorem I. — *When the different terms of series* (3) *are functions of the same variable z and are continuous with respect to this variable in the neighborhood of a particular value for which this series is convergent, the sum s of this series is also a continuous function of z in the neighborhood of this particular value.*[2]

Proof. — Indeed, in the neighborhood of the particular value attributed to the variable z, series (3) cannot be convergent and have continuous functions of z for its different terms unless the real series (1) and (2) both enjoy the same properties. Now under this hypothesis, because both of the sums

$$p_0 + p_1 + p_2 + \dots \quad \text{and}$$
$$q_0 + q_1 + q_2 + \dots$$

are continuous functions of the variable z (by virtue of theorem I, Chapter VI, § I) it follows that the sum of series (3), namely

$$s = (p_0 + p_1 + p_2 + \dots) + (q_0 + q_1 + q_2 + \dots)\sqrt{-1}$$

is also a continuous function of this variable.

Now suppose that we denote by

$$\rho_0, \quad \rho_1, \quad \rho_2, \quad \dots$$

the moduli of the various terms of series (3), and by

$$\cos\theta_0 + \sqrt{-1}\sin\theta_0, \ \cos\theta_1 + \sqrt{-1}\sin\theta_1, \ \cos\theta_2 + \sqrt{-1}\sin\theta_2, \ \dots$$

the corresponding reduced expressions, so that in general we have

---

[2] This theorem as stated is incorrect. Cauchy's proof depends on his incorrect theorem I of Chapter VI, § I. See the footnote on p. 90.

$$\rho_n = \left(p_n^2 + q_n^2\right)^{\frac{1}{2}} \quad \text{and}$$
$$p_n + q_n\sqrt{-1} = \rho_n\left(\cos\theta_n + \sqrt{-1}\sin\theta_n\right).$$

[235] Series (3) becomes

$$(12) \qquad \begin{cases} \rho_0\left(\cos\theta_0 + \sqrt{-1}\sin\theta_0\right), \\ \rho_1\left(\cos\theta_1 + \sqrt{-1}\sin\theta_1\right), \\ \rho_2\left(\cos\theta_2 + \sqrt{-1}\sin\theta_2\right), \\ \dots\dots\dots\dots\dots\dots, \\ \rho_n\left(\cos\theta_n + \sqrt{-1}\sin\theta_n\right), \\ \dots\dots\dots\dots\dots\dots, \end{cases}$$

and we can ordinarily decide if this series is convergent or divergent with the aid of the theorem I am about to state.

Theorem II.[3] — *To find the limit or limits towards which the expression* $(\rho_n)^{\frac{1}{n}}$ *converges as n grows indefinitely. Series* (3) *converges or diverges according to whether the largest of these limits is less than or greater than 1.*

Proof. — First consider the case where the largest values of the expression $(\rho_n)^{\frac{1}{n}}$ converge towards a limit less than 1 as n grows indefinitely. In this case, because the series

$$(13) \qquad \rho_0, \quad \rho_1, \quad \rho_2, \quad \dots, \quad \rho_n, \quad \dots$$

is convergent (Chapter VI, § II, theorem I), the two series

$$(14) \qquad \begin{cases} \rho_0\cos\theta_0, & \rho_1\cos\theta_1, & \rho_2\cos\theta_2, & \dots, & \rho_n\cos\theta_n, & \dots, \\ \rho_0\sin\theta_0, & \rho_1\sin\theta_1, & \rho_2\sin\theta_2, & \dots, & \rho_n\sin\theta_n, & \dots \end{cases}$$

are convergent as well (Chapter VI, § III, theorem IV), and the convergence of these last series entails that of series (12), which is nothing but series (3) presented in a different form.

In the second place, suppose that for increasing values of n, the largest values of $(\rho_n)^{\frac{1}{n}}$ converge towards a limit greater than 1. Under this hypothesis and using reasoning similar to that which we used in Chapter VI (§ II, theorem I), we prove that the largest values of the modulus

$$\rho_n = \left(p_n^2 + q_n^2\right)^{\frac{1}{2}}$$

[236] increase with n beyond all limits, which cannot be true unless the largest values of the two quantities p_n and q_n, or at least one of them, likewise grows indefinitely. Now as these two quantities are the general terms of series (1) and (2), we

[3] This is the Root Test adapted to complex numbers.

must conclude that at least one of these two series must be divergent, which suffices to assure the divergence of series (3).

Scholium I. — The theorem that we have just established leaves no doubt about the convergence or divergence of an imaginary series except in the particular case where the limit of the largest values of $(\rho_n)^{\frac{1}{n}}$ becomes equal to 1. In this particular case, it is not always easy to decide the issue. Nevertheless, we can affirm that if series (13) is convergent, then series (14) and consequently series (12) are as well. The converse is not true, and it can turn out that although series (12) remains convergent, series (13) is divergent. Thus, for example, if we take

$$\rho_n = \frac{1}{n+1} \quad \text{and} \quad \theta_n = \left(n + \frac{1}{2}\right)\pi,$$

then we get for series (12) and (13) the two following ones:

$$\begin{array}{lllll} \sqrt{-1}, & -\frac{1}{2}\sqrt{-1}, & +\frac{1}{3}\sqrt{-1}, & -\frac{1}{4}\sqrt{-1}, & \ldots, \\ 1, & \frac{1}{2}, & \frac{1}{3}, & \frac{1}{4}, & \ldots, \end{array}$$

where the second is divergent, while the first remains convergent and has for its sum

$$\sqrt{-1}\ln 2,$$

where ln denotes the characteristic of Napierian logarithms.

Scholium II. — Whenever the ratio

$$\frac{\rho_{n+1}}{\rho_n}$$

indefinitely approaches a fixed limit for increasing values of n, this limit is the same
[237] as the limit towards which the largest values of the expression $(\rho_n)^{\frac{1}{n}}$ converge.

Theorem V of § III (Chapter VI) is evidently applicable to imaginary series as well as to real series. As for theorem VI of the same section, when it is a question of imaginary series, we ought to replace it with the following:

Theorem III. — *Let*

$$(15) \qquad \begin{cases} u_0, & u_1, & u_2, & \ldots, & u_n, & \ldots, \\ v_0, & v_1, & v_2, & \ldots, & v_n, & \ldots \end{cases}$$

be two convergent imaginary series that have s and s′, respectively, as their sums. If each of these series remains convergent when we reduce its terms to their respective moduli, then

$$(16) \qquad \begin{cases} u_0v_0, \quad u_0v_1 + u_1v_0, \quad u_0v_2 + u_1v_1 + u_2v_0, \quad \ldots, \\ \ldots, \quad u_0v_n + u_1v_{n-1} + \ldots + u_{n-1}v_1 + u_nv_0, \quad \ldots \end{cases}$$

is a new convergent imaginary series that has ss' as its sum.

Proof. – Denote, respectively, by s_n and s'_n the sums of the first n terms of the two series (15), and by s''_n the sum of the first n terms of series (16). We find that

$$\begin{aligned} s_n s'_n - s''_n = u_{n-1}v_{n-1} + (u_{n-1}v_{n-2} + u_{n-2}v_{n-1}) + \dots \\ + (u_{n-1}v_1 + u_{n-2}v_2 + \dots + u_n v_{n-2} + u_1 v_{n-1}). \end{aligned}$$

Again denote by ρ_n and ρ'_n the moduli of the imaginary expressions u_n and v_n, so that these expressions are determined by equations of the form

$$\begin{aligned} u_n &= \rho_n \left(\cos\theta_n + \sqrt{-1}\sin\theta_n\right) \quad \text{and} \\ v_n &= \rho'_n \left(\cos\theta'_n + \sqrt{-1}\sin\theta'_n\right). \end{aligned}$$

Because the real series

$$\begin{aligned} &\rho_0, \rho_1, \rho_2, \dots, \rho_n, \dots \quad \text{and} \\ &\rho'_0, \rho'_1, \rho'_2, \dots, \rho'_n, \dots \end{aligned}$$

[238] are convergent by hypothesis, we conclude, as in Chapter VI (§ III, theorem VI) that the sum

$$\begin{aligned} \rho_{n-1}\rho'_{n-1} + (\rho_{n-1}\rho'_{n-2} + \rho_{n-2}\rho'_{n-1}) + \dots \\ + (\rho_{n-1}\rho'_1 + \rho_{n-2}\rho'_2 + \dots + \rho_2\rho'_{n-2} + \rho_1\rho'_{n-1}) \end{aligned}$$

converges towards the limit zero for increasing values of n. It is the same, *a fortiori*, for the two sums

$$\begin{aligned} &\rho_{n-1}\rho'_{n-1}\cos\left(\theta_{n-1} + \theta'_{n-1}\right) \\ &\quad + \left[\rho_{n-1}\rho'_{n-2}\cos\left(\theta_{n-1} + \theta'_{n-2}\right) + \rho_{n-2}\rho'_{n-1}\cos\left(\theta_{n-2} + \theta'_{n-1}\right)\right] \\ &\quad + \dots\dots\dots\dots\dots\dots\dots\dots\dots\dots\dots\dots \\ &\quad + \left[\rho_{n-1}\rho'_1\cos(\theta_{n-1} + \theta'_1) + \rho_{n-2}\rho'_2\cos(\theta_{n-2} + \theta'_2) + \dots \right. \\ &\qquad \left. + \rho_2\rho'_{n-2}\cos\left(\theta_2 + \theta'_{n-2}\right) + \rho_1\rho'_{n-1}\cos\left(\theta_1 + \theta'_{n-1}\right)\right] \end{aligned}$$

and

$$\begin{aligned} &\rho_{n-1}\rho'_{n-1}\sin\left(\theta_{n-1} + \theta'_{n-1}\right) \\ &\quad + \left[\rho_{n-1}\rho'_{n-2}\sin\left(\theta_{n-1} + \theta'_{n-2}\right) + \rho_{n-2}\rho'_{n-1}\sin\left(\theta_{n-2} + \theta'_{n-1}\right)\right] \\ &\quad + \dots\dots\dots\dots\dots\dots\dots\dots\dots\dots\dots\dots \\ &\quad + \left[\rho_{n-1}\rho'_1\sin(\theta_{n-1} + \theta'_1) + \rho_{n-2}\rho'_2\sin(\theta_{n-2} + \theta'_2) + \dots \right. \\ &\qquad \left. + \rho_2\rho'_{n-2}\sin\left(\theta_2 + \theta'_{n-2}\right) + \rho_1\rho'_{n-1}\sin\left(\theta_1 + \theta'_{n-1}\right)\right], \end{aligned}$$

where the first series evidently represents the real part of the imaginary expression

$$s_n s'_n - s''_n,$$

while the second series represents the coefficient of $\sqrt{-1}$ in this expression. Consequently, $s_n s'_n - s''_n$ also converges towards the limit zero for increasing values of n. Because $s_n s'_n$ indefinitely approaches the limit ss', it is certainly necessary that the expression s''_n, that is to say the sum of the first n terms of series (16), itself indefinitely approaches this last limit. It follows that: 1° series (16) is convergent; and 2° that this convergent series has as its sum ss'.

9.2 On imaginary series ordered according to the ascending integer powers of a single variable.

[239] Let x be an imaginary variable. Any imaginary series ordered according to the ascending integer powers of the variable x is of the form

$$a_0+b_0\sqrt{-1},\quad (a_1+b_1\sqrt{-1})\,x,\quad (a_2+b_2\sqrt{-1})\,x^2,\quad \ldots,$$
$$\ldots,\quad (a_n+b_n\sqrt{-1})\,x^n,\quad \ldots,$$

where $a_0,\ a_1,\ a_2,\ \ldots,\ a_n,\ \ldots$ and $b_0,\ b_1,\ b_2,\ \ldots,\ b_n,\ \ldots$ denote two sequences of constant quantities. In the case where the constants in the second sequence vanish, the preceding series reduces to

$$(1)\qquad a_0,\quad a_1x,\quad a_2x^2,\quad \ldots,\quad a_nx^n,\ldots.$$

In this section, we consider in particular series of this last kind. If, for simplicity, we put

$$(2)\qquad x=z\left(\cos\theta+\sqrt{-1}\sin\theta\right),$$

where z denotes a real variable and θ denotes a real arc, then series (1) becomes

$$(3)\qquad \begin{cases} a_0,\ a_1z(\cos\theta+\sqrt{-1}\sin\theta),\ a_2z^2(\cos2\theta+\sqrt{-1}\sin2\theta),\ \ldots,\\ \qquad\ldots,\quad a_nz^n(\cos n\theta+\sqrt{-1}\sin n\theta),\quad \ldots. \end{cases}$$

Now, as in Chapter VI (§ IV), let A be the largest of the limits towards which the nth root of the numerical value of a_n converges as n increases indefinitely. Under the same hypothesis, the largest of the limits towards which the nth root of the modulus of the imaginary expression

$$a_nx^n=a_nz^n\left(\cos n\theta+\sqrt{-1}\sin n\theta\right)$$

converges is equivalent to the numerical value of the product

$$Az.$$

Consequently (see above, § I, theorem II), series (3) is [240] convergent or divergent according to whether the product Az has a value less than or greater than 1. We deduce the following proposition immediately from this remark:

Theorem I. — *Series* (3) *is convergent for all values of z contained between the limits*

$$z = -\frac{1}{A} \quad \text{and} \quad z = +\frac{1}{A},$$

and divergent for all values of z located outside these same limits. In other words, series (1) *is convergent or divergent depending on whether the modulus of the imaginary expression x is less than or greater than $\frac{1}{A}$.*

Scholium. — When the numerical value of the ratio $\frac{a_{n+1}}{a_n}$ converges to a fixed limit for increasing values of n, this limit is precisely the value of the positive quantity denoted by A.

Corollary I. — In comparing the preceding theorem to theorem I of Chapter VI (§ IV), we recognize that if series (1) is convergent for a certain real value of the variable x, it remains convergent for every imaginary value that has this real value as its modulus, up to sign. Consequently, if series (1) is convergent for all real values of the variable x, it remains convergent for whatever imaginary value we may attribute to this variable.

Corollary II. — To apply theorem I and the preceding corollary, consider the following four series:

$$(4) \qquad 1,\ x,\quad x^2,\quad \ldots,\quad x^n,\quad \ldots,$$

$$(5) \qquad 1,\ \tfrac{\mu}{1}x,\ \tfrac{\mu(\mu-1)}{1\cdot 2}x^2,\ \ldots,\ \tfrac{\mu(\mu-1)\ldots(\mu-n+1)}{1\cdot 2\cdot 3\ldots n}x^n,\ \ldots,$$

$$(6) \qquad 1,\ \frac{x}{1},\quad \frac{x^2}{1\cdot 2},\quad \ldots,\quad \frac{x^n}{1\cdot 2\cdot 3\ldots n},\quad \ldots,$$

$$(7) \qquad x,\quad -\frac{x^2}{2},\quad \ldots,\quad \pm\frac{x^n}{n},\quad \ldots,$$

[241] where in the second series μ denotes any quantity whatsoever. Of these four series, the first two, as well as the last, remain convergent for all real values of x contained between the limits

$$x = -1 \quad \text{and} \quad x = +1,$$

and the third remains convergent for all real values of the variable x. However, instead of giving x a real value, if we suppose that

$$x = z\left(\cos\theta + \sqrt{-1}\sin\theta\right),$$

then instead of these four series, we get the following ones:

$$(8)\quad \begin{cases} 1,\ z(\cos\theta+\sqrt{-1}\sin\theta),\ z^2(\cos 2\theta+\sqrt{-1}\sin 2\theta),\ \ldots, \\ \qquad \ldots,\quad z^n(\cos n\theta+\sqrt{-1}\sin n\theta),\quad \ldots; \end{cases}$$

$$(9)\quad \begin{cases} 1,\ \frac{\mu}{1}z(\cos\theta+\sqrt{-1}\sin\theta), \\ \frac{\mu(\mu-1)}{1\cdot 2}z^2(\cos 2\theta+\sqrt{-1}\sin 2\theta),\quad \ldots, \\ \frac{\mu(\mu-1)\ldots(\mu-n+1)}{1\cdot 2\cdot 3\ldots n}z^n(\cos n\theta+\sqrt{-1}\sin n\theta),\quad \ldots; \end{cases}$$

$$(10)\quad \begin{cases} 1,\ \dfrac{z(\cos\theta+\sqrt{-1}\sin\theta)}{1},\ \dfrac{z^2(\cos 2\theta+\sqrt{-1}\sin 2\theta)}{1\cdot 2},\ \ldots, \\ \qquad \ldots,\quad \dfrac{z^n(\cos n\theta+\sqrt{-1}\sin n\theta)}{1\cdot 2\cdot 3\ldots n},\quad \ldots;\quad \text{and} \end{cases}$$

$$(11)\quad \begin{cases} \dfrac{z(\cos\theta+\sqrt{-1}\sin\theta)}{1},\ -\dfrac{z^2(\cos 2\theta+\sqrt{-1}\sin 2\theta)}{2},\ \ldots, \\ \qquad \ldots,\quad \pm\dfrac{z^n(\cos n\theta+\sqrt{-1}\sin n\theta)}{n},\quad \ldots, \end{cases}$$

where the first two and the last one remain convergent for all values of z contained between the limits

$$z=-1 \quad \text{and} \quad z=+1,$$

while the remaining one is always convergent, whatever the real value of z may be.

Having fixed the limits between which z must be contained in order to render series (3) convergent, we make the remark that, by virtue [242] of the principles established in the preceding section, theorems III, IV and V of Chapter VI (§ IV), with their corollaries, can be extended to the case where the variable x becomes imaginary. We need only assume that in the statement of theorem IV, each of the series

$$a_0,\quad a_1x,\quad a_2x^2,\quad \ldots \quad \text{and}$$
$$b_0,\quad b_1x,\quad b_2x^2,\quad \ldots$$

remains convergent when we reduce the terms not just to their numerical values but to their respective moduli. Given this, if we denote by $\varpi(\mu)$ what the right-hand side of equation (15) (Chapter VI § IV) becomes when we give to x the imaginary value

$$z\left(\cos\theta+\sqrt{-1}\sin\theta\right),$$

or in other words, if we make

$$(12)\qquad \varpi(\mu)=1+\frac{\mu}{1}z\left(\cos\theta+\sqrt{-1}\sin\theta\right) + \frac{\mu(\mu-1)}{1\cdot 2}z^2\left(\cos 2\theta+\sqrt{-1}\sin 2\theta\right)+\ldots,$$

we find, in place of formula (16) (Chapter VI, § IV), the following:

$$\varpi(\mu)\,\varpi(\mu') = \varpi(\mu+\mu'). \tag{13}$$

It is essential to remark that this last formula remains true only for values of z contained between the limits $z=-1$ and $z=+1$, and that between these limits, the imaginary function $\varpi(\mu)$, that is to say the sum of series (9), is at the same time continuous with respect to z and with respect to μ (see above, § I, theorem I).

Imagine for the time being that instead of series (9) we consider more generally series (3), and that in this last series we make the value of z vary by insensible degrees. As long as series (3) is convergent, that is to say as long as the value of z remains contained between the limits

$$-\frac{1}{A} \quad \text{and} \quad +\frac{1}{A},$$

the sum of the series is a continuous imaginary function of the [243] variable z. Let $\varpi(z)$ be this continuous function. The equation

$$\varpi(z) = a_0 + a_1 z\left(\cos\theta + \sqrt{-1}\sin\theta\right) + a_2 z^2\left(\cos 2\theta + \sqrt{-1}\sin 2\theta\right) + \ldots$$

remains true for all values of z contained between the limits $-\frac{1}{A}$ and $+\frac{1}{A}$, which we indicate by writing these limits beside the series,[4] as we see here:

$$\left\{\begin{aligned} \varpi(z) = a_0 &+ a_1 z\,\left(\cos\theta + \sqrt{-1}\sin\theta\right) \\ &+ a_2 z^2\left(\cos 2\theta + \sqrt{-1}\sin 2\theta\right) + \ldots \\ &\left(z = -\tfrac{1}{A},\quad z = +\tfrac{1}{A}\right). \end{aligned}\right. \tag{14}$$

We ought to observe that the preceding equation is always equivalent to two real equations. Indeed, if we set

$$\varpi(z) = \varphi(z) + \chi(z)\sqrt{-1}, \tag{15}$$

where $\varphi(z)$ and $\chi(z)$ denote two real functions, we get from equation (14) that

$$\left\{\begin{aligned} \varphi(z) &= a_0 + a_1 z\cos\theta + a_2 z^2\cos 2\theta + \ldots, \\ \chi(z) &= \qquad a_1 z\sin\theta + a_2 z^2\sin 2\theta + \ldots \\ &\left(z = -\tfrac{1}{A},\quad z = +\tfrac{1}{A}\right). \end{aligned}\right. \tag{16}$$

When series (3) is given, we can sometimes deduce the value of the function $\varpi(x)$ in a finite form, and to do this is called *summing* the series. In § I, we have already resolved this question for series (8). We will now try to resolve it for series (9), (10) and (11), and as a consequence, we will treat the three problems that follow, one after another.

---

[4] In [Cauchy 1821, p. 290 ff], the limits really are written beside the series. However, in [Cauchy 1897, p. 243 ff], they are written below. (tr.)

Problem I. — *To find the sum of the series*

$$(9) \qquad 1,\ \tfrac{\mu}{1}z(\cos\theta+\sqrt{-1}\sin\theta),\ \tfrac{\mu(\mu-1)}{1\cdot 2}z^2(\cos 2\theta+\sqrt{-1}\sin 2\theta),\ \ldots,$$

in the case where we attribute to the variable z a value contained between the limits

$$z=-1 \quad \textit{and} \quad z=+1.$$

[244] *Solution.* — Let $\varpi(\mu)$ be the sum being sought. Let μ' denote a real quantity different from μ. We find

$$(13) \qquad \varpi(\mu)\,\varpi(\mu') = \varpi(\mu+\mu').$$

The preceding equation, being similar to equation (2) of Chapter VIII (§ V), is solved in the same manner, and we thus conclude that

$$\varpi(\mu) = r^{\mu}\left(\cos\mu t+\sqrt{-1}\sin\mu t\right),$$

where the modulus r and the angle t are two quantities constant with respect to μ, but which necessarily depend on z and θ. Thus, between the limits $z=-1$ and $z=+1$, we have

$$(17) \qquad \begin{cases} 1+\tfrac{\mu}{1}z\left(\cos\theta+\sqrt{-1}\sin\theta\right)+\tfrac{\mu(\mu-1)}{1\cdot 2}z^2\left(\cos 2\theta+\sqrt{-1}\sin 2\theta\right)+\ldots \\ \quad = r^{\mu}\left(\cos\mu t+\sqrt{-1}\sin\mu t\right). \end{cases}$$

To determine the unknown values of r and t, we set $\mu=1$ in equation (17), and then we get

$$1+z\cos\theta+z\sin\theta\sqrt{-1} = r\cos t+r\sin t\sqrt{-1},$$

or what amounts to the same thing,

$$\begin{aligned} 1+z\cos\theta &= r\cos t \\ z\sin\theta &= r\sin t. \end{aligned}$$

Consequently we find

$$r=\left(1+2z\cos\theta+z^2\right)^{\frac{1}{2}}.$$

Then, by observing that $\cos t=\frac{1+z\cos\theta}{r}$ remains postive for every numerical value of z less than 1, and denoting by k any integer number, we also find

$$t=\arctan\frac{z\sin\theta}{1+z\cos\theta}\pm 2k\pi.$$

Given this, if for brevity we make

$$(18) \qquad s=\arctan\frac{z\sin\theta}{1+z\cos\theta},$$

[245] then equation (17) becomes

$$(19)\quad \begin{cases} 1+\frac{\mu}{1}z\left(\cos\theta+\sqrt{-1}\sin\theta\right)+\frac{\mu(\mu-1)}{1\cdot 2}z^2\left(\cos 2\theta+\sqrt{-1}\sin 2\theta\right)+\dots \\ \quad =\left(1+2z\cos\theta+z^2\right)^{\frac{1}{2}\mu}\left(\cos\mu t+\sqrt{-1}\sin\mu t\right) \end{cases}$$

$$(z=-1,\quad z=+1),$$

where the value of t is determined by the formula

$$(20)\qquad t=s\pm 2k\pi,$$

in which the integer number k depends only on the quantities z and θ.

We remark now that, between the limits $z=-1$ and $z=+1$, the left-hand side of equation (19) is a continuous function of z that varies with z by insensible degrees, whatever the value of μ. The right-hand side of the equation thus ought to enjoy the same property. In other words, the quantities

$$\left(1+2z\cos\theta+z^2\right)^{\frac{\mu}{2}}\cos\mu t,\quad \text{and}$$
$$\left(1+2z\cos\theta+z^2\right)^{\frac{\mu}{2}}\sin\mu t,$$

and consequently

$$\cos\mu t\quad \text{and}\quad \sin\mu t$$

ought to vary with z by insensible degrees for all possible values of μ. Now, this condition cannot be satisfied except in the case where t itself varies with z by insensible degrees. Indeed, if an infinitely small increase in z produces a finite increase in t in such a way as to change t into $t+a$, where a denotes a finite quantity, the sines and cosines of the two arcs

$$\mu t\quad \text{and}\quad \mu(t+a)$$

could not remain sensibly equal, except when the numerical value of the product μa is very close to a multiple of the [246] circumference, which cannot be true except for particular values of the coefficient μ, and not generally for any finite values of this coefficient. Thus we must conclude that the arc $t=s\pm 2k\pi$ is a continuous function of z. Because the first of the two quantities s and k, determined by equation (18), varies with z in a continuous way between the limits $z=-1$ and $z=+1$, while the second, which must always be an integer, admits only finite variations that are multiples of 1, it is clear that to satisfy the stated condition, the quantity s must be the only one to vary and the quantity k must remain constant. Thus this last quantity is independent of z, and to know its value in all possible cases, it suffices to find it by supposing that $z=0$. Because under this hypothesis, we have $s=0$ and $t=s\pm 2k\pi$, we get

$$1=\cos(2k\mu\pi)\pm\sqrt{-1}\sin(2k\mu\pi)$$

from equation (19), whatever the value of μ may be. Consequently

$$k = 0.$$

Given this, in general, formula (20) gives

$$t = s,$$

and equation (19) is found to reduce to

$$(21)\quad \begin{cases} 1 + \frac{\mu}{1} z\left(\cos\theta + \sqrt{-1}\sin\theta\right) + \frac{\mu(\mu-1)}{1\cdot 2} z^2\left(\cos 2\theta + \sqrt{-1}\sin 2\theta\right) + \dots \\ \quad = \left(1 + 2z\cos\theta + z^2\right)^{\frac{1}{2}\mu}\left(\cos\mu s + \sqrt{-1}\sin\mu s\right) \end{cases}$$
$$(z = -1, \quad z = +1).$$

Moreover, if we consider formula (27) of Chapter VII (§ IV), we easily recognize that the right-hand side of equation (21) can be represented by the notation

$$\left[1 + z\left(\cos\theta + \sqrt{-1}\sin\theta\right)\right]^{\mu}.$$

Thus, always supposing that the value of z is contained between the [247] limits $+1$ and -1, we have

$$(22)\quad \begin{cases} 1 + \frac{\mu}{1} z\left(\cos\theta + \sqrt{-1}\sin\theta\right) + \frac{\mu(\mu-1)}{1\cdot 2} z^2\left(\cos 2\theta + \sqrt{-1}\sin 2\theta\right) + \dots \\ \quad = \left[1 + z\left(\cos\theta + \sqrt{-1}\sin\theta\right)\right]^{\mu} \end{cases}$$
$$(z = -1, \quad z = +1).$$

In other words, equation (20) of Chapter VI (§ IV), namely

$$1 + \frac{\mu}{1} x + \frac{\mu(\mu-1)}{1\cdot 2} x^2 + \dots = (1+x)^{\mu},$$

remains true not only if we attribute to the variable x real values contained between the limits -1 and $+1$ but also if we let

$$x = z\left(\cos\theta + \sqrt{-1}\sin\theta\right),$$

the numerical value of z being less than 1.

Corollary I. — Formula (21), as with all imaginary equations, is equivalent to two real equations, which we obtain by equating on both sides the real parts and the coefficients of $\sqrt{-1}$. In this way we find

$$(23)\quad \begin{cases} 1 + \frac{\mu}{1} z\cos\theta + \frac{\mu(\mu-1)}{1\cdot 2} z^2\cos 2\theta + \dots = \left(1 + 2z\cos\theta + z^2\right)^{\frac{1}{2}\mu}\cos\mu s, \\ \frac{\mu}{1} z\sin\theta + \frac{\mu(\mu-1)}{1\cdot 2} z^2\sin 2\theta + \dots = \left(1 + 2z\cos\theta + z^2\right)^{\frac{1}{2}\mu}\sin\mu s \end{cases}$$
$$(z = -1, \quad z = +1),$$

where the value of s is still determined by equation (18).

Corollary II. — If in formulas (22) and (23) we set $\mu = -1$, and if we then replace z by $-z$, we get equations (8) and (10) of § I.

Corollary III. — If we set $\theta = \frac{\pi}{2}$, or what amounts to the same thing,

$$\cos\theta = 0 \quad \text{and} \quad \sin\theta = 1,$$

[248] the value of s given by formula (18) becomes

$$s = \arctan z,$$

and remains contained between the limits $-\frac{\pi}{4}$ and $+\frac{\pi}{4}$ for any numerical value of z less than 1. Under the same hypothesis, we evidently have

$$z = \tan s = \frac{\sin s}{\cos s} \quad \text{and}$$
$$\left(1+2z\cos\theta+z^2\right)^{\frac{\mu}{2}} = (\sec s)^{\mu} = \frac{1}{(\cos s)^{\mu}},$$

and we get from equations (23), but only for values between the given limits, that

$$(24)\qquad \begin{cases} \cos\mu s = \cos^{\mu} s - \frac{\mu(\mu-1)}{1\cdot 2}\cos^{\mu-2} s\sin^2 s \\ \qquad + \frac{\mu(\mu-1)(\mu-2)(\mu-3)}{1\cdot 2\cdot 3\cdot 4}\cos^{\mu-4}\sin^4 s - \ldots, \\ \sin\mu s = \frac{\mu}{1}\cos^{\mu-1} s\sin s \\ \qquad - \frac{\mu(\mu-1)(\mu-2)}{1\cdot 2\cdot 3}\cos^{\mu-3} s\sin^3 s + \ldots \end{cases}$$
$$\left(s = -\tfrac{\pi}{4}, \quad s = +\tfrac{\pi}{4}\right).$$

Consequently, if in formulas (12) of Chapter VII (§ II), we replace the integer number m by any quantity μ, these formulas, which held for all possible real values of the arc z, are not generally true except for numerical values of this arc less than $\frac{\pi}{4}$.

Problem II. — *To find the sum of the series*

$$(10)\qquad 1,\ \frac{z}{1}\left(\cos\theta+\sqrt{-1}\sin\theta\right),\ \frac{z^2}{1\cdot 2}\left(\cos 2\theta+\sqrt{-1}\sin 2\theta\right),\ \ldots,$$

whatever the numerical value of z might be.

Solution. — If in equations (18) and (21), we replace z by αz and μ by $\frac{1}{\alpha}$, where α denotes an infinitely small quantity, we find that [249] for all values of αz contained between the limits -1 and $+1$, or what amounts to the same thing, for all values of z contained between the limits $-\frac{1}{\alpha}$ and $+\frac{1}{\alpha}$,

$$(25)\quad \begin{cases} 1+\frac{z}{1}\left(\cos\theta+\sqrt{-1}\sin\theta\right) \\ \quad +\frac{z^2}{1\cdot 2}\left(\cos 2\theta+\sqrt{-1}\sin 2\theta\right)(1-\alpha) \\ \quad +\frac{z^3}{1\cdot 2\cdot 3}\left(\cos 3\theta+\sqrt{-1}\sin 3\theta\right)(1-\alpha)(1-2\alpha)+\dots \\ \quad\quad =\left(1+2\alpha z\cos\theta+\alpha^2z^2\right)^{\frac{1}{2\alpha}}\left(\cos\frac{s}{\alpha}+\sqrt{-1}\sin\frac{s}{\alpha}\right) \end{cases}$$

$$\left(z=-\tfrac{1}{\alpha},\quad z=+\tfrac{1}{\alpha}\right),$$

where the arc s is determined by the formula

$$(26)\qquad s=\arctan\frac{\alpha z\sin\theta}{1+\alpha z\cos\theta}.$$

Now if we let the numerical value of α in equation (25) decrease indefinitely, then by passing to the limit we find

$$(27)\quad \begin{cases} 1+\frac{z}{1}\left(\cos\theta+\sqrt{-1}\sin\theta\right)+\frac{z^2}{1\cdot 2}\left(\cos 2\theta+\sqrt{-1}\sin 2\theta\right) \\ \quad\quad +\frac{z^3}{1\cdot 2\cdot 3}\left(\cos 3\theta+\sqrt{-1}\sin 3\theta\right)+\dots \\ \quad =\lim\left[\left(1+2\alpha z\cos\theta+\alpha^2z^2\right)^{\frac{1}{2\alpha}}\left(\cos\frac{s}{\alpha}+\sqrt{-1}\sin\frac{s}{\alpha}\right)\right] \end{cases}$$

$$\left(z=-\infty,\quad z=+\infty\right).$$

It remains to find the limit of the product

$$\left(1+2\alpha z\cos\theta+\alpha^2z^2\right)^{\frac{1}{2\alpha}}\left(\cos\frac{s}{\alpha}+\sqrt{-1}\sin\frac{s}{\alpha}\right),$$

and consequently, the limits of each of the quantities

$$\left(1+2\alpha z\cos\theta+\alpha^2z^2\right)^{\frac{1}{2\alpha}}\quad\text{and}\quad\frac{s}{\alpha}.$$

Now, in the first place, if we make

$$2\alpha z\cos\theta+\alpha^2z^2=\beta,$$

[250] we conclude that[5]

$$\left(1+2\alpha z\cos\theta+\alpha^2z^2\right)^{\frac{1}{2\alpha}}=(1+\beta)^{\frac{z\cos\theta+\frac{\alpha z^2}{2}}{\beta}}$$

and consequently

[5] In [Cauchy 1897, p. 250] the numerator of the exponent on the right-hand side reads $s\cos\theta+\frac{\alpha s^2}{2}$. In [Cauchy 1821, p. 299] the numerator is given correctly, with z in place of s. (tr.)

$$\lim\left(1+2\alpha z\cos\theta+\alpha^2z^2\right)^{\frac{1}{2\alpha}} = \left[\lim\left(1+\beta\right)^{\frac{1}{\beta}}\right]^{\lim\left(z\cos\theta+\frac{\alpha z^2}{2}\right)}$$
$$= e^{z\cos\theta}.$$

Moreover, because the value of s given by equation (26) is infinitely small, the ratio

$$\frac{s}{\tan s} = \frac{1}{\frac{\sin s}{s}}\cos s$$

has 1 as its limit. Also, we get from equation (26) that

$$\frac{\tan s}{\alpha} = \frac{z\sin\theta}{1+\alpha z\cos\theta} \quad \text{and}$$
$$\frac{s}{\alpha} = \frac{s}{\tan s}\frac{z\sin\theta}{1+\alpha z\cos\theta}.$$

Thus we find, by passing to the limit, that

$$\lim\left(\frac{s}{\alpha}\right) = z\sin\theta.$$

Given this, it is clear that the right-hand side of equation (25) has as its limit the imaginary expression

$$e^{z\cos\theta}\left[\cos\left(z\sin\theta\right)+\sqrt{-1}\sin\left(z\sin\theta\right)\right],$$

so that formula (27) becomes

$$(28) \qquad \begin{cases} 1+\frac{z}{1}\left(\cos\theta+\sqrt{-1}\sin\theta\right)+\frac{z^2}{1\cdot 2}\left(\cos 2\theta+\sqrt{-1}\sin 2\theta\right)+\dots \\ \qquad = e^{z\cos\theta}\left[\cos\left(z\sin\theta\right)+\sqrt{-1}\sin\left(z\sin\theta\right)\right] \end{cases}$$
$$\left(z=-\infty, \quad z=+\infty\right).$$

The value of the real variable z is completely arbitrary because it can be chosen at will between the extreme values $z=-\infty$ and $z=+\infty$.

[251] *Corollary I.* — If, in comparing the two sides of equation (28), we equate: 1° the real parts; and 2° the coefficients of $\sqrt{-1}$, we obtain the two real equations

$$(29) \qquad \begin{cases} 1+\frac{z}{1}\cos\theta+\frac{z^2}{1\cdot 2}\cos 2\theta+\dots = e^{z\cos\theta}\cos\left(z\sin\theta\right), \\ \frac{z}{1}\sin\theta+\frac{z^2}{1\cdot 2}\sin 2\theta+\dots = e^{z\cos\theta}\sin\left(z\sin\theta\right) \end{cases}$$
$$\left(z=-\infty, \quad z=+\infty\right).$$

Corollary II. — If we suppose that $\theta=\frac{\pi}{2}$, or what amounts to the same thing

$$\cos\theta=0 \quad \text{and} \quad \sin\theta=1,$$

then equations (29) become

$$(30)\qquad \begin{cases} 1-\frac{z^2}{1\cdot 2}+\frac{z^4}{1\cdot 2\cdot 3\cdot 4}-\ldots=\cos z, \\ \frac{z}{1}-\frac{z^3}{1\cdot 2\cdot 3}+\ldots=\sin z \end{cases}$$

$$(z=-\infty, \quad z=+\infty).$$

These last equations, as well as equations (29), remain true for any real values of z, and it follows that the functions $\sin z$ and $\cos z$ can always be expanded into series ordered by the ascending powers of the variables they contain. As this proposition is noteworthy, I will prove it here directly.

Because the series

$$1, \quad \frac{x}{1}, \quad \frac{x^2}{1\cdot 2}, \quad \ldots$$

is convergent for all possible real values of the variable x, it remains convergent (by virtue of theorem I, corollary I) for all imaginary values of the same variable. If we multiply the sum of this series by the sum of [252] the similar series

$$1, \quad \frac{y}{1}, \quad \frac{y^2}{1\cdot 2}, \quad \ldots,$$

and we take into consideration both theorem II of § I and formula (6) of Chapter VIII (§ IV), we find that for all possible values, real and imaginary, attributed to x and y,

$$(31)\qquad \begin{cases} \left(1+\frac{x}{1}+\frac{x^2}{1\cdot 2}+\ldots\right)\left(1+\frac{y}{1}+\frac{y^2}{1\cdot 2}+\ldots\right) \\ \quad =1+\frac{x+y}{1}+\frac{(x+y)^2}{1\cdot 2}+\ldots. \end{cases}$$

In the preceding equation, when we replace x by $x\sqrt{-1}$ and y by $y\sqrt{-1}$, we obtain the following

$$(32)\qquad \begin{cases} \left(1+\frac{x\sqrt{-1}}{1}-\frac{x^2}{1\cdot 2}-\frac{x^3\sqrt{-1}}{1\cdot 2\cdot 3}+\ldots\right) \\ \quad \times\left(1+\frac{y\sqrt{-1}}{1}-\frac{y^2}{1\cdot 2}-\frac{y^3\sqrt{-1}}{1\cdot 2\cdot 3}+\ldots\right) \\ \quad =1+\frac{(x+y)\sqrt{-1}}{1}-\frac{(x+y)^2}{1\cdot 2}-\ldots, \end{cases}$$

in which we may, if we wish, assume that the variables x and y are real. Under this hypothesis, take

$$\varpi(x)=1+\frac{x\sqrt{-1}}{1}-\frac{x^2}{1\cdot 2}-\frac{x^3\sqrt{-1}}{1\cdot 2\cdot 3}+\ldots.$$

Equation (32) becomes

$$\varpi(x)\,\varpi(y)=\varpi(x+y),$$

and we conclude that (see Chapter VIII, § V, equation (12))

$$\varpi(x) = A^x \left(\cos bx + \sqrt{-1}\sin bx\right),$$

or what amounts to the same thing,

$$(33)\qquad \begin{cases} 1 + \frac{x\sqrt{-1}}{1} - \frac{x^2}{1\cdot 2} - \frac{x^3\sqrt{-1}}{1\cdot 2\cdot 3} + \frac{x^4}{1\cdot 2\cdot 3\cdot 4} + \dots \\ \qquad = A^x \left(\cos bx + \sqrt{-1}\sin bx\right) \end{cases}$$

$$(x = -\infty, \quad x = +\infty),$$

[253] where the letters A and b denote two unknown constants, the first one of which is necessarily positive. Consequently, we have

$$(34)\qquad \begin{cases} 1 - \frac{x^2}{1\cdot 2} + \frac{x^4}{1\cdot 2\cdot 3\cdot 4} - \dots = A^x \cos bx, \\ \quad \frac{x}{1} - \frac{x^3}{1\cdot 2\cdot 3} + \dots = A^x \sin bx \end{cases}$$

$$(x = -\infty, \quad x = +\infty).$$

To determine the unknown constants A and b, it suffices to observe: 1° that formulas (34) must remain true when we change x to $-x$, and that to fulfill this condition it is necessary to suppose that

$$A^x = A^{-x},$$

and consequently

$$A = 1;$$

and 2° to observe that if we divide both sides of the second of formulas (34) by x, and then we let the variable x converge towards the limit zero, then the left-hand side converges towards the limit 1, and the right-hand side, namely

$$A^x \frac{\sin bx}{x} = A^x \frac{\sin bx}{bx} \times b,$$

converges towards the limit b. From this it follows that

$$b = 1.$$

Given this, formulas (33) and (34) become, respectively,

$$(35)\qquad \begin{cases} 1 + \frac{x\sqrt{-1}}{1} - \frac{x^2}{1\cdot 2} - \frac{x^3\sqrt{-1}}{1\cdot 2\cdot 3} + \frac{x^4}{1\cdot 2\cdot 3\cdot 4} + \dots \\ \qquad = \cos x + \sqrt{-1}\sin x \end{cases}$$

$$(x = -\infty, \quad x = +\infty)$$

and[6]

---

[6] The denominator of the x^2 term is incorrectly given as 1 instead of $1 \cdot 2$ in [Cauchy 1821, p. 304, Cauchy 1897, p. 253]. (tr.)

$$(36)\qquad \begin{cases} 1-\frac{x^2}{1\cdot 2}+\frac{x^4}{1\cdot 2\cdot 3\cdot 4}-\ldots=\cos x, \\ \frac{x}{1}-\frac{x^3}{1\cdot 2\cdot 3}+\ldots=\sin x \end{cases}$$

$$(x=-\infty, \quad x=+\infty).$$

[254] If in these last two formulas we replace the variable x with the variable z, we rediscover formulas (30).

It is essential to observe that when we suppose that $x = z\sin\theta$, equation (35) gives the expansion of

$$\cos(z\sin\theta)+\sqrt{-1}\sin(z\sin\theta)$$

according to the ascending powers of z. If we multiply this expansion by that of

$$e^{z\cos\theta},$$

and take into consideration formula (31), which remains true for all values, real and imaginary, of the variables it contains, then we get precisely equation (28).

Problem III. — *To find the sum of the series*

$$(11)\qquad \begin{cases} \frac{z}{1}\left(\cos\theta+\sqrt{-1}\sin\theta\right)-\frac{z^2}{2}\left(\cos 2\theta+\sqrt{-1}\sin 2\theta\right) \\ +\frac{z^3}{3}\left(\cos 3\theta+\sqrt{-1}\sin 3\theta\right)-\ldots \end{cases}$$

in the case where we attribute to the variable z a value contained between the limits

$$z=-1 \quad \textit{and} \quad z=+1.$$

Solution. — If we use the notation ln for the characteristic of the Napierian logarithms, then we have

$$\left(1+2z\cos\theta+z^2\right)^{\frac{1}{2}\mu}=e^{\frac{1}{2}\mu\ln\left(1+2z\cos\theta+z^2\right)},$$

and consequently equation (21) can be put into the form

$$\begin{gathered} 1+\frac{\mu}{1}z\left(\cos\theta+\sqrt{-1}\sin\theta\right)+\frac{\mu(\mu-1)}{1\cdot 2}z^2\left(\cos 2\theta+\sqrt{-1}\sin 2\theta\right)+\ldots \\ =e^{\frac{1}{2}\mu\ln\left(1+2z\cos\theta+z^2\right)}\left(\cos\mu s+\sqrt{-1}\sin\mu s\right) \\ (z=-1, \quad z=+1), \end{gathered}$$

[255] where the value of s is still given by formula (18). If we expand the two factors of the right-hand side of the preceding equation into convergent series ordered according to the ascending powers of μ, then, if we form the product of these two expansions with the aid of formula (31), we find

$$\begin{aligned}1+\tfrac{\mu}{1}z\left(\cos\theta+\sqrt{-1}\sin\theta\right)+\tfrac{\mu(\mu-1)}{1\cdot 2}z^2\left(\cos 2\theta+\sqrt{-1}\sin 2\theta\right)+\dots\\ =1+\tfrac{\mu}{1}\left[\tfrac{1}{2}\ln\left(1+2z\cos\theta+z^2\right)+s\sqrt{-1}\right]\\ +\tfrac{\mu^2}{1\cdot 2}\left[\tfrac{1}{2}\ln\left(1+2z\cos\theta+z^2\right)+s\sqrt{-1}\right]^2+\dots\\ (z=-1,\quad z=+1).\end{aligned}$$

Finally, after subtracting 1 from both sides, then dividing them by μ, if we let μ converge towards the limit zero, we obtain the equation

(37) $$\left\{\begin{aligned}&\tfrac{z}{1}\left(\cos\theta+\sqrt{-1}\sin\theta\right)-\tfrac{z^2}{2}\left(\cos 2\theta+\sqrt{-1}\sin 2\theta\right)+\dots\\ &\quad=\tfrac{1}{2}\ln\left(1+2z\cos\theta+z^2\right)+s\sqrt{-1}\end{aligned}\right.$$

$$(z=-1,\quad z=+1).$$

Corollary I. — If in the two sides of equation (37), we equate: 1° the real parts; and 2° the coefficients of $\sqrt{-1}$, then if we substitute for s its value determined by formula (18), we obtain the two real equations

(38) $$\left\{\begin{aligned}&\tfrac{z}{1}\cos\theta-\tfrac{z^2}{2}\cos 2\theta+\tfrac{z^3}{3}\cos 3\theta-\dots=\tfrac{1}{2}\ln\left(1+2z\cos\theta+z^2\right),\\ &\tfrac{z}{1}\sin\theta-\tfrac{z^2}{2}\sin 2\theta+\tfrac{z^3}{3}\sin 3\theta-\dots=\arctan\tfrac{z\sin\theta}{1+z\cos\theta}\end{aligned}\right.$$

$$(z=-1,\quad z=+1).$$

[256] *Corollary II.* — If we suppose that $\theta=\frac{\pi}{2}$, or what amounts to the same thing,

$$\cos\theta=0\quad\text{and}\quad\sin\theta=1,$$

the second of equations (38) becomes

(39) $$z-\frac{z^3}{3}+\frac{z^5}{5}-\dots=\arctan z\qquad(z=-1,\quad z=+1).$$

The series that forms the left-hand side of this last equation is convergent, not only for any numerical value of z less than 1, but also when we suppose that $z=1$ (see Chapter VI, § III, theorem III), and as a result the equation remains true in this last hypothesis. Moreover, because we have

$$\arctan(1)=\frac{\pi}{4},$$

we conclude that

(40) $$1-\frac{1}{3}+\frac{1}{5}-\dots=\frac{\pi}{4}.$$

Formula (40)[7] can be used to calculate an approximation of the value of π, that is to say the ratio if the circumference to the diameter.

[7] This series was originally discovered independently by James Gregory (1638–1675) and Gottfried Wilhelm von Leibniz (1646–1716).

9.3 Notations used to represent various imaginary functions which arise from the summation of convergent series. Properties of these same functions.

Consider the six notations

$$A^x, \quad \sin x, \quad \cos x,$$
$$\log x, \ \arcsin x \ \text{and} \ \arccos x.$$

As we know, if we give the variable x a real value, these six notations represent as many real functions of x, which, taken two by two, are *inverses* of each other, that is to say [257] given by inverse operations, provided, however, that, where A denotes a number, log expresses the characteristic of the logarithms in the system for which the base is A. It remains to clarify the sense of these same notations in the case where the variable x becomes imaginary. We will do this here, starting with the first three.

We have proved that in the case where the variable x is taken to be real, the three functions represented by

$$A^x, \quad \sin x \quad \text{and} \quad \cos x$$

can always be expanded into series ordered according to the ascending integer powers of this variable. Indeed, under this hypothesis we have

$$(1) \qquad \begin{cases} A^x &= 1 + \frac{x \ln A}{1} + \frac{x^2 (\ln A)^2}{1 \cdot 2} + \frac{x^3 (\ln A)^3}{1 \cdot 2 \cdot 3} + \dots, \\ \cos x &= 1 - \frac{x^2}{1 \cdot 2} + \frac{x^4}{1 \cdot 2 \cdot 3 \cdot 4} - \dots, \\ \sin x &= \frac{x}{1} - \frac{x^3}{1 \cdot 2 \cdot 3} + \dots, \end{cases}$$

where the characteristic ln denotes a Napierian logarithm. Moreover (by virtue of theorem I, corollary I, § II), the above series remain convergent for all values, real and imaginary, of the variable x, so we agree to extend equations (1) to all possible cases and to consider them as clarifying the meanings of the three notations

$$A^x, \quad \sin x \quad \text{and} \quad \cos x,$$

even when the variable becomes imaginary.

Now we observe that if we make

$$A = e$$

in the first of equations (1), where e denotes the base of the Napierian logarithms, then we find that

$$(2) \qquad e^x = 1 + \frac{x}{1} + \frac{x^2}{1 \cdot 2} + \dots.$$

[258] Then in place of x, we can successively write $x \ln A$, $x\sqrt{-1}$ and $-x\sqrt{-1}$ to get

$$(3)\qquad \begin{cases} e^{x\ln A} = 1 + \frac{x\ln A}{1} + \frac{x^2(\ln A)^2}{1\cdot 2} + \frac{x^3(\ln A)^3}{1\cdot 2\cdot 3} + \dots, \\ e^{x\sqrt{-1}} = 1 + \frac{x}{1}\sqrt{-1} - \frac{x^2}{1\cdot 2} - \frac{x^3}{1\cdot 2\cdot 3}\sqrt{-1} + \dots, \\ e^{-x\sqrt{-1}} = 1 - \frac{x}{1}\sqrt{-1} - \frac{x^2}{1\cdot 2} + \frac{x^3}{1\cdot 2\cdot 3}\sqrt{-1} + \dots. \end{cases}$$

As a consequence we have[8]

$$(4)\qquad \begin{cases} e^{x\ln A} = A^x, \\ e^{x\sqrt{-1}} = \cos x + \sqrt{-1}\sin x, \\ e^{-x\sqrt{-1}} = \cos x - \sqrt{-1}\sin x, \end{cases}$$

where the variable x may be either real or imaginary. Moreover, whatever x and y may be, equation (31) (§ II) gives

$$(5)\qquad e^x e^y = e^{x+y}.$$

Given this, it becomes easy to find in finite form the values of A^x, $\sin x$ and $\cos x$ corresponding to the imaginary values of the variable x. Indeed, if we suppose that

$$(6)\qquad x = \alpha + \beta\sqrt{-1},$$

where α and β represent real quantities, then we conclude from the first two of equations (4) together with equation (5) that

$$(7)\qquad \begin{cases} A^x = e^{x\ln A} = e^{\left(\alpha+\beta\sqrt{-1}\right)\ln A} = e^{\alpha\ln A} e^{\beta\ln(A)\sqrt{-1}} \\ \quad = A^\alpha\left(\cos\beta\ln A + \sqrt{-1}\sin\beta\ln A\right). \end{cases}$$

From the last two of equations (4), we conclude that

$$(8)\qquad \begin{cases} \cos x = \dfrac{e^{x\sqrt{-1}} + e^{-x\sqrt{-1}}}{2}, \\ \sin x = \dfrac{e^{x\sqrt{-1}} - e^{-x\sqrt{-1}}}{2\sqrt{-1}}. \end{cases}$$

Then, by substituting the value $\alpha+\beta\sqrt{-1}$ for x and expanding the [259] right-hand sides, we get

$$(9)\qquad \begin{cases} \cos x = \frac{e^\beta + e^{-\beta}}{2}\cos\alpha - \frac{e^\beta - e^{-\beta}}{2}\sin\alpha\sqrt{-1}, \\ \sin x = \frac{e^\beta + e^{-\beta}}{2}\sin\alpha + \frac{e^\beta - e^{-\beta}}{2}\cos\alpha\sqrt{-1} \\ \quad = \cos\left(\frac{\pi}{2} - \alpha - \beta\sqrt{-1}\right). \end{cases}$$

And so, under the given hypotheses, the three notations

[8] The second of formulas (4) is Euler's Identity, extended to complex numbers.

$$A^x, \quad \sin x \quad \text{and} \quad \cos x,$$

respectively, denote the three imaginary expressions

$$\begin{aligned}
&A^\alpha \left(\cos \ln A + \sqrt{-1} \sin \ln A\right), \\
&\tfrac{e^\beta + e^{-\beta}}{2} \sin \alpha + \tfrac{e^\beta - e^{-\beta}}{2} \cos \alpha \sqrt{-1} \quad \text{and} \\
&\tfrac{e^\beta + e^{-\beta}}{2} \cos \alpha - \tfrac{e^\beta - e^{-\beta}}{2} \sin \alpha \sqrt{-1}.
\end{aligned}$$

Under the same hypothesis, if we make

$$A = e$$

then equation (7) gives the following value

$$e^\alpha \left(\cos \beta + \sqrt{-1} \sin \beta\right)$$

for the notation

$$e^x.$$

Now that we have determined the values of the three functions

$$A^x, \quad \sin x \quad \text{and} \quad \cos x$$

in the case where the variable x becomes imaginary, we still have to look for which definitions to give in the same case for the inverse functions

$$\log x, \quad \arcsin x \quad \text{and} \quad \arccos x,$$

[260] or more generally, what meaning to give to the notations

$$\log((x)), \quad \arcsin((x)) \quad \text{and} \quad \arccos((x)).$$

We continue to suppose that

$$x = \alpha + \beta \sqrt{-1} = \rho \left(\cos \theta + \sqrt{-1} \sin \theta\right),$$

where α and β denote two real quantities which can be replaced by the modulus ρ and the real arc θ. Every imaginary expression $u + \sqrt{-1} v$ that satisfies the equation

$$A^{u + v\sqrt{-1}} = \alpha + \beta \sqrt{-1} = x \tag{10}$$

is what we call an *imaginary logarithm* of x taken in the system where the base is A. As we will see below, equation (10) gives several values of $u + v\sqrt{-1}$, even in the case where β is zero. It follows that any expression, imaginary or real, has several imaginary logarithms. Whenever we wish to designate indistinctly any one of these logarithms (among which we ought to include the real one, if there is one), we use the characteristic log or ln followed by double parentheses, taking care to state the base of the system in the narrative. We prefer to use the characteristic ln when it is a question of Napierian logarithms taken in the system for which the base is e. By virtue of these conventions, the various logarithms of the real and imaginary quantities

$$1, \quad -1, \quad \alpha+\beta\sqrt{-1} \quad \text{and} \quad x$$

are respectively denoted, in the system for which the base is A by

$$\log((1)), \quad \log((-1)), \quad \log\left(\left(\alpha+\beta\sqrt{-1}\right)\right) \quad \text{and} \quad \log((x)),$$

and in the Napierian system for which the base is e by

$$\ln((1)), \quad \ln((-1)), \quad \ln\left(\left(\alpha+\beta\sqrt{-1}\right)\right) \quad \text{and} \quad \ln((x)).$$

Given this, to determine these various logarithms it suffices to solve the following problems.[9]

[261] **Problem I.** — *To find the various values, real and imaginary, of the expression*

$$\ln((1)).$$

Solution. — Let $u+v\sqrt{-1}$ be one of these values, where u and v denote two real quantities. From the definition itself of the expression $\ln((1))$, we have

$$e^{u+v\sqrt{-1}} = 1, \tag{11}$$

or what amounts to the same thing,

$$e^u\left(\cos v+\sqrt{-1}\sin v\right) = 1.$$

From this last equation we get

$$e^u = 1 \quad \text{and}$$
$$\cos v+\sqrt{-1}\sin v = 1,$$

and consequently

$$u = 0,$$
$$\cos v = 1, \quad \sin v = 0 \quad \text{and} \quad v = \pm 2k\pi,$$

where k represents any integer number. With the quantities u and v determined in this way, the various values of $u+v\sqrt{-1}$ satisfying equation (11) are evidently contained in the formula

---

[9] Euler was the first to resolve these problems; see [Euler 1751].

$$u+v\sqrt{-1}=\pm 2k\pi\sqrt{-1}.$$

In other words, the various values of $\ln((1))$ are given by the equation

(12) $$\ln((1))=\pm 2k\pi\sqrt{-1}.$$

Only one of these values is real, namely the one that we obtain by setting $k=0$, which reduces the value itself to zero. To represent this real value, we commonly use the simple notation

$$\ln(1) \quad \text{or} \quad \ln 1.$$

There are evidently an infinite number of imaginary values of $\ln((1))$.

[262] **Problem II.** — *To find the various values of the expression*

$$\ln((-1)).$$

Solution. — Let $u+v\sqrt{-1}$ be one of these values, where u and v denote two real quantities. From the definition itself of the expression $\ln((-1))$, we have

(13) $$e^{u+v\sqrt{-1}}=-1,$$

or what amounts to the same thing,

$$e^u\left(\cos v+\sqrt{-1}\sin v\right)=-1.$$

From this last equation we get

$$e^u=1 \quad \text{and}$$
$$\cos v+\sqrt{-1}\sin v=-1,$$

and consequently

$$u=0,$$
$$\cos v=-1, \quad \sin v=0 \quad \text{and} \quad v=\pm(2k+1)\pi,$$

where k represents any integer number. With the quantities u and v determined in this way, the various values of $u+v\sqrt{-1}$ satisfying equation (13) are evidently contained in the formula

$$u+v\sqrt{-1}=\pm(2k+1)\pi\sqrt{-1}.$$

In other words, the various values of $\ln((-1))$ are given by the equation

(14) $$\ln((-1))=\pm(2k+1)\pi\sqrt{-1}.$$

Consequently, there are infinitely many such values and they are all imaginary.

Problem III. — *To find the various values of the expression*

$$\ln\left(\left(\alpha+\beta\sqrt{-1}\right)\right).$$

Solution. — Let $u+v\sqrt{-1}$ be one of these values. From the [263] definition itself of the expression $\ln\left(\left(\alpha+\beta\sqrt{-1}\right)\right)$ we have

(15) $$e^{u+v\sqrt{-1}}=\alpha+\beta\sqrt{-1}=\rho\left(\cos\theta+\sqrt{-1}\sin\theta\right),$$

or what amounts to the same thing,

$$e^{u}\left(\cos v+\sqrt{-1}\sin v\right)=\rho\left(\cos\theta+\sqrt{-1}\sin\theta\right),$$

where ρ denotes the modulus of $\alpha+\beta\sqrt{-1}$. From the preceding equation, we get

$$e^{u}=\rho \quad \text{and}$$
$$\cos v+\sqrt{-1}\sin v=\cos\theta+\sqrt{-1}\sin\theta,$$

and consequently,

$$u=\ln(\rho),$$
$$\cos v=\cos\theta,\quad \sin v=\sin\theta \quad \text{and} \quad v=\theta\pm 2k\pi,$$

where k represents any integer number. With the quantities u and v determined in this way, the various values of $u+v\sqrt{-1}$ are contained in the formula

$$u+v\sqrt{-1}=\ln(\rho)+\theta\sqrt{-1}\pm 2k\pi\sqrt{-1}.$$

In other words, the various values of

$$\ln\left(\left(\alpha+\beta\sqrt{-1}\right)\right)$$

are given by the equation

(16) $$\ln\left(\left(\alpha+\beta\sqrt{-1}\right)\right)=\ln(\rho)+\theta\sqrt{-1}+\ln((1)).$$

It is worth observing that in this last equation, the value of ρ is completely determined and is equal to

$$\sqrt{\alpha^2+\beta^2},$$

while θ can be any arc which has $\frac{\alpha}{\sqrt{\alpha^2+\beta^2}}$ as its cosine and $\frac{\beta}{\sqrt{\alpha^2+\beta^2}}$ as its sine.

[264] *Corollary I.* — If we make

$$(17) \qquad \zeta = \arctan \frac{\beta}{\alpha}$$

for greater convenience, then it is easy to substutute the arc ζ in place of the arc θ in formula (16). Indeed, we may suppose that

$$\theta = \zeta$$

if α is positive, and

$$\theta = \zeta + \pi$$

if α is negative. In the first case, we find that

$$(18) \qquad \ln\left(\left(\alpha+\beta\sqrt{-1}\right)\right) = \ln(\rho) + \zeta\sqrt{-1} + \ln((1)),$$

and in the second case that

$$(19) \qquad \ln\left(\left(\alpha+\beta\sqrt{-1}\right)\right) = \ln(\rho) + \zeta\sqrt{-1} + \pi\sqrt{-1} + \ln((1)).$$

In particular, if in this last equation we make

$$\alpha + \beta\sqrt{-1} = -1,$$

that is to say

$$\alpha = -1 \quad \text{and} \quad \beta = 0,$$

and consequently

$$\rho = 1 \quad \text{and} \quad \zeta = 0,$$

we obtain

$$(20) \qquad \ln((-1)) = \pi\sqrt{-1} + \ln((1)).$$

In general, it follows that for negative values of α we have

$$(21) \qquad \ln\left(\left(\alpha+\beta\sqrt{-1}\right)\right) = \ln(\rho) + \zeta\sqrt{-1} + \ln((-1)).$$

Now suppose that we substitute the values

$$\left(\alpha^2+\beta^2\right)^{\frac{1}{2}} \quad \text{and} \quad \arctan\frac{\beta}{\alpha}$$

for ρ and ζ in formulas (18) and (21). We find for the various values of

$$\ln\left(\left(\alpha+\beta\sqrt{-1}\right)\right):$$

[265] 1° if α is positive, that

$$(22) \qquad \ln\left(\left(\alpha+\beta\sqrt{-1}\right)\right) = \frac{1}{2}\ln\left(\alpha^2+\beta^2\right) + \left(\arctan\frac{\beta}{\alpha}\right)\sqrt{-1} + \ln((1));$$

and 2° if α is negative, that

$$(23)\quad \ln\left(\left(\alpha+\beta\sqrt{-1}\right)\right)=\frac{1}{2}\ln\left(\alpha^2+\beta^2\right)+\left(\arctan\frac{\beta}{\alpha}\right)\sqrt{-1}+\ln\left((-1)\right).$$

Corollary II. — If we suppose that $\beta=0$ in equations (22) and (23), then for positive values of α we have

$$(24)\qquad \ln\left((\alpha)\right)=\ln(\alpha)+\ln\left((1)\right)=\ln(\alpha)\pm 2k\pi\sqrt{-1},$$

and for negative values of α we have

$$(25)\qquad \ln\left((\alpha)\right)=\ln(-\alpha)+\ln\left((-1)\right)=\ln(-\alpha)\pm(2k+1)\pi\sqrt{-1},$$

where k as always is an integer number. It follows from these last formulas that a real quantity α has an infinity of imaginary logarithms, among which, in the case where α is positive, we find just one real logarithm. We obtain this real logarithm, denoted by the simple notation $\ln(\alpha)$ or $\ln\alpha$, by setting $k=0$ in equation (24).

Scholium I. — Among the various values of $\ln((1))$, as we have just remarked, there is one that is equal to zero, and which we indicate by the notation $\ln(1)$ or $\ln 1$, making use of the simple parentheses or suppressing them altogether. If we substitute this particular value in equation (22), we obtain a corresponding value

$$\ln\left(\left(\alpha+\beta\sqrt{-1}\right)\right),$$

which analogy leads us to indicate, with the aid of simple parentheses, by the notation

$$\ln\left(\alpha+\beta\sqrt{-1}\right).$$

We will do so from now on. Consequently, supposing that α is positive, we have

$$(26)\qquad \ln\left(\alpha+\beta\sqrt{-1}\right)=\frac{1}{2}\ln\left(\alpha^2+\beta^2\right)+\left(\arctan\frac{\beta}{\alpha}\right)\sqrt{-1}.$$

[266] On the other hand, if α becomes negative, then $-\alpha$ is positive and we find that

$$\ln\left(-\alpha-\beta\sqrt{-1}\right)=\frac{1}{2}\ln\left(\alpha^2+\beta^2\right)+\left(\arctan\frac{-\beta}{-\alpha}\right)\sqrt{-1},$$

or what amounts to the same thing,

$$(27)\qquad \ln\left(-\alpha-\beta\sqrt{-1}\right)=\frac{1}{2}\ln\left(\alpha^2+\beta^2\right)+\left(\arctan\frac{\beta}{\alpha}\right)\sqrt{-1}.$$

By making use of the preceding notations, we can reduce equations (22) and (23) to the following

(28) $$\ln\left(\left(\alpha+\beta\sqrt{-1}\right)\right)=\ln\left(\alpha+\beta\sqrt{-1}\right)+\ln((1)) \quad \text{and}$$

(29) $$\ln\left(\left(\alpha+\beta\sqrt{-1}\right)\right)=\ln\left(-\alpha-\beta\sqrt{-1}\right)+\ln((-1)),$$

where the first equation applies for positive values of α, while the second applies for negative values of that quantity. In other words, depending on whether the real part of an imaginary expression x is positive or negative, we have

(30) $$\ln((x))=\ln(x)+\ln((1))$$

or else

(31) $$\ln((x))=\ln(-x)+\ln((-1)).$$

To summarize what we have just said, we see that the notation

$$\ln(x)$$

has a precise meaning determined by equation (26) only in the first case, where the real part of the imaginary expression x is positive, while in all possible cases the notation

$$\ln((x))$$

has infinitely many values determined by one of equations (28) or (29).

[267] **Problem IV.** — *To find the various values of the expression*

$$\log\left(\left(\alpha+\beta\sqrt{-1}\right)\right),$$

where the characteristic log *indicates a logarithm taken in the system where the base is A.*

Solution. — Let $u+v\sqrt{-1}$ still denote one of the values of the expression we are considering. From the definition itself of this expression, we have

(32) $$A^{u+v\sqrt{-1}}=\alpha+\beta\sqrt{-1},$$

or what amounts to the same thing,

$$e^{\left(u+v\sqrt{-1}\right)\ln A}=\alpha+\beta\sqrt{-1},$$

where ln is the characteristic of the Napierian logarithms. Then we conclude that

$$\left(u+v\sqrt{-1}\right)\ln A=\ln\left(\left(\alpha+\beta\sqrt{-1}\right)\right),$$

and consequently

$$u+v\sqrt{-1}=\frac{\ln\left(\left(\alpha+\beta\sqrt{-1}\right)\right)}{\ln A},$$

or in other words,

$$\log\left(\left(\alpha+\beta\sqrt{-1}\right)\right)=\frac{\ln\left(\left(\alpha+\beta\sqrt{-1}\right)\right)}{\ln A}. \tag{33}$$

This last equation remains true in the case where β vanishes, that is to say when the imaginary expression $\alpha+\beta\sqrt{-1}$ reduces to a real quantity.

Scholium. — If we suppose that the quantity α is positive, then the particular value of $\ln\left(\left(\alpha+\beta\sqrt{-1}\right)\right)$ represented by $\ln\left(\alpha+\beta\sqrt{-1}\right)$ corresponds to a particular value of $\log\left(\left(\alpha+\beta\sqrt{-1}\right)\right)$, which analogy leads us to indicate with the aid of simple parentheses by the notation

$$\log\left(\alpha+\beta\sqrt{-1}\right).$$

[268] Given this, for positive values of α we have

$$\begin{cases}\log\left(\alpha+\beta\sqrt{-1}\right)=\dfrac{\ln\left(\alpha+\beta\sqrt{-1}\right)}{\ln A}\\ \qquad\qquad =\frac{1}{2}\log\left(\alpha^2+\beta^2\right)+\dfrac{\arctan\frac{\beta}{\alpha}}{\ln A}\sqrt{-1}.\end{cases} \tag{34}$$

Moreover, if in equation (33) we substitute for $\ln\left(\left(\alpha+\beta\sqrt{-1}\right)\right)$ its value given successively in formulas (28) and (29), we find that for positive values of the quantity α,

$$\begin{cases}\log\left(\left(\alpha+\beta\sqrt{-1}\right)\right)=\dfrac{\ln\left(\alpha+\beta\sqrt{-1}\right)}{\ln A}+\dfrac{\ln((1))}{\ln A}\\ \qquad\qquad =\log\left(\alpha+\beta\sqrt{-1}\right)+\log((1)),\end{cases} \tag{35}$$

and for negative values of the same quantity,[10]

$$\begin{cases}\log\left(\left(\alpha+\beta\sqrt{-1}\right)\right)=\dfrac{\ln\left(-\alpha-\beta\sqrt{-1}\right)}{\ln A}+\dfrac{\ln((-1))}{\ln A}\\ \qquad\qquad =\log\left(-\alpha-\beta\sqrt{-1}\right)+\log((-1)).\end{cases} \tag{36}$$

In other words, according to whether the real part of an imaginary expression x is positive or negative, we have, respectively,

$$\log((x))=\log x+\log((1))=\log(x)\pm\frac{2k\pi\sqrt{-1}}{\ln A} \tag{37}$$

or else

$$\log((x))=\log(-x)+\log((-1))=\log(-x)\pm\frac{(2k+1)\pi\sqrt{-1}}{\ln A}, \tag{38}$$

[10] In [Cauchy 1821, p. 323, Cauchy 1897, p. 268], single parentheses were used on the left-hand side of equation (36). (tr.)

where k denotes any integer number. We can add that of the two preceding formulas, the first remains true for all positive real values of x and the second for all negative real values of the same variable.

After having calculated the various logarithms of the imaginary expression

$$x = \alpha + \beta\sqrt{-1},$$

we propose to find the imaginary arcs for which the cosine is equal to x. If we denote any one of these arcs by

$$\arccos((x)) = u + v\sqrt{-1},$$

[269] then to determine $u + v\sqrt{-1}$, we have the equation

$$\cos\left(u + v\sqrt{-1}\right) = \alpha + \beta\sqrt{-1},$$

or what amounts to the same thing,

$$\frac{e^v + e^{-v}}{2}\cos u - \frac{e^v - e^{-v}}{2}\sin u\sqrt{-1} = \alpha + \beta\sqrt{-1}. \tag{39}$$

This separates into two other equations, namely

$$\frac{e^v + e^{-v}}{2}\cos u = \alpha \quad \text{and} \quad \frac{e^v - e^{-v}}{2}\sin u = -\beta. \tag{40}$$

For these last two equations, we can substitute the equivalent system of two formulas

$$e^v = \frac{\alpha}{\cos u} - \frac{\beta}{\sin u} \quad \text{and} \quad e^{-v} = \frac{\alpha}{\cos u} + \frac{\beta}{\sin u}. \tag{41}$$

Moreover, if we eliminate v from formulas (41), it follows that

$$\frac{\alpha^2}{\cos^2 u} - \frac{\beta^2}{\sin^2 u} = 1 \quad \text{and}$$
$$\sin^4 u - (1 - \alpha^2 - \beta^2)\sin^2 u - \beta^2 = 0.$$

Then, by observing that $\sin^2 u$ is necessarily a positive quantity, we have

$$\sin^2 u = \frac{1 - \alpha^2 - \beta^2}{2} + \sqrt{\left(\frac{1 - \alpha^2 - \beta^2}{2}\right)^2 + \beta^2}.$$

Consequently we have

$$\cos^2 u = \frac{1 + \alpha^2 + \beta^2}{2} - \sqrt{\left(\frac{1 + \alpha^2 + \beta^2}{2}\right)^2 - \alpha^2}$$

$$= \frac{\alpha^2}{\frac{1+\alpha^2+\beta^2}{2} + \sqrt{\left(\frac{1+\alpha^2+\beta^2}{2}\right)^2 - \alpha^2}},$$

and because (by virtue of the first of equations (40)) $\cos u$ and α [270] must have the same sign, we have, by extracting square roots,

(42) $$\cos u = \frac{\alpha}{\left[\frac{1+\alpha^2+\beta^2}{2} + \sqrt{\left(\frac{1+\alpha^2+\beta^2}{2}\right)^2 - \alpha^2}\right]^{\frac{1}{2}}}.$$

Given this, if for convenience we make

(43) $$\begin{cases} U = \arccos \dfrac{\alpha}{\left[\frac{1+\alpha^2+\beta^2}{2} + \sqrt{\left(\frac{1+\alpha^2+\beta^2}{2}\right)^2 - \alpha^2}\right]^{\frac{1}{2}}} & \text{and} \\ V = \ln\left(\frac{\alpha}{\cos U} - \frac{\beta}{\sin U}\right), & \end{cases}$$

we conclude from equations (41) and (42) that

(44) $$u = \pm U \pm 2k\pi \quad \text{and} \quad v = \pm V,$$

where k denotes any integer number and the two letters U and V must have the same sign. Thus, we finally have

(45) $$\arccos((x)) = \pm 2k\pi \pm \left(U + V\sqrt{-1}\right).$$

Among the various values of $\arccos((x))$ given by the preceding equation, the simplest is the one obtained by setting $k = 0$ in the first term of the right-hand side, and giving a $+$ sign to the other term. We denote this particular value with the aid of simple parentheses, and consequently we write

$$\arccos(x) = U + V\sqrt{-1},$$

or even, by suppressing the parentheses entirely,

(46) $$\arccos x = U + V\sqrt{-1}.$$

In the particular case where β is zero, the quantity α remains contained between the limits -1 and $+1$, and formula (46) reduces, as we [271] should expect, to the identity

$$\arccos \alpha = \arccos \alpha.$$

On the other hand, if we note that $\pm 2k\pi$ represents any of the arcs that have 1 for their cosines, we recognize that equation (45) can be put into the form

$$\arccos((x)) = \pm \arccos x + \arccos((1)). \tag{47}$$

Yet, it is essential to remark that in the case where we suppose that $\beta = 0$ and the numerical value of α is greater than 1, the expression

$$\arccos \alpha$$

always takes an imaginary value. This imaginary value is given by the equation

$$\arccos \alpha = \ln(\alpha)\sqrt{-1} \tag{48}$$

if α is positive, and by

$$\arccos \alpha = \pi + \ln(-\alpha)\sqrt{-1} = \left[\ln(-\alpha) - \pi\sqrt{-1}\right]\sqrt{-1} \tag{49}$$

if α is negative.

Now consider the imaginary arcs for which the sine is $x = \alpha + \beta\sqrt{-1}$. If we denote any one of these arcs by

$$\arcsin((x)) = u + v\sqrt{-1},$$

then by taking into consideration the second of equations (9), we find

$$x = \sin\left(u + v\sqrt{-1}\right) = \cos\left(\frac{\pi}{2} - u - v\sqrt{-1}\right),$$

and we conclude

$$\arcsin((x)) = u + v\sqrt{-1} = \frac{\pi}{2} \arccos((x)). \tag{50}$$

In the previous formula, if we substitute the various values of $\arccos((x))$, one of which is distinguished by the notation $\arccos(x)$ or $\arccos x$, we obtain the various values of $\arcsin((x))$, one of which [272] is distinguished by the notation $\arcsin(x)$ or $\arcsin x$, and determined by the equation

$$\arcsin x = \frac{\pi}{2} - \arccos x. \tag{51}$$

With the aid of the principles that we have just established, it is easy to recognize the most essential properties that are enjoyed by those functions of the imaginary variable x represented by the notations

$$A^x, \quad \cos x, \quad \sin x,$$
$$\log x, \ \arccos x \text{ and } \arcsin x.$$

To obtain these properties, it suffices to extend the formulas that these functions satisfy in the case where the variable x is real to the case where the variable becomes imaginary. This extension is ordinarily carried out without difficulty for each of the three functions

$$A^x, \quad \cos x \quad \text{and} \quad \sin x.$$

Thus, for example, if A, B, C, ... denote several numbers, we can easily prove that the equations

(52)
$$\begin{cases} A^x A^y A^z \ldots = A^{x+y+z+\ldots}, \\ A^x B^x C^x \ldots = (ABC\ldots)^x \end{cases}$$

and

(53)
$$\begin{cases} \cos(x+y) = \cos x \cos y - \sin x \sin y, \\ \sin(x+y) = \sin x \cos y + \sin y \cos x \end{cases}$$

remain equally true for any values, real or imaginary, of the variables x, y, z, ... But if we consider formulas that involve the inverse functions

$$\log x, \quad \arccos x \quad \text{or} \quad \arcsin x,$$

we usually find that these formulas, extended to the case where the variables become imaginary, remain true only with considerable restrictions, and only for certain values of the variables that they involve. For example, if we make

$$x = \alpha + \beta\sqrt{-1}, \quad y = \alpha' + \beta'\sqrt{-1}, \quad z = \alpha'' + \beta''\sqrt{-1}, \quad \ldots,$$

[273] and if we denote by μ any real quantity, we recognize that the formula

(54)
$$\log(x) + \log(y) + \log(z) + \ldots = \log(xyz\ldots)$$

remains true only in the case where α, α', α'', ... are positive and the sum

$$\arctan\frac{\beta}{\alpha} + \arctan\frac{\beta'}{\alpha'} + \arctan\frac{\beta''}{\alpha''} + \ldots$$

remains contained between the limits $-\frac{\pi}{2}$ and $+\frac{\pi}{2}$. The formula

(55)
$$\log(x^\mu) = \mu \log(x)$$

remains true only in the case where α is positive and the product

$$\mu \arctan\frac{\beta}{\alpha}$$

remains contained between the same limits.

Chapter 10
On real or imaginary roots of algebraic equations for which the left-hand side is a rational and integer function of one variable. The solution of equations of this kind by algebra or trigonometry.

10.1 We can satisfy any equation for which the left-hand side is a rational and integer function of the variable x by real or imaginary values of that variable. Decomposition of polynomials into factors of the first and second degree. Geometric representation of real factors of the second degree.

[274] Consider an algebraic equation for which the left-hand side is a rational and integer function of the variable x. Such an equation can be put into the form

$$(1) \qquad a_0x^n + a_1x^{n-1} + a_2x^{n-2} + \ldots + a_{n-1}x + a_n = 0,$$

where n represents the degree of this equation and a_0, a_1, a_2, ..., a_{n-1}, a_n, are constant coefficients, real or imaginary. A *root* of this equation is any expression, real or imaginary, that when substituted in place of the unknown value x, makes the left-hand side equal to zero. First, to clarify the ideas, suppose that the constants $a_0, a_1, a_2, \ldots, a_n$, reduce to real quantities. Then if two real values of x substituted into the left-hand side of equation (1) give two results containing zero between them, that is to say, results with opposite signs, we conclude from Chapter II (§ II, theorem IV)[1] that equation (1) admits one or more real roots contained between these two values. It follows that any equation of odd degree has at least one real root. Indeed, if n is an odd number, the left-hand side [275] of equation (1) changes signs, with its first term a_0x^n, whenever, by giving the variable x very large numerical values, we make this variable pass from positive to negative (see theorem VIII of Chapter II, § I).

When n is an even number, the quantity x^n remains positive as long as the variable x is real. Thus, for very large numerical values of x, the left-hand side of equation

[1] This is the Intermediate Value Theorem, which was proven intuitively in Chapter II, and will be proven rigorously in Note III.

R.E. Bradley, C.E. Sandifer, *Cauchy's Cours d'analyse*, Sources and Studies in the History of Mathematics and Physical Sciences, DOI 10.1007/978-1-4419-0549-9_10,

(1) will eventually always be the same sign as a_0. If, under the same hypothesis, a_n and a_0 are of opposite signs, the left-hand side evidently changes signs as it passes from a very large numerical value of x to a very small one, while remaining either always positive or always negative. Then equation (1) has at least two real roots, one positive and the other negative.

When n is an even number and a_0 and a_n have the same sign, it can happen that the left-hand side of equation (1) remains of the same sign as a_0 for all real values of x, without ever vanishing. This is what happens, for example, for each of the binomial equations

$$x^2+1=0, \quad x^4+1=0, \quad x^6+1=0, \quad \ldots.$$

In such a case, equation (1) no longer has real roots, but we satisfy the equation by taking for x an imaginary expression

$$u+v\sqrt{-1},$$

where u and v denote two finite real quantities. This proposition and the ones that we have just established are found contained in the following theorem:

Theorem I. — *Whatever the values, real or imaginary, of the constants a_0, a_1, ..., a_{n-1}, a_n may be, the equation*

$$a_0x^n+a_1x^{n-1}+a_2x^{n-2}+\ldots+a_{n-1}x+a_n=0, \tag{1}$$

in which n denotes an integer number greater than or equal to 1, always has real or imaginary roots.

[276] *Proof.* — For brevity, denote the left-hand side of equation (1) by $f(x)$. Then $f(x)$ is a function, real or imaginary, but always integer, of the variable x. Because any real expression u is contained as a particular case of some imaginary expression $u+v\sqrt{-1}$, to establish the stated theorem it suffices to prove in general that we can satisfy the equation

$$f(x)=0 \tag{1}$$

by taking

$$x=u+v\sqrt{-1},$$

then giving the new variables u and v real values. Now, if we substitute the preceding value of x in the function $f(x)$, the result is of the form

$$\varphi(u,v)+\sqrt{-1}\chi(u,v),$$

where $\varphi(u,v)$ and $\chi(u,v)$ denote two real integer functions of the variables u and v. Given this, equation (1) becomes

$$\varphi(u,v)+\sqrt{-1}\chi(u,v)=0.$$

To satisfy this equation, it suffices to satisfy the two real equations

(2) $$\begin{cases} \varphi(u,v) = 0 \quad \text{and} \\ \chi(u,v) = 0, \end{cases}$$

or what amounts to the same thing, the single equation

(3) $$[\varphi(u,v)]^2 + [\chi(u,v)]^2 = 0.$$

Thus, if for convenience we set

(4) $$F(u,v) = [\varphi(u,v)]^2 + [\chi(u,v)]^2,$$

it remains only to show that we can find real values of u and v that make the function

$$F(u,v)$$

vanish. We can easily do this with the aid of the following considerations.

[277] First, to determine the general value of the function $F(u,v)$, we represent each of the real or imaginary constants a_0, a_1, ..., a_{n-1}, a_n, as well as the imaginary variable $u+v\sqrt{-1}$, by the product of a modulus and a reduced expression. Consequently, we write

(5) $$\begin{cases} a_0 \quad = \rho_0\left(\cos\theta_0 + \sqrt{-1}\sin\theta_0\right), \\ a_1 \quad = \rho_1\left(\cos\theta_1 + \sqrt{-1}\sin\theta_1\right), \\ \dots\dots\dots\dots\dots\dots\dots\dots\dots\dots, \\ a_{n-1} = \rho_{n-1}\left(\cos\theta_{n-1} + \sqrt{-1}\sin\theta_{n-1}\right), \\ a_n \quad = \rho_n\left(\cos\theta_n + \sqrt{-1}\sin\theta_n\right) \end{cases}$$

and

(6) $$u + v\sqrt{-1} = r\left(\cos t + \sqrt{-1}\sin t\right).$$

Consequently we have

(7) $$\begin{cases} f\left(u+v\sqrt{-1}\right) \\ \quad = \rho_0 r^n\left[\cos(nt+\theta_0) + \sqrt{-1}\sin(nt+\theta_0)\right] \\ \qquad + \rho_1 r^{n-1}\left[\cos\left(\overline{n-1}\cdot t+\theta_1\right) + \sqrt{-1}\sin\left(\overline{n-1}\cdot t+\theta_1\right)\right] \\ \qquad + \ldots + \rho_{n-1} r\left[\cos(t+\theta_{n-1}) + \sqrt{-1}\sin(t+\theta_{n-1})\right] \\ \qquad + \rho_n\left(\cos\theta_0 + \sqrt{-1}\sin\theta_0\right). \end{cases}$$

From this we deduce that

$$(8)\quad \begin{cases} \varphi(u,v) = \rho_0 r^n \cos(nt+\theta_0) \\ \qquad +\rho_1 r^{n-1}\cos(\overline{n-1}\cdot t+\theta_1)+\dots \\ \qquad \dots+\rho_{n-1} r\cos(t+\theta_{n-1})+\rho_n\cos\theta_n, \\ \chi(u,v) = \rho_0 r^n \sin(nt+\theta_0) \\ \qquad +\rho_1 r^{n-1}\sin(\overline{n-1}\cdot t+\theta_1)+\dots \\ \qquad \dots+\rho_{n-1} r\sin(t+\theta_{n-1})+\rho_n\sin\theta_n \end{cases}$$

and

$$(9)\quad \begin{cases} F(u,v) = \quad [\rho_0 r^n \cos(nt+\theta_0) \\ \qquad +\rho_1 r^{n-1}\cos(\overline{n-1}\cdot t+\theta_1)+\dots \\ \qquad \dots+\rho_{n-1} r\cos(t+\theta_{n-1})+\rho_n\cos\theta_n]^2 \\ \quad +[\rho_0 r^n \sin(nt+\theta_0) \\ \qquad +\rho_1 r^{n-1}\sin(\overline{n-1}\cdot t+\theta_1)+\dots \\ \qquad \dots+\rho_{n-1} r\sin(t+\theta_{n-1})+\rho_n\sin\theta_n]^2 \\ \quad = r^{2n}\left[\rho_0^2+\dfrac{2\rho_0\rho_1\cos(t+\theta_0-\theta_1)}{r}\right. \\ \qquad \left.+\dfrac{\rho_1^2+2\rho_0\rho_2\cos(2t+\theta_0-\theta_2)}{r^2}+\dots\right]. \end{cases}$$

[278] It follows from this last formula that the function $F(u,v)$, which is evidently always positive, is the product of two factors, of which one, namely

$$r^{2n} = \left(u^2+v^2\right)^n,$$

grows indefinitely if we give one or both of the variables u and v larger and larger numerical values, while under the same hypothesis, the other factor converges towards the limit ρ_0^2, that is to say towards a finite limit different from zero. Thus we conclude that the function $F(u,v)$ cannot retain a finite value except when both of the two quantities u and v receive values of this kind, and it becomes infinitely large when either of the two quantities grows indefinitely. Moreover, equation (4) gives an integer function for $F(u,v)$, and consequently a continuous function of the variables u and v. Thus, it is clear that $F(u,v)$ varies with u and v by insensible degrees and cannot drop below zero, and so it attains, one or several times, a certain lower limit below which it never descends. Denote this limit by A, and by u_0 and v_0 one of the systems of finite values of u and v for which $F(u,v)$ reduces to A. Consequently, we have identically

$$(10)\qquad F(u_0,v_0) = A.$$

The difference $F(u,v)-F(u_0,v_0)$ can never fall below zero. As a consequence, if we make

$$(11)\qquad u = u_0+\alpha h \quad\text{and}\quad v = v_0+\alpha k$$

where α denotes an infinitely small quantity and h and k denote two finite quantities, then the expression

$$F\left(u_0+\alpha h, v_0+\alpha k\right)-F\left(u_0, v_0\right)$$

is never negative. On the basis of this principle, it is easy to determine the value of the constant A, as we shall see.

In the imaginary expression $f\left(u+v\sqrt{-1}\right)$, if we substitute for u and v their values given in formulas (11), this expression becomes [279] an imaginary and integer function of the product

$$\alpha\left(h+k\sqrt{-1}\right),$$

and it can be expanded according to the ascending integer powers of this product. If we denote the imaginary coefficients of these powers by

$$\begin{gathered}R\left(\cos T+\sqrt{-1}\sin T\right),\\ R_1\left(\cos T_1+\sqrt{-1}\sin T_1\right),\\ \dots\dots\dots\dots\dots\dots\dots\dots,\\ R_n\left(\cos T_n+\sqrt{-1}\sin T_n\right),\end{gathered}$$

some of which may be reduced to zero, and if we make, for convenience

$$h+k\sqrt{-1}=\rho\left(\cos\theta+\sqrt{-1}\sin\theta\right), \tag{12}$$

we obtain the equation

$$\left\{\begin{aligned} & f\left[u_0+v_0\sqrt{-1}+\alpha\left(h+k\sqrt{-1}\right)\right]\\ &\quad=R\left(\cos T+\sqrt{-1}\sin T\right)\\ &\qquad+\alpha R_1\rho\left[\cos\left(T_1+\theta\right)+\sqrt{-1}\sin\left(T_1+\theta\right)\right]+\dots\\ &\qquad\dots+\alpha^n R_n\rho^n\left[\cos\left(T_n+n\theta\right)+\sqrt{-1}\sin\left(T_n+n\theta\right)\right],\end{aligned}\right. \tag{13}$$

in which the terms on the right-hand side, and thus the moduli

$$R_1,\quad R_2,\quad \dots,\quad R_n,$$

do not all vanish at the same time. Moreover, because we have

$$\left\{\begin{aligned} & f\left[u_0+\alpha h+\left(v_0+\alpha k\right)\sqrt{-1}\right]\\ &\quad=\varphi\left(u_0+\alpha h, v_0+\alpha k\right)+\sqrt{-1}\chi\left(u_0+\alpha h, v_0+\alpha k\right),\end{aligned}\right. \tag{14}$$

we conclude from equation (13) that

$$(15)\quad \begin{cases} \varphi(u_0+\alpha h, v_0+\alpha k) \\ \quad = R\cos T+\alpha R_1\rho\cos(T_1+\theta)+\ldots+\alpha^n R_n\rho^n\cos(T_n+n\theta), \\ \chi(u_0+\alpha h, v_0+\alpha k) \\ \quad = R\sin T+\alpha R_1\rho\sin(T_1+\theta)+\ldots+\alpha^n R_n\rho^n\sin(T_n+n\theta), \end{cases}$$

[280] and as a consequence,

$$(16)\quad \begin{cases} F(u_0+\alpha h, v_0+\alpha k) \\ \quad = \quad [R\cos T+\alpha R_1\rho\cos(T_1+\theta)+\alpha^n R_n\rho^n\cos(T_n+n\theta)]^2 \\ \qquad +[R\sin T+\alpha R_1\rho\sin(T_1+\theta)+\alpha^n R_n\rho^n\sin(T_n+n\theta)]^2. \end{cases}$$

If we set $\alpha=0$ in this last formula, we get

$$F(u_0,v_0)=R^2.$$

Moreover, $R^2=A$ and so $R=A^{\frac{1}{2}}$. If we now expand the right-hand side of equation (16) according to the descending powers of R and then replace R by $A^{\frac{1}{2}}$, this equation becomes

$$(17)\quad \begin{cases} F(u_0+\alpha h, v_0+\alpha k) \\ \quad = A+2A^{\frac{1}{2}}\alpha\rho\,[R_1\cos(T_1-T+\theta)+\ldots \\ \qquad\qquad \ldots+\alpha^{n-1}\rho^{n-1}R_n\cos(T_n-T+n\theta)] \\ \qquad +\alpha^2\rho^2\Big\{[R_1\cos(T_1+\theta)+\ldots+\alpha^{n-1}\rho^{n-1}R_n\cos(T_n+n\theta)]^2 \\ \qquad\qquad +[R_1\sin(T_1+\theta)+\ldots+\alpha^{n-1}\rho^{n-1}R_n\sin(T_n+n\theta)]^2\Big\}. \end{cases}$$

If we move the quantity $A=F(u_0,v_0)$ to the left-hand side, we finally find that

$$(18)\quad \begin{cases} F(u_0+\alpha h, v_0+\alpha k)-F(u_0,v_0) \\ \quad = 2A^{\frac{1}{2}}\alpha\rho\,[R_1\cos(T_1-T+\theta)+\ldots \\ \qquad\qquad \ldots+\alpha^{n-1}\rho^{n-1}R_n\cos(T_n-T+n\theta)] \\ \qquad +\alpha^2\rho^2\Big\{[R_1\cos(T_1+\theta)+\ldots+\alpha^{n-1}\rho^{n-1}R_n\cos(T_n+n\theta)]^2 \\ \qquad\qquad +[R_1\sin(T_1+\theta)+\ldots+\alpha^{n-1}\rho^{n-1}R_n\sin(T_n+n\theta)]^2\Big\}. \end{cases}$$

Given this, because the difference

$$F(u_0+\alpha h, v_0+\alpha k)-F(u_0,v_0)$$

ought never fall below the limit zero, it is absolutely necessary that, for very small numerical values of α, the right-hand side of the preceding equation, and hence the first term of the right-hand side, that is to say, the term which contains the smallest power of α, can never become negative. Now, denoting by R_m the first of the quantities

$$R_1, \quad R_2, \quad \ldots, \quad R_n$$

[281] which has a value different from zero, we find that the term in question is,

$$2A^{\frac{1}{2}}\alpha^m\rho^m R_m \cos\left(T_m - T + m\theta\right),$$

if A is not zero, and

$$\alpha^{2m}\rho^{2m}R_m^2$$

otherwise. Moreover, the value of the arc θ is entirely indeterminate, so we can choose it in such a way as to give the factor

$$\cos\left(T_m - T + m\theta\right),$$

and hence the product

$$2A^{\frac{1}{2}}\alpha^m\rho^m R_m \cos\left(T_m - T + m\theta\right),$$

whichever sign we wish. Thus it is clear that only the second hypothesis remains admissible. Thus we necessarily have

(19) $$A = 0,$$

which reduces equation (10) to

(20) $$F\left(u_0, v_0\right) = 0.$$

It follows that the function $F\left(u, v\right)$ vanishes if we attribute to the variables u and v the real values u_0 and v_0, and consequently that the equation

(1) $$f\left(x\right) = 0$$

is satisfied by taking

$$x = u_0 + v_0\sqrt{-1}.$$

In other words, $u_0 + v_0\sqrt{-1}$ is a root of the equation

(1) $$a_0x^n + a_1x^{n-1} + \ldots + a_{n-1}x + a_n = 0.$$

The preceding proof of theorem I, while different in several points from that given by M. Legendre (*Théorie des Nombres*, 1st Part, § XIV),[2] is based on the same principles.

[282] *Corollary.* — The polynomial

---

[2] See [Legendre 1808].

$$f(x) = a_0x^n + a_1x^{n-1} + \ldots + a_{n-1}x + a_n,$$

which vanishes, as we have just said, for

$$x = u_0 + v_0\sqrt{-1},$$

is algebraically divisible by the factor

$$x - u_0 - v_0\sqrt{-1},$$

by virtue of theorem I (Chapter VII, § IV). Because the quotient is just a new polynomial of degree $n-1$ with respect to x, it is again necessarily divisible by a new factor of the same form as the previous one, that is to say, of first degree with respect to x. Denote this new factor by

$$x - u_1 - v_1\sqrt{-1}.$$

The polynomial $f(x)$ is equivalent to the product of the two factors

$$x - u_0 - v_0\sqrt{-1} \quad \text{and} \quad x - u_1 - v_1\sqrt{-1}$$

and a third polynomial of degree $n-2$. We can prove that this third polynomial is divisible by a third factor similar to the two others, and by continuing to operate in the same manner, we eventually obtain n linear factors of the polynomial $f(x)$. Let these factors be

$$x - u_0 - v_0\sqrt{-1}, \quad x - u_1 - v_1\sqrt{-1}, \quad \ldots, \quad x - u_{n-1} - v_{n-1}\sqrt{-1},$$

respectively. By dividing the polynomial $f(x)$ by their product, we find the quotient to be a constant, evidently equal to the coefficient a_0, of the greatest power of x in $f(x)$. Consequently we have

$$(21) \qquad f(x) = a_0\left(x - u_0 - v_0\sqrt{-1}\right)\left(x - u_1 - v_1\sqrt{-1}\right)\ldots$$
$$\ldots\left(x - u_{n-1} - v_{n-1}\sqrt{-1}\right).$$

This last equation contains a theorem that we may state as follows:

[283] **Theorem II.**[3] — *Whatever the values, real or imaginary, of the constants* $a_0, a_1, \ldots, a_{n-1}, a_n$ *may be, the polynomial*

$$a_0x^n + a_1x^{n-1} + \ldots + a_{n-1}x + a_n = f(x)$$

is equivalent to the product of the constant a_0 *by* n *linear factors of the form*

$$x - \alpha - \beta\sqrt{-1}.$$

[3] This is the Fundamental Theorem of Algebra.

To determine the factors in question here is called to *decompose* the polynomial $f(x)$ into its linear factors. There is only one way to carry out this decomposition. To demonstrate this, suppose that there were two different ways of forming the two equations[4]

$$(22)\qquad \begin{cases} f(x) = a_0\left(x-u_0-v_0\sqrt{-1}\right)\left(x-u_1-v_1\sqrt{-1}\right)\dots \\ \qquad \dots\left(x-u_{n-1}-v_{n-1}\sqrt{-1}\right) \quad \text{and} \\ f(x) = a_0\left(x-\alpha_0-\beta_0\sqrt{-1}\right)\left(x-\alpha_1-\beta_1\sqrt{-1}\right)\dots \\ \qquad \dots\left(x-\alpha_{n-1}-\beta_{n-1}\sqrt{-1}\right). \end{cases}$$

We get that

$$(23)\qquad \begin{cases} \left(x-\alpha_0-\beta_0\sqrt{-1}\right)\left(x-\alpha_1-\beta_1\sqrt{-1}\right)\dots \\ \qquad \dots\left(x-\alpha_{n-1}-\beta_{n-1}\sqrt{-1}\right) \\ = \left(x-u_0-v_0\sqrt{-1}\right)\left(x-u_1-v_1\sqrt{-1}\right)\dots \\ \qquad \dots\left(x-u_{n-1}-v_{n-1}\sqrt{-1}\right). \end{cases}$$

Because the right-hand side of the preceding formula vanishes when we give the variable x the particular value $u_0+v_0\sqrt{-1}$, it is necessary that, for this value of x, the left-hand side, and hence one of its factors (see Chapter VII, § II, theorem VII, corollary II), reduces to zero. Let

$$x-\alpha_0-\beta_0\sqrt{-1}$$

be that factor. We have identically

$$\alpha_0+\beta_0\sqrt{-1} = u_0+v_0\sqrt{-1},$$

and consequently,

$$x-\alpha_0-\beta_0\sqrt{-1} = x-u_0-v_0\sqrt{-1}.$$

[284] Given this, formula (23) can be replaced by the following:

$$\begin{aligned} &\left(x-\alpha_1-\beta_1\sqrt{-1}\right)\dots\left(x-\alpha_{n-1}-\beta_{n-1}\sqrt{-1}\right) \\ &\quad = \left(x-u_1-v_1\sqrt{-1}\right)\dots\left(x-u_{n-1}-v_{n-1}\sqrt{-1}\right). \end{aligned}$$

Because the right-hand side of this vanishes when we suppose that

$$x = u_1+v_1\sqrt{-1},$$

one of the factors of the left-hand side, for example,

$$x-\alpha_1-\beta_1\sqrt{-1},$$

[4] In [Cauchy 1897, p. 283], the last term of the second line of (22) has a_{n-1} in place of α_{n-1}. The equation is given correctly in [Cauchy 1821, p. 341]. (tr.)

must to vanish under the same hypotheses, and this entails two new identity equations of the form

$$\alpha_1 + \beta_1\sqrt{-1} = u_1 + v_1\sqrt{-1} \quad \text{and}$$
$$x - \alpha_1 - \beta_1\sqrt{-1} = x - u_1 - v_1\sqrt{-1}.$$

By repeating the same reasoning several times, we prove that the different linear factors that comprise the right-hand sides of equations (22) are absolutely the same as each other. It is essential to add that each imaginary factor of the form

$$x - \alpha - \beta\sqrt{-1}$$

is changed into a real factor $x - \alpha$ any time that the quantity β is reduced to zero.

Because, as we have just said, the left-hand side of equation (1) is decomposable into linear factors in just one way, it cannot vanish except when one of these factors vanishes. Thus if we successively make them equal to zero, we obtain all the possible values of x that satisfy equation (1), that is to say, all the roots of this equation. The number of these roots, like the number of linear factors, is equal to n. Moreover, each real factor of the form $x - \alpha$ corresponds to one real root α, and each imaginary factor of the form

$$x - \alpha - \beta\sqrt{-1}$$

[285] corresponds to an imaginary root

$$\alpha + \beta\sqrt{-1}.$$

These remarks suffice to establish the following proposition:

Theorem III. — *Whatever the values, real or imaginary, of the constants a_0, a_1, ..., a_{n-1}, a_n may be, the equation*

$$a_0x^n + a_1x^{n-1} + \ldots + a_{n-1}x + a_n = 0 \tag{1}$$

always has n roots, real or imaginary, and it will never have a greater number.

It can happen that several of the roots of equation (1) are equal to each other. In this case, the number of different values of the variable that satisfy this equation necessarily becomes less than n. Thus, for example, because the second-degree equation

$$x^2 - 2ax + a^2 = 0,$$

has two equal roots, it cannot be satisfied except by a single value of x, namely

$$x = a.$$

Whenever the constants $a_0, a_1, \ldots, a_{n-1}, a_n$ are all real, the imaginary expression

$$\alpha + \beta\sqrt{-1}$$

evidently cannot be a root of equation (1) except when the conjugate expression

$$\alpha - \beta\sqrt{-1}$$

is also a root of the same equation. Consequently, under this hypothesis, the imaginary linear factors of the polynomial that form the left-hand side of equation (1) are pairwise conjugate and of the form[5]

$$x - \alpha - \beta\sqrt{-1} \quad \text{and} \quad x - \alpha + \beta\sqrt{-1}.$$

The product of two such factors is always a real polynomial of the second degree, namely

$$(x-\alpha)^2 + \beta^2,$$

[286] and so we deduce the following theorem immediately from the observation that we have just made:

Theorem IV. — *When a_0, a_1, ..., a_{n-1}, a_n denote real constants, the polynomial*

(24) $$a_0 x^n + a_1 x^{n-1} + \ldots + a_{n-1} x + a_n$$

is decomposable into real factors of the first and second degree.

In the preceding, we have presented the imaginary roots of equation (1) in the form

$$\alpha \pm \beta\sqrt{-1}.$$

Then for polynomial (24), a real factor of the second degree corresponding to two conjugate imaginary roots

$$\alpha + \beta\sqrt{-1} \quad \text{and} \quad \alpha - \beta\sqrt{-1}$$

is of the form

$$(x-\alpha)^2 + \beta^2.$$

For convenience, if we make

$$\alpha \pm \beta\sqrt{-1} = \rho\left(\cos\theta \pm \sqrt{-1}\sin\theta\right),$$

(where ρ denotes a positive quantity and θ denotes an angle that we can assume is contained between the limits 0 and π), then the same real factor of the second degree becomes

$$(x - \rho\cos\theta)^2 + (\rho\sin\theta)^2 = x^2 - 2\rho\cos\theta + \rho^2.$$

[5] In [Cauchy 1897, p. 285], the second of the factors below is given as $x - \alpha + \sqrt{-1}$. The coeffeicient β is correctly included in [Cauchy 1821, p. 344]. (tr.)

It is easy to construct this last expression geometrically in the case where we give the variable x a real value. Indeed, if we trace a triangle in which one angle is equal to θ and the two adjacent sides are first the numerical value of x and second the modulus ρ, then the square of the third side is (from a well-known theorem of Trigonometry)[6] the value of the trinomial

$$x^2 - 2\rho x \cos\theta + \rho^2,$$

[287] whenever the value of the variable x is positive. If the value of x becomes negative, it suffices to replace the given angle θ in the construction by its supplement.

The third side of the triangle in question cannot vanish unless the two other sides fall on the same straight line and their extremities coincide, and this requires: 1° that the angle θ reduces to zero or to π; and 2° that the numerical value of x is equal to ρ. Consequently, the factor

$$x^2 - 2\rho\cos\theta + \rho^2$$

cannot become zero for a real value of x, at least when we do not suppose that

$$\cos\theta = 1 \quad \text{or} \quad \cos\theta = -1,$$

and the only value of x that makes this factor vanish is, in the first case,

$$x = \rho,$$

and in the second,

$$x = -\rho.$$

We arrive directly at the same conclusion by observing that the equation

$$x^2 - 2\rho\cos\theta + \rho^2$$

has two roots,

$$\rho\left(\cos\theta + \sqrt{-1}\sin\theta\right) \quad \text{and} \quad \rho\left(\cos\theta - \sqrt{-1}\sin\theta\right),$$

which cannot cease to be imaginary without becoming equal, and that the only values of θ capable of producing this effect are those which satisfy the formula

$$\sin\theta = 0.$$

From this we get

$$\cos\theta = \pm 1,$$

and consequently

$$x = \pm\rho$$

for the common value of the two roots.

---

[6] This is the Law of Cosines.

Up to now, we have been limited to determining the number [288] of roots of equation (1), along with the form of these roots and of their corresponding factors. In the following sections, we will review some particular cases in which we are able to solve similar equations without being required to imagine their coefficients converted into numbers, and to express the roots of these coefficients as algebraic or trigonometric functions of the coefficients. On this matter, we observe here that in every algebraic equation for which the left-hand side is a rational and integer function of the variable x, we can reduce the coefficient of the highest power of x to 1 by division, and the coefficient of the next-highest power of x to zero by a change of variable. Indeed, if a_0 is not equal to 1 in the equation

$$a_0x^n + a_1x^{n-1} + \ldots + a_{n-1}x + a_n = 0,$$

it suffices to divide the equation by a_0 to reduce the coefficient of x^n to 1. If an equation has been put into the form

$$x^n + a_1x^{n-1} + \ldots + a_{n-1}x + a_n = 0$$

and a_1 is not zero, then it suffices to set

$$x = z - \frac{a_1}{n}$$

to obtain a transformation into z of degree n which no longer has the second term, that is to say, a transformation in which the coefficient of z^{n-1} vanishes.

10.2 Algebraic or trigonometric solution of binomial equations and of some trinomial equations. The theorems of de Moivre and of Cotes.

Consider the binomial equation

$$(1) \qquad x^n + p = 0,$$

where p denotes a constant quantity. We get that

$$x^n = -p$$

[289] or, if ρ denotes the numerical value of p, then

$$x^n = \pm\rho.$$

Thus we have to solve the equation

$$(2) \qquad x^n = \rho,$$

if $-p$ is positive, and the following,

$$x^n = -\rho, \tag{3}$$

if $-p$ is negative. We satisfy the first one by taking

$$x = ((\rho))^{\frac{1}{n}} = \rho^{\frac{1}{n}}((1))^{\frac{1}{n}}, \tag{4}$$

and the second one by taking

$$x = ((-\rho))^{\frac{1}{n}} = \rho^{\frac{1}{n}}((-1))^{\frac{1}{n}}. \tag{5}$$

As for the various values of each of the two expressions $((1))^{\frac{1}{n}}$ and $((-1))^{\frac{1}{n}}$, there are always n of them (see Chapter VII, § III), and they are deduced from these two formulas:

$$\left\{\begin{array}{l} ((1))^{\frac{1}{n}} = \cos\frac{2k\pi}{n} \pm \sqrt{-1}\sin\frac{2k\pi}{n} \quad \text{and} \\ ((-1))^{\frac{1}{n}} = \cos\frac{(2k+1)\pi}{n} \pm \sqrt{-1}\sin\frac{(2k+1)\pi}{n}, \end{array}\right. \tag{6}$$

in which it suffices to give k successively all the integer values which do not surpass $\frac{n}{2}$. When n is an even number, the first of equations (6) gives two real values of $((1))^{\frac{1}{n}}$, namely $+1$ and -1, the first of which corresponds to $k=0$ and the second to $k=\frac{n}{2}$. Under the same hypothesis, all of the values of $((-1))^{\frac{1}{n}}$ are imaginary. When n is an odd number, the expression $((1))^{\frac{1}{n}}$ has a single real value, $+1$, corresponding to $k=0$, and the expression [290] $((-1))^{\frac{1}{n}}$ has a single real value, -1, corresponding to $k=\frac{n-1}{2}$. Consequently, when n is an even number equation (1) either admits two real roots or it admits none at all, and in the contrary case the same equation admits a single real root. Moreover, we recognize immediately by inspection of formulas (6) that the imaginary roots form conjugate pairs, as we ought to expect.

Now consider the trinomial equation

$$x^{2n} + px^n + q = 0, \tag{7}$$

where p and q denote two constant quantities chosen at will. We get

$$x^{2n} + px^n = -q,$$

and consequently

$$\left(x^n + \frac{p}{2}\right)^2 = \frac{p^2}{4} - q. \tag{8}$$

If $\frac{p^2}{4} - q$ is positive, the preceding equation will lead to one of the two following ones:

$$x^n + \tfrac{p}{2} = +\sqrt{\tfrac{p^2}{4} - q} \quad \text{or}$$

$$x^n + \tfrac{p}{2} = -\sqrt{\tfrac{p^2}{4} - q},$$

so that x^n admits two real values contained in the formula[7]

(9) $$x^n = -\frac{p}{2} \pm \sqrt{\frac{p^2}{4} - q}.$$

When the number n reduces to 1, formula (9) immediately gives the two real roots of the trinomial equation of the second degree

(10) $$x^2 + px + q.$$

When n is not equal to 1, then by substituting the formula under consideration into equation [291] (7), we have only to solve two binomial equations similar to those we have treated above.

Now suppose that the quantity $\frac{p^2}{4} - q$ is negative. Then equation (8) leads to one of the two following ones:

$$x^n + \tfrac{p}{2} = +\sqrt{q - \tfrac{p^2}{4}}\sqrt{-1} \quad \text{or}$$
$$x^n + \tfrac{p}{2} = -\sqrt{q - \tfrac{p^2}{4}}\sqrt{-1}.$$

Consequently, x^n admits two imaginary values contained in the formula

(11) $$x^n = -\frac{p}{2} \pm \sqrt{q - \frac{p^2}{4}}\sqrt{-1}.$$

If the number n reduces to 1, these values will be the imaginary roots of equation (10). However, if we suppose that $n > 1$, it still remains to deduce the values of x from the known values of x^n. Under this hypothesis, denote by ρ the modulus of the imaginary expression that serves as the right-hand side of formula (11). We evidently have

(12) $$\rho = q^{\frac{1}{2}}.$$

Moreover, for convenience make

(13) $$\zeta = \arctan \frac{\sqrt{q - \frac{p^2}{4}}}{-\frac{p}{2}}.$$

When p is negative, the two values of x^n given by formula (11) become

(14) $$x^n = \rho\left(\cos\zeta \pm \sqrt{-1}\sin\zeta\right),$$

and thus we conclude that

---

[7] Readers in North America may not be aware that in Europe the most commonly taught version of the quadratic formula gives the roots of a monic quadratic $x^2 + px + q$ as $-\frac{p}{2} \pm \sqrt{\frac{p^2}{4} - q}$. To Cauchy's readers, the version in formula (9) would have been very familiar.

$$x = \rho^{\frac{1}{n}} \left(\cos \frac{\zeta}{n} \pm \sqrt{-1} \sin \frac{\zeta}{n} \right) ((1))^{\frac{1}{n}}. \tag{15}$$

[292] On the other hand, if p is positive we find that

$$x^n = -\rho \left(\cos \zeta \pm \sqrt{-1} \sin \zeta \right), \tag{16}$$

and consequently

$$x = \rho^{\frac{1}{n}} \left(\cos \frac{\zeta}{n} \pm \sqrt{-1} \sin \frac{\zeta}{n} \right) ((-1))^{\frac{1}{n}}. \tag{17}$$

In the particular case where we have

$$\frac{p^2}{4} - q = 0,$$

ζ becomes zero, so that equations (15) and (17) take the form of equations (4) and (5).

If for brevity we denote $\rho^{\frac{1}{n}}$ by r, then by supposing that the quantity p is negative, we get from equations (12) and (13) that

$$p = -2r^n \cos \zeta, \quad q = r^{2n} \quad \text{and}$$
$$x^{2n} + px^n + q = x^{2n} - 2r^n x^n \cos \zeta + r^{2n}.$$

Under the same hypothesis, formula (15) gives

$$\begin{aligned} x &= r \left(\cos \tfrac{\zeta}{n} \pm \sqrt{-1} \sin \tfrac{\zeta}{n} \right) \left(\cos \tfrac{2k\pi}{n} \pm \sqrt{-1} \sin \tfrac{2k\pi}{n} \right) \\ &= r \left(\cos \tfrac{\zeta \pm 2k\pi}{n} \pm \sin \tfrac{\zeta \pm 2k\pi}{n} \right), \end{aligned}$$

where k represents a whole number. Thus we conclude that the trinomial

$$x^{2n} - 2r^n x^n \cos \zeta + r^{2n}$$

is decomposable into real factors of the second degree of the form

$$x^2 - 2rx \cos \frac{\zeta \pm 2k\pi}{n} + r^2.$$

On the other hand, if we suppose that the quantity p is positive, the trinomial

$$x^{2n} + px^n + q$$

becomes

$$x^{2n} + 2r^n x^n \cos \zeta + r^{2n},$$

[293] and its real factors of the second degree are of the form

$$x^2 - 2rx\cos\frac{\zeta \pm (2k+1)\pi}{n} + r^2.$$

Under both hypotheses, whenever we give real values to the variable x, we can construct the real factors of the second degree geometrically by the method indicated above (see § I). If we take the numerical value of the variable x as the common base of all the triangles that correspond to the different factors, and in each triangle we always join to the same end of this base the known side represented by r, we find that the vertices of these various triangles coincide with points that divide the circumference of a circle of radius of r into equal parts. Consequently, *if we multiply together the squares of the lines taken from the second extremity of the base to the points in question, the product of these squares will be the value of the trinomial*

$$x^{2n} + px^n + q = x^{2n} \pm 2r^n x^n \cos\zeta + r^{2n}.$$

In the particular case where $\zeta = 0$, *the product of the lines themselves represents the numerical value of the binomial*

$$x^n \pm r^n,$$

which corresponds to the positive square root of the trinomial

$$x^{2n} \pm 2r^n x^n + r^{2n}.$$

Of the two propositions that we have just stated, the first is the theorem of *de Moivre* and the second that of *Cotes.*

10.3 Algebraic or trigonometric solution of equations of the third and fourth degree.

Consider the general equation of the third degree. By making the second term of this equation vanish, we can always [294] reduce it to the form

(1) $$x^3 + px + q = 0,$$

where p and q denote two constant quantities. Moreover, if we set

$$x = u + v,$$

where u and v are two new variables, we conclude that

$$x^3 = (u+v)^3 = u^3 + v^3 + 3uvx,$$

or

(2) $$x^3 - 3uvx - \left(u^3 + v^3\right) = 0.$$

To make equation (2) identical to the given equation, it suffices to subject the unknowns u and v to the two conditions

$$(3) \qquad u^3+v^3=-q$$

and

$$(4) \qquad uv=-\frac{p}{3}.$$

Thus we find that the solution of equation (1) is reduced to the simultaneous solution of equations (3) and (4).

First, let us seek the values of u^3 and v^3. If we make

$$(5) \qquad u^3=z_1 \quad \text{and} \quad v^3=z_2,$$

then we have, by virtue of equations (3) and (4), that

$$z_1+z_2=-q \quad \text{and} \quad z_1z_2=-\frac{p^3}{27},$$

and consequently, by naming a new variable z,

$$(z-z_1)(z-z_2)=z^2+qz-\frac{p^3}{27}.$$

As a result, z_1 and z_2 are the two roots of the equation

$$(6) \qquad z^2+qz-\frac{p^3}{27}=0.$$

Knowing these two roots, we deduce from formulas (5) the three values of u and of v that correspond, two by two, [295] in a way that satisfies formula (4). Let U be any one of the three values of u, and let V be the corresponding value of v, so that we have

$$UV=-\frac{p}{3}.$$

Moreover, denote the imaginary expression

$$\cos\frac{2\pi}{3}+\sqrt{-1}\sin\frac{2\pi}{3}$$

by α. Then the three values of the expression $((1))^{\frac{1}{3}}$ are, respectively,

$$\alpha^0=1,$$

$$\alpha=\cos\tfrac{2\pi}{3}+\sqrt{-1}\sin\tfrac{2\pi}{3}=-\tfrac{1}{2}+\tfrac{3^{\frac{1}{2}}}{2}\sqrt{-1} \quad \text{and}$$

$$\alpha^2=\cos\tfrac{2\pi}{3}-\sqrt{-1}\sin\tfrac{2\pi}{3}=-\tfrac{1}{2}-\tfrac{3^{\frac{1}{2}}}{2}\sqrt{-1},$$

and the three values of u, evidently contained in the general formula $((1))^{\frac{1}{3}}\,U$, must be

$$U, \quad \alpha U \quad \text{and} \quad \alpha^2 U.$$

We find that the corresponding values of v are

$$V, \quad \frac{V}{\alpha} \quad \text{and} \quad \frac{V}{\alpha^2},$$

or what amounts to the same thing,

$$V, \quad \alpha^2 V \quad \text{and} \quad \alpha V.$$

Consequently, if we name the three roots of equation (1) x_0, x_1 and x_2, we have

$$\text{(7)} \qquad \begin{cases} x_0 = U + V, \\ x_1 = \alpha U + \alpha^2 V \quad \text{and} \\ x_2 = \alpha^2 U + \alpha V. \end{cases}$$

It is essential to observe that because U, αU and $\alpha^2 U$ are the three values of $u = ((z_1))^{\frac{1}{3}}$, and that because V, $\alpha^2 V$ and αV are the corresponding values of $v = -\frac{p}{3((z_1))^{\frac{1}{3}}}$, the roots x_0, x_1 and x_2 determined by equations (7) are, respectively, equal to the three values of x given by the formula[8] [296]

$$\text{(8)} \qquad x = ((z_1))^{\frac{1}{3}} - \frac{p}{3\,((z_1))^{\frac{1}{3}}}.$$

Whenever equation (6) has all real roots, formulas (5) give a system of real values of u and v that correspond in a way that satisfies equation (4). If we take these same values for U and V, we recognize immediately that of the three roots x_0, x_1 and x_2, the first is necessarily real and the two others may be real or imaginary, according to whether the quantity

$$\frac{q^2}{4} + \frac{p^3}{27}$$

is zero or positive, that is to say according to whether equation (6) has roots that are equal or unequal. In the first case, we find that

$$x_0 = 2U \quad \text{and} \quad x_1 = x_2 = -U.$$

Whenever the roots of equation (6) become imaginary, we can present them in the form

$$z_1 = \rho\left(\cos\theta + \sqrt{-1}\sin\theta\right) \quad \text{and} \quad z_2 = \rho\left(\cos\theta - \sqrt{-1}\sin\theta\right),$$

[8] Cauchy neglects to remind us here that it is necessary to use the same particular value of $((z_1))^{\frac{1}{3}}$ in each term of the right-hand side.

where the modulus ρ is determined by the equation

$$\rho^2 = -\frac{p^3}{27}.$$

Because under this hypothesis we have

$$((z_1))^{\frac{1}{3}} = \rho^{\frac{1}{3}}\left(\cos\frac{\theta}{3} + \sqrt{-1}\sin\frac{\theta}{3}\right)((1))^{\frac{1}{3}},$$

we find that formula (8) reduces to

$$\text{(9)} \qquad \begin{cases} x = \rho^{\frac{1}{3}}\left[\left(\cos\frac{\theta}{3} + \sqrt{-1}\sin\frac{\theta}{3}\right)((1))^{\frac{1}{3}}\right. \\ \qquad\qquad \left. + \left(\cos\frac{\theta}{3} - \sqrt{-1}\sin\frac{\theta}{3}\right)\dfrac{1}{((1))^{\frac{1}{3}}}\right]. \end{cases}$$

[297] Moreover, by taking for U the imaginary expression

$$\rho^{\frac{1}{3}}\left(\cos\frac{\theta}{3} + \sqrt{-1}\sin\frac{\theta}{3}\right),$$

we conclude from equations (7) that

$$\text{(10)} \qquad \begin{cases} x_0 = 2\rho^{\frac{1}{3}}\cos\frac{\theta}{3}, \\ x_1 = 2\rho^{\frac{1}{3}}\cos\frac{\theta+2\pi}{3} \quad \text{and} \\ x_2 = 2\rho^{\frac{1}{3}}\cos\frac{\theta-2\pi}{3}. \end{cases}$$

These last three values of x are all real and coincide with those which are given by formula (9).

In the preceding calculations, equation (6), the solution of which leads to that of equation (1), is what we call the *reduced* equation. Its roots z_1 and z_2 are necessarily equivalent to certain functions of the required roots x_0, x_1 and x_2. To determine these functions, it suffices to observe that, by virtue of formulas (5), we have

$$z_1 = U^3 \quad \text{and} \quad z_2 = V^3,$$

where U and V denote particular values of u and v. Moreover, from equations (7) we get that

$$\begin{aligned} 3U &= x_0 + \alpha x_2 + \alpha^2 x_1 \\ &= \alpha \left(x_2 + \alpha x_1 + \alpha^2 x_0 \right) \\ &= \alpha^2 \left(x_1 + \alpha x_0 + \alpha^2 x_2 \right) \quad \text{and} \\ 3V &= x_0 + \alpha x_1 + \alpha^2 x_2 \\ &= \alpha \left(x_1 + \alpha x_2 + \alpha^2 x_0 \right) \\ &= \alpha^2 \left(x_2 + \alpha x_0 + \alpha^2 x_1 \right). \end{aligned}$$

Consequently we find that

$$\text{(11)} \qquad \begin{cases} 27z_1 = \left(x_0 + \alpha x_2 + \alpha^2 x_1 \right)^3 \\ \qquad = \left(x_2 + \alpha x_1 + \alpha^2 x_0 \right)^3 \\ \qquad = \left(x_1 + \alpha x_0 + \alpha^2 x_2 \right)^3 \quad \text{and} \\ 27z_2 = \left(x_0 + \alpha x_1 + \alpha^2 x_2 \right)^3 \\ \qquad = \left(x_1 + \alpha x_2 + \alpha^2 x_0 \right)^3 \\ \qquad = \left(x_2 + \alpha x_0 + \alpha^2 x_1 \right)^3. \end{cases}$$

It follows that z_1 and z_2 are, respectively, equal (except for a numerical coefficient) to the only two distinct values which arise as the cube of the linear function

$$x_0 + \alpha x_1 + \alpha^2 x_2,$$

[298] when we interchange the roots, x_0, x_1 and x_2 of this function in every manner possible. The numerical coefficient is evidently $\frac{1}{27}$, or the cube of the fraction $\frac{1}{3}$.[9]

Now consider the general equation of the fourth degree. By making the second term disappear, we can reduce it to the form

$$\text{(12)} \qquad x^4 + px^2 + qx + r = 0,$$

where p, q and r denote constant quantities. Moreover, if we set

$$x = u + v + w,$$

where u, v and w are three new variables, we then conclude that

$$x^2 = u^2 + v^2 + w^2 + 2\left(uv + uw + vw\right),$$

and consequently,

$$\left[x^2 - \left(u^2 + v^2 + w^2\right)\right]^2 = 4\left(u^2v^2 + u^2w^2 + v^2w^2\right) + 8uvw \cdot x,$$

or what amounts to the same thing,

---

[9] What Cauchy has derived in this first part of § III is sometimes called the Cardano Formula for the cubic.

(13)
$$\begin{cases} x^4 - 2\left(u^2+v^2+w^2\right)x^2 - 8uvw\cdot x \\ \quad + \left(u^2+v^2+w^2\right)^2 - 4\left(u^2v^2+u^2w^2+v^2w^2\right) = 0. \end{cases}$$

To make this last equation identical to the given one, it suffices to subject the unknowns u, v and w to the conditions

(14)
$$\begin{cases} 4\left(u^2+v^2+w^2\right) = -2p, \\ 8uvw = -q \quad \text{and} \\ 16\left(u^2v^2+u^2w^2+v^2w^2\right) = p^2-4r. \end{cases}$$

Thus we find that the solution of equation (12) reduces to the simultaneous solution of equations (14).

First, we seek the values of $4u^2$, $4v^2$ and $4w^2$. If we make

(15)
$$4u^2 = z_1, \quad 4v^2 = z_2 \quad \text{and} \quad 4w^2 = z_3,$$

[299] we have, by virtue of formulas (14),

$$z_1+z_2+z_3 = -2p, \quad z_1z_2+z_1z_3+z_2z_3 = p^2-4r \quad \text{and} \quad z_1z_2z_3 = q^2.$$

Consequently, letting z be a new variable, we have

$$(z-z_1)(z-z_2)(z-z_3) = z^3+2pz^2+\left(p^2-4r\right)z-q^2.$$

It follows that z_1, z_2 and z_3 are the three roots of the equation

(16)
$$z^3+2pz^2+\left(p^2-4r\right)z-q^2 = 0,$$

and because these three roots must satisfy the formula $z_1z_2z_3 = q^2$, we can be sure that at least one of the roots will be positive and that the other two will be either both positive, both negative or both imaginary. When we have determined these roots, the first two of equations (15) give two equal values for each of the variables u and v, up to sign. Let

$$u = \pm U \quad \text{and} \quad v = \pm V$$

be the values, real or imaginary, in question, and let W be a real quantity or an imaginary expression determined by the equation

$$8UVW = -q.$$

If we suppose that in the second of formulas (14)

$$u = +U \quad \text{and} \quad v = +V$$

or else

$$u = -U \quad \text{and} \quad v = -V,$$

we get

$$w = +W.$$

On the other hand, if we make

$$u = +U \quad \text{and} \quad v = -V$$

or else[10]

$$u = -U \quad \text{and} \quad v = +V,$$

we find[11]

$$w = -W.$$

In this way, we obtain for the variables u, v and w four systems [300] of values that satisfy equations (14). If we represent by x_0, x_1, x_2 and x_3 the four values corresponding to the unknown

$$x = u + v + w,$$

then we have

$$\text{(17)} \qquad \begin{cases} x_0 = U + V + W, \\ x_1 = -U - V + W, \\ x_2 = U - V - W \quad \text{and} \\ x_3 = -U + V - W. \end{cases}$$

It is easy to recognize that if equation (16) has three positive roots, then these four values of x are all real; if equation (16) has two distinct negative roots, then they are all imaginary; while if equation (16) has two equal negative roots or two imaginary roots, then two values will be real and two will be imaginary.

By the method that we have just described, the solution of equation (12) is reduced to that of equation (16). This last equation, which we call the *reduced* equation, necessarily has for its roots certain functions of the roots of the given equation. If we wish to determine these functions, that is to say, to express z_1, z_2 and z_3 in terms of x_0, x_1, x_2 and x_3, it suffices to observe that because U, V and W are particular values of u, v and w, we have, by virtue of formulas (15), that

$$z_1 = 4U^2, \quad z_2 = 4V^2 \quad \text{and} \quad z_3 = 4W^2.$$

Moreover, we get from equations (17) that

$$\begin{aligned} 4U &= x_0 - x_1 + x_2 - x_3, \\ 4V &= x_0 - x_1 + x_3 - x_2 \quad \text{and} \\ 4W &= x_0 - x_2 + x_1 - x_3. \end{aligned}$$

As a consequence, we find

---

[10] In [Cauchy 1897, p. 299], this is written $u = -U$ and $w = +V$. It is $v = +V$ in [Cauchy 1821, p. 362]. (tr.)

[11] In [Cauchy 1897, p. 299], this is written $u = -W$. It is $w = -W$ in [Cauchy 1821, p. 362]. (tr.)

$$(18)\qquad \begin{cases} 4z_1 = (x_0 - x_1 + x_2 - x_3)^2 = (x_1 - x_0 + x_3 - x_2)^2, \\ 4z_2 = (x_0 - x_1 + x_3 - x_2)^2 = (x_1 - x_0 + x_2 - x_3)^2 \text{ and} \\ 4z_3 = (x_0 - x_2 + x_1 - x_3)^2 = (x_2 - x_0 + x_3 - x_1)^2. \end{cases}$$

[301] It follows that z_1, z_2 and z_3 are, if we ignore the numerical coefficient $\frac{1}{4} = \left(\frac{1}{2}\right)^2$, respectively equal to the three distinct values that are given by the square of the linear function

$$x_0 - x_1 + x_2 - x_3,$$

when we interchange the roots x_0, x_1, x_2 and x_3 in this function in all possible ways. This same linear function can thus be written as follows:

$$x_0 + (-1)x_1 + (-1)^2 x_2 + (-1)^3 x_3,$$

which is evidently a particular case of the general formula

$$x_0 + \alpha x_1 + \alpha^2 x_2 + \alpha^3 x_3,$$

when we denote by α one of the values of the expression $((1))^{\frac{1}{4}}$.

Chapter 11
Decomposition of rational fractions.

11.1 Decomposition of a rational fraction into two other fractions of the same kind.

[302] Let $f(x)$ and $F(x)$ be two integer functions of the variable x. Then

$$\frac{f(x)}{F(x)}$$

is what we call a *rational function*. If we denote the degree of the denominator $F(x)$ by m, then the equation

(1) $$F(x) = 0$$

admits m roots, real or imaginary, equal or not equal to each other. Supposing them to be distinct, if we represent them by

$$x_0, \quad x_1, \quad x_2, \quad \ldots, \quad x_{m-1},$$

then the linear factors of the polynomial $F(x)$ are, respectively,

$$x - x_0, \quad x - x_1, \quad x - x_2, \quad \ldots, \quad x - x_{m-1}.$$

Given this, make

(2) $$F(x) = (x - x_0)\,\varphi(x)$$

and

(3) $$\frac{f(x_0)}{\varphi(x_0)} = A.$$

[303] Because $\varphi(x_0)$ is not zero, the constant A is finite and the difference

$$\frac{f(x)}{\varphi(x)} - A = \frac{f(x) - A\varphi(x)}{\varphi(x)}$$

R.E. Bradley, C.E. Sandifer, *Cauchy's Cours d'analyse*, Sources and Studies in the History of Mathematics and Physical Sciences, DOI 10.1007/978-1-4419-0549-9_11,

vanishes for $x = x_0$. Consequently, the same is true of the polynomial

$$f(x) - A\varphi(x)$$

and this polynomial is algebraically divisible by $x - x_0$. Thus we have

$$f(x) - A\varphi(x) = (x - x_0)\chi(x)$$

or

(4) $$f(x) = A\varphi(x) + (x - x_0)\chi(x),$$

where $\chi(x)$ denotes a new integer function of the variable x. If we divide the two sides of this last equation by $F(x)$ and take into account formula (2), we conclude that

(5) $$\frac{f(x)}{F(x)} = \frac{A}{x - x_0} + \frac{\chi(x)}{\varphi(x)}.$$

Thus, if we separate the polynomial $F(x)$ into two factors, one of which is linear, we can decompose the rational fraction $\frac{f(x)}{F(x)}$ into two others which have as their respective denominators the two factors in question, and for which the simpler one has a constant numerator.

Imagine now that we separate the function $F(x)$ into two factors where the first, instead of being linear, corresponds to several roots of the equation $F(x) = 0$. For example, take for the first factor the factor of second degree

$$(x - x_0)(x - x_1).$$

As a consequence, we have

(6) $$F(x) = (x - x_0)(x - x_1)\varphi(x).$$

The fraction $\frac{f(x)}{\varphi(x)}$ still has a finite value, not only for $x = x_0$, but also for $x = x_1$. If we denote by u a polynomial [304] which, under both hypotheses is equal to $\frac{f(x)}{\varphi(x)}$, we find (Chapter IV, § I)

(7) $$u = \frac{f(x_0)}{\varphi(x_0)} \frac{x - x_1}{x_0 - x_1} + \frac{f(x_1)}{\varphi(x_1)} \frac{x - x_0}{x_1 - x_0}.$$

Because the polynomial u is determined, as we have just said, the equation

$$\frac{f(x)}{\varphi(x)} - u = 0$$

or

$$f(x) - u\varphi(x) = 0$$

includes x_0 and x_1 among its roots and consequently the polynomial

$$f(x) - u\varphi(x)$$

is divisible by the product

$$(x - x_0)(x - x_1).$$

Thus we have

$$f(x) - u\varphi(x) = (x - x_0)(x - x_1)\chi(x),$$

or

(8) $$f(x) = u\varphi(x) + (x - x_0)(x - x_1)\chi(x),$$

where $\chi(x)$ denotes a new integer function of the variable x. If we divide the last equation by $F(x)$ and take into account formula (6), we conclude

(9) $$\frac{f(x)}{F(x)} = \frac{u}{(x - x_0)(x - x_1)} + \frac{\chi(x)}{\varphi(x)}.$$

Likewise, we could prove that it suffices to set

(10) $$F(x) = (x - x_0)(x - x_1)(x - x_2)\varphi(x)$$

and

(11) $$\begin{cases} u = & \dfrac{f(x_0)}{\varphi(x_0)} \dfrac{(x - x_1)(x - x_2)}{(x_0 - x_1)(x_0 - x_2)} \\ & + \dfrac{f(x_1)}{\varphi(x_1)} \dfrac{(x - x_0)(x - x_2)}{(x_1 - x_0)(x_1 - x_2)} \\ & + \dfrac{f(x_2)}{\varphi(x_2)} \dfrac{(x - x_0)(x - x_1)}{(x_2 - x_0)(x_2 - x_1)} \end{cases}$$

[305] to obtain an equation of the form

(12) $$\frac{f(x)}{F(x)} = \frac{u}{(x - x_0)(x - x_1)(x - x_2)} + \frac{\chi(x)}{\varphi(x)},$$

etc.

Thus, in general, whenever the equation $F(x) = 0$ does not have equal roots, if we separate the polynomial $F(x)$ into two factors of which the first is the product of several linear factors, then the rational fraction $\frac{f(x)}{F(x)}$ is decomposable into two other fractions of the same kind which have as their respective denominators the two factors mentioned above, and of which the first has a numerator of a degree less than that of its denominator.

I move on to the case where we suppose that the equation $F(x) = 0$ has equal roots. Under this second hypothesis, let

$$a, \quad b, \quad c, \quad \ldots$$

be the various roots of this same equation, and denote by m' the number of roots equal to a, by m'' the number of roots equal to b, by m''' the number of roots equal to c, etc. The function $F(x)$ is equal to the product

$$(x-a)^{m'}(x-b)^{m''}(x-c)^{m'''}\ldots$$

or to this product multiplied by a constant coefficient, and we have

$$m'+m''+m'''+\ldots=m.$$

Given this, make

$$F(x)=(x-a)^{m'}\varphi(x) \tag{13}$$

and

$$\frac{f(a)}{\varphi(a)}=A. \tag{14}$$

Because $\varphi(a)$ is not zero, the constant A remains finite and the difference

$$\frac{f(x)}{\varphi(x)}-A$$

[306] vanishes for $x=a$. Thus we conclude that the polynomial

$$f(x)-A\varphi(x)$$

is divisible by $x-a$, and consequently we have

$$f(x)=A\varphi(x)+(x-a)\chi(x), \tag{15}$$

where $\chi(x)$ denotes a new integer function of the variable x. Finally, if we divide both sides of equation (15) by $F(x)$ and take into consideration formula (13), we find

$$\frac{f(x)}{F(x)}=\frac{A}{(x-a)^{m'}}+\frac{\chi(x)}{(x-a)^{m'-1}\varphi(x)}. \tag{16}$$

By reasoning in the same way, we could prove that it suffices to take

$$F(x)=(x-a)^{m'}(x-b)^{m''}\varphi(x) \tag{17}$$

and

$$u=\frac{f(a)}{\varphi(a)}\frac{x-b}{a-b}+\frac{f(b)}{\varphi(b)}\frac{x-a}{b-a} \tag{18}$$

to obtain an equation of the form

$$\frac{f(x)}{F(x)}=\frac{u}{(x-a)^{m'}(x-b)^{m''}}+\frac{\chi(x)}{(x-a)^{m'-1}(x-b)^{m''-1}\varphi(x)}, \tag{19}$$

etc.

11.2 Decomposition of a rational fraction for which the denominator is the product of several unequal factors into simple fractions which have for their respective denominators these same linear factors and have constant numerators.

Let

$$\frac{f(x)}{F(x)}$$

be the rational fraction under consideration, m be the degree of the function $F(x)$ and

$$x_0, \quad x_1, \quad x_2, \quad \ldots, \quad x_{m-1}$$

[307] the roots, assumed to be unequal, of the equation

(1) $$F(x) = 0.$$

If k denotes a constant coefficient, we have

(2) $$F(x) = k(x-x_0)(x-x_1)\ldots(x-x_{m-1}),$$

and by virtue of the principles established in the preceding section, the rational fraction $\frac{f(x)}{F(x)}$ can be decomposed into two others, of which the first is of the form

$$\frac{A_0}{x-x_0},$$

where A_0 represents a constant, while the second has as its denominator

$$\frac{F(x)}{x-x_0} = k(x-x_1)(x-x_2)\ldots(x-x_{m-1}).$$

By decomposing this second rational fraction by the same method, we obtain:

1° A new simple fraction of the form

$$\frac{A_1}{x-x_1}; \quad \text{and}$$

2° A fraction which has as its denominator

$$k(x-x_2)\ldots(x-x_{m-1}).$$

By continuing in this way, we make all the linear factors contained in the polynomial

$$F(x) = k(x-x_0)(x-x_1)\ldots(x-x_{m-1})$$

successively disappear. Consequently, we finally reduce the polynomial to the constant k. Thus, when by a series of such partial decompositions like those we have just indicated, we have extracted from the fraction $\frac{f(x)}{F(x)}$ a series of simple fractions [308] of the form

$$\frac{A_0}{x-x_0}, \quad \frac{A_1}{x-x_1}, \quad \frac{A_2}{x-x_2}, \quad \ldots, \quad \frac{A_{m-1}}{x-x_{m-1}},$$

where the remainder is just a rational fraction with a constant denominator, that is to say an integer function of the variable x. Denoting this integer function by R, we find

(3) $$\frac{f(x)}{F(x)} = R + \frac{A_0}{x-x_0} + \frac{A_1}{x-x_1} + \frac{A_2}{x-x_2} + \ldots + \frac{A_{m-1}}{x-x_{m-1}}.$$

Now it remains to find the values of the constants

$$A_0, \quad A_1, \quad A_2, \quad \ldots, \quad A_{m-1}.$$

These values are deduced without difficulty by the method of decomposition indicated in § I. However, we arrive more directly at their determination with the aid of the following considerations:

If we multiply the two sides of equation (3) by $F(x)$, we get

(4) $$\begin{cases} f(x) = RF(x) + A_0 \dfrac{F(x)}{x-x_0} + A_1 \dfrac{F(x)}{x-x_1} \\ \qquad\qquad + A_2 \dfrac{F(x)}{x-x_2} + \ldots + A_{m-1} \dfrac{F(x)}{x-x_{m-1}}. \end{cases}$$

If we make

$$x = x_0 + z$$

in both sides of this last formula, then the sum

$$RF(x) + A_1 \frac{F(x)}{x-x_1} + A_2 \frac{F(x)}{x-x_2} + \ldots + A_{m-1} \frac{F(x)}{x-x_{m-1}},$$

which is evidently a polynomial in x divisible by $x - x_0$, takes the form

$$zZ,$$

where Z denotes an integer function of z. It follows that we have

(5) $$f(x_0+z) = A_0 \frac{F(x_0+z)}{z} + zZ.$$

[309] Now suppose that the substitution of $x+z$ in place of x in the function $F(x)$ gives generally

(6) $$F(x+z) = F(x) + zF_1(x) + z^2 F_2(x) + \ldots.$$

We then deduce that

$$F(x_0+z) = zF_1(x_0) + z^2F_2(x_0) + \ldots,$$

and equation (5) becomes

$$(x_0+z) = A_0\left[F_1(x_0) + zF_2(x_0) + \ldots\right] + zZ.$$

When we make $z=0$ in this last equation, it reduces to

$$f(x_0) = A_0F_1(x_0),$$

and we conclude that

(7) $$A_0 = \frac{f(x_0)}{F_1(x_0)}.$$

By an entirely similar calculation, we find that

(8) $$\begin{cases} A_1 & = \dfrac{f(x_1)}{F_1(x_1)}, \\ A_2 & = \dfrac{f(x_2)}{F_1(x_2)}, \\ \ldots\ldots\ldots\ldots\ldots\ldots, \\ A_{m-1} & = \dfrac{f(x_{m-1})}{F_1(x_{m-1})}. \end{cases}$$

The values that we have just obtained for

$$A_0, \quad A_1, \quad A_2, \quad \ldots, \quad A_{m-1}$$

are evidently independent of the method used for the decomposition of the rational fraction $\frac{f(x)}{F(x)}$. From this it follows that this fraction can be decomposed in only one way into simple fractions which have as denominators linear factors of the polynomial $F(x)$ with constant numerators.

It is easy to see how equation (7) and formula (3) of the preceding section [310] agree with each other. Indeed, $F_1(x_0)$ is what the polynomial

$$F_1(x_0) + zF_2(x_0) + \ldots = \frac{F(x_0+z)}{z} = \frac{F(x)}{x-x_0}$$

becomes when we make $z=0$ or $x=x_0$. Consequently, if we set

(9) $$F(x) = (x-x_0)\,\varphi(x),$$

we have

$$F_1(x_0) = \varphi(x_0)$$

and

$$A_0 = \frac{f(x_0)}{\varphi(x_0)}. \tag{10}$$

To show an application of the formulas established above, suppose that it is a question of decomposing the rational fraction

$$\frac{x^n}{x^m - 1}$$

into simple fractions, where n denotes an integer number less than m. In this particular case, we have

$$f(x) = x^n, \quad F(x) = x^m - 1 \quad \text{and} \quad k = 1.$$

If we represent an integer number which does not surpass $\frac{m}{2}$ by h, then the various roots of the equation $F(x) = 0$, all unequal to each other, are contained in the formula

$$\cos\frac{2h\pi}{m} \pm \sqrt{-1}\sin\frac{2h\pi}{m}.$$

Let a be one of these roots. We seek the numerator A of the simple fraction that has $x - a$ as its denominator. This numerator is

$$A = \frac{f(a)}{F_1(a)} = \frac{a^n}{F_1(a)},$$

where the value of $F_1(a)$ is determined by the equation

$$\begin{aligned} F(a) + zF_1(a) + \ldots = F(a+z) &= (a+z)^m - 1 \\ &= a^m - 1 + ma^{m-1}z + \ldots, \end{aligned}$$

[311] and as a consequence is equal to ma^{m-1}. Thus we find that

$$A = \frac{a^n}{ma^{m-1}} = \frac{1}{m}a^{n+1-m}.$$

Moreover, because we have

$$\left(\cos\frac{2h\pi}{m} \pm \sqrt{-1}\sin\frac{2h\pi}{m}\right)^{n+1-m} = \cos\frac{2h(n+1)\pi}{m} \pm \sqrt{-1}\sin\frac{2h(n+1)\pi}{m},$$

and taking

$$\frac{(n+1)\pi}{m} = \theta \tag{11}$$

for brevity, we conclude from the preceding, that

(12)
$$\left\{\begin{aligned}\frac{x^n}{x^m-1}=\frac{1}{m}\Bigg(&\frac{1}{x-1}+\frac{\cos 2\theta+\sqrt{-1}\sin 2\theta}{x-\cos\frac{2\pi}{m}-\sqrt{-1}\sin\frac{2\pi}{m}}\\&+\frac{\cos 2\theta-\sqrt{-1}\sin 2\theta}{x-\cos\frac{2\pi}{m}+\sqrt{-1}\sin\frac{2\pi}{m}}\\&+\frac{\cos 4\theta+\sqrt{-1}\sin 4\theta}{x-\cos\frac{4\pi}{m}-\sqrt{-1}\sin\frac{4\pi}{m}}\\&+\frac{\cos 4\theta-\sqrt{-1}\sin 4\theta}{x-\cos\frac{4\pi}{m}+\sqrt{-1}\sin\frac{4\pi}{m}}+\cdots\Bigg).\end{aligned}\right.$$

By reasoning in the same manner, we find that

(13)
$$\left\{\begin{aligned}\frac{x^n}{x^m+1}=-\frac{1}{m}\Bigg(&\frac{1}{x-1}+\frac{\cos\theta+\sqrt{-1}\sin\theta}{x-\cos\frac{\pi}{m}-\sqrt{-1}\sin\frac{\pi}{m}}\\&+\frac{\cos\theta-\sqrt{-1}\sin\theta}{x-\cos\frac{\pi}{m}+\sqrt{-1}\sin\frac{\pi}{m}}\\&+\frac{\cos 3\theta+\sqrt{-1}\sin 3\theta}{x-\cos\frac{3\pi}{m}-\sqrt{-1}\sin\frac{3\pi}{m}}\\&+\frac{\cos 3\theta-\sqrt{-1}\sin 3\theta}{x-\cos\frac{3\pi}{m}+\sqrt{-1}\sin\frac{3\pi}{m}}+\cdots\Bigg).\end{aligned}\right.$$

It is essential to observe that, in equation (12) for even values of m and in equation (13) for odd values of m, the last of the simple fractions contained in the right-hand side of the equation is [312]

$$\frac{\cos m\theta}{x+1}=\frac{\cos(n+1)\pi}{x+1}=\frac{(-1)^{n+1}}{x+1}.$$

Thus, for example, we have

(14)
$$\frac{1}{x^2-1}=\frac{1}{2}\left(\frac{1}{x-1}-\frac{1}{x+1}\right),$$

(15)
$$\frac{x}{x^2-1}=\frac{1}{2}\left(\frac{1}{x-1}+\frac{1}{x+1}\right),$$

(16)
$$\begin{aligned}\frac{1}{x^3+1}=-\frac{1}{3}\Bigg(&\frac{\cos\frac{\pi}{3}+\sqrt{-1}\sin\frac{\pi}{3}}{x-\cos\frac{\pi}{3}-\sqrt{-1}\sin\frac{\pi}{3}}\\&+\frac{\cos\frac{\pi}{3}-\sqrt{-1}\sin\frac{\pi}{3}}{x-\cos\frac{\pi}{3}+\sqrt{-1}\sin\frac{\pi}{3}}-\frac{1}{x+1}\Bigg),\end{aligned}$$

....

We could also remark that if in the right-hand sides of equations of (12) and (13) we combine by addition two simple fractions corresponding to conjugate linear factors of the binomial $x^m \pm 1$, then the sum is a new fraction which has as denominator a real factor of the second degree and for its numerator a real linear function of the variable x. For example, by taking $n = 0$ and $m = 3$, we find

$$(17) \qquad \begin{cases} \dfrac{1}{x^3+1} = -\dfrac{1}{3}\left(\dfrac{2x\cos\frac{\pi}{3}-2}{x^2-2x\cos\frac{\pi}{3}+1} - \dfrac{1}{x+1}\right) \\ \qquad\quad\; = \dfrac{1}{3}\left(\dfrac{2-x}{x^2-x+1} + \dfrac{1}{x+1}\right). \end{cases}$$

It is easy to generalize this remark as follows.

Because the integer functions $f(x)$ and $F(x)$ are real, suppose that we denote two conjugate imaginary roots of equation (1) by

$$\alpha+\beta\sqrt{-1} \quad \text{and} \quad \alpha-\beta\sqrt{-1}$$

and take [313] A and B to be two real quantities that satisfy the formula

$$(18) \qquad \frac{f\left(\alpha+\beta\sqrt{-1}\right)}{F_1\left(\alpha+\beta\sqrt{-1}\right)} = A - B\sqrt{-1},$$

where $F_1(x)$ still represents the coefficient of z in the expansion of $F(x+z)$. We necessarily have

$$(19) \qquad \frac{f\left(\alpha-\beta\sqrt{-1}\right)}{F_1\left(\alpha-\beta\sqrt{-1}\right)} = A + B\sqrt{-1}.$$

As a consequence, if we decompose the rational fraction $\frac{f(x)}{F(x)}$, then the two simple fractions corresponding to the conjugate linear factors

$$x-\alpha-\beta\sqrt{-1} \quad \text{and} \quad x-\alpha+\beta\sqrt{-1}$$

are, respectively,

$$(20) \qquad \frac{A-B\sqrt{-1}}{x-\alpha-\beta\sqrt{-1}} \quad \text{and} \quad \frac{A+B\sqrt{-1}}{x-\alpha+\beta\sqrt{-1}}.$$

By adding these two fractions we obtain the following:

$$(21) \qquad \frac{2A(x-\alpha)+2B\beta}{(x-\alpha)^2+\beta^2}.$$

This last formula, which has as its numerator a real linear function of the variable x and as its denominator a real factor of the second degree of the polynomial $F(x)$, does not differ from the fraction

$$\frac{u}{(x-x_0)(x-x_1)},$$

which in formula (9) of section I contains in the case where we suppose

$$x_0 = \alpha + \beta\sqrt{-1} \quad \text{and} \quad x_1 = \alpha - \beta\sqrt{-1}.$$

11.3 Decomposition of a given rational fraction into other simpler ones which have for their respective denominators the linear factors of the first rational fraction, or of the powers of these same factors, and constants as their numerators.

[314] Let

$$\frac{f(x)}{F(x)}$$

be the rational fraction under consideration, m be the degree of the polynomial $F(x)$, and

$$a, \quad b, \quad c, \quad \ldots$$

the various roots of the equation

$$F(x) = 0. \tag{1}$$

Denote by k a constant coefficient and by m', m'', m''', ... several integer numbers for which the sum is equal to m. Then we have

$$F(x) = k(x-a)^{m'}(x-b)^{m''}(x-c)^{m'''}\ldots. \tag{2}$$

Given this, if we make use of the method explained in section I, we decompose the rational fraction $\frac{f(x)}{F(x)}$ into two others for which the first one is of the form

$$\frac{A}{(x-a)^{m'}},$$

while the second has as its denominator

$$\frac{F(x)}{x-a} = k(x-a)^{m'-1}(x-b)^{m''}(x-c)^{m'''}\ldots.$$

By decomposing this second rational fraction by the same method, we obtain: 1° a new simple fraction

$$\frac{A_1}{(x-a)^{m'-1}},$$

[315] in which A_1 represents a constant; and 2° a fraction which has as its denominator

$$k(x-a)^{m'-2}(x-b)^{m''}(x-c)^{m'''}\dots.$$

By continuing like this, we successively make the different linear factors composing the power $(x-a)^{m'}$ of the polynomial $F(x)$ disappear. When we have extracted from $\frac{f(x)}{F(x)}$ a sequence of simple fractions of the form

$$\frac{A}{(x-a)^{m'}},\quad \frac{A_1}{(x-a)^{m'-1}},\quad \frac{A_2}{(x-a)^{m'-2}},\quad \dots,\quad \frac{A_{m'-1}}{x-a},$$

what remains is a new rational fraction for which the denominator is reduced to

$$k(x-b)^{m''}(x-c)^{m'''}\dots.$$

If we extract a second sequence of simple fractions of the form

$$\frac{B}{(x-b)^{m''}},\quad \frac{B_1}{(x-b)^{m''-1}},\quad \frac{B_2}{(x-b)^{m''-2}},\quad \dots,\quad \frac{B_{m''-1}}{x-b}$$

from what remains, we obtain a second remainder for which the denominator is

$$k(x-c)^{m'''}\dots.$$

Finally, if we extend these operations until the polynomial $F(x)$ is reduced to the constant k, the last of all the remainders is a rational function with a constant denominator, that is to say an integer function of the variable x. Call this integer function R. Finally, we have as the value of $\frac{f(x)}{F(x)}$ decomposed into simple fractions

$$(3)\qquad \left\{\begin{aligned} \frac{f(x)}{F(x)} = R &+ \frac{A}{(x-a)^{m'}} + \frac{A_1}{(x-a)^{m'-1}} + \dots + \frac{A_{m'-1}}{x-a}\\ &+ \frac{B}{(x-b)^{m''}} + \frac{B_1}{(x-b)^{m''-1}} + \dots + \frac{B_{m''-1}}{x-b}\\ &+ \frac{C}{(x-c)^{m'''}} + \frac{C_1}{(x-c)^{m'''-1}} + \dots + \frac{C_{m'''-1}}{x-c}\\ &+ \dots\dots\dots\dots\dots\dots\dots\dots\dots\dots\dots\dots, \end{aligned}\right.$$

[316] where $A, A_1, \dots, A_{m'-1}$; $B, B_1, \dots, B_{m''-1}$; $C, C_1, \dots, C_{m'''-1}$; ... denote constants which we can easily deduce from the principles described in section I, or calculate directly with the aid of the following considerations.

For convenience, make

$$(4)\qquad \begin{cases} R+\dfrac{B}{(x-b)^{m''}}+\dfrac{B_1}{(x-b)^{m''-1}}+\ldots+\dfrac{B_{m''-1}}{x-b} \\ +\dfrac{C}{(x-c)^{m'''}}+\dfrac{C_1}{(x-c)^{m'''-1}}+\ldots+\dfrac{C_{m'''-1}}{x-c} \\ +\ldots\ldots\ldots\ldots\ldots\ldots\ldots\ldots \\ =\dfrac{Q}{(x-b)^{m''}(x-c)^{m'''}\ldots}, \end{cases}$$

where Q is a new integer function of the variable x. Equation (3) then becomes

$$\frac{f(x)}{F(x)}=\frac{A}{(x-a)^{m'}}+\frac{A_1}{(x-a)^{m'-1}}+\ldots+\frac{A_{m'-1}}{x-a}$$
$$+\frac{Q}{(x-b)^{m''}(x-c)^{m'''}\ldots}.$$

If we multiply both sides of this last formula by

$$F(x)=k(x-a)^{m'}(x-b)^{m''}(x-c)^{m'''}\ldots$$

we then conclude that

$$(5)\qquad \begin{cases} f(x)=[A+A_1(x-a)+\ldots \\ \qquad \ldots+A_{m'-1}(x-a)^{m'-1}\Big]\dfrac{F(x)}{(x-a)^{m'}}+kQ(x-a)^{m'}. \end{cases}$$

Consequently, by making

$$x=a+z,$$

we find that

$$(6)\qquad f(a+z)=\left(A+A_1z+\ldots+A_{m'-1}z^{m'+1}\right)\frac{F(a+z)}{z^{m'}}+Zz^{m'},$$

where Z denotes the value of the polynomial kQ expressed as a function of z. Now suppose that the substitution of $x+z$ in place of x in the functions $f(x)$ and $F(x)$ gives in general

$$(7)\qquad \begin{cases} f(x+z)=f(x)\ +zf_1(x)+z^2f_2(x)+\ldots, \\ F(x+z)=F(x)\ +zF_1(x)+z^2F_2(x)+\ldots \\ \qquad\qquad +z^{m'}F_{m'}(x)+z^{m'+1}F_{m'+1}(x)+\ldots. \end{cases}$$

[317] By taking $x=a+z$ and observing that the expansion of the function

$$F(x)=F(a+z)$$

ought to be divisible by $(x-a)^{m'} = z^{m'}$, we have that

$$(8) \qquad \begin{cases} f(a+z) = f(a) + zf_1(x) + z^2 f_2(x) + \ldots, \\ F(a+z) = \left[F_{m'}(a) + zF_{m'+1}(a) + z^2 F_{m'+2}(a) + \ldots\right] z^{m'} \end{cases}$$

and

$$(9) \qquad F(a) = 0, \quad F_1(a) = 0, \quad \ldots, \quad F_{m'-1}(a) = 0.$$

Given this, formula (6) is found to reduce to

$$(10) \qquad \begin{cases} f(a) + zf_1(a) + z^2 f_2(a) + \ldots \\ \quad = \left(A + A_1 z + A_2 z^2 + \ldots\right) \\ \qquad \times \left[F_{m'}(a) + zF_{m'+1}(a) + z^2 F_{m'+2}(a) + \ldots\right] + z^{m''} Z. \end{cases}$$

By equating the coefficients of similar powers of z on the two sides of the equation, we derive from this that

$$(11) \qquad \begin{cases} f(a) = AF_{m'}(a), \\ f_1(a) = A_1 F_{m'}(a) + AF_{m'+1}(a), \\ f_2(a) = A_2 F_{m'}(a) + A_1 F_{m'+1}(a) + AF_{m'+2}(a), \\ \ldots\ldots\ldots\ldots\ldots\ldots\ldots\ldots\ldots\ldots\ldots\ldots \end{cases}$$

By an entirely similar calculation we find

$$(12) \qquad \begin{cases} f(b) = BF_{m''}(b), \; f_1(b) = B_1 F_{m''}(b) + BF_{m''+1}(b), \; f_2(b) = \ldots, \\ f(c) = CF_{m'''}(c), \; f_1(c) = C_1 F_{m'''}(c) + CF_{m'''+1}(c), \; f_2(c) = \ldots, \\ \ldots\ldots\ldots\ldots, \; \ldots\ldots\ldots\ldots\ldots\ldots\ldots\ldots, \; \ldots\ldots\ldots \end{cases}$$

These various equations suffice to determine completely the values of the constants $A, A_1, A_2, \ldots, B, B_1, B_2, \ldots, C, C_1, C_2, \ldots$. They give, for example,

$$(13) \qquad \begin{cases} A = \dfrac{f(a)}{F_{m'}(a)}, \\ A_1 = \dfrac{f_1(a) - AF_{m'+1}(a)}{F_{m'}(a)}, \\ A_2 = \dfrac{f_2(a) - A_1 F_{m'+1}(a) - AF_{m'+2}(a)}{F_{m'}(a)}, \\ \ldots\ldots\ldots\ldots\ldots\ldots\ldots\ldots\ldots\ldots \end{cases}$$

[318] Because the constants thus determined are evidently independent of the method used for the decomposition of the rational fraction $\frac{f(x)}{F(x)}$, it follows that this fraction is decomposable into simple fractions of the form of those on the right-hand side of equation (3) in only one way.

It is easy to see that the first of equations (13) agrees with formula (14) of section I. Indeed, the quantity $F_{m'}(a)$ is what becomes of the polynomial

$$F_{m'}(a) + zF_{m'+1}(a) + z^2 F_{m'+2}(a) + \ldots = \frac{F(a+z)}{z^{m'}} = \frac{F(x)}{(x-a)^{m'}}$$

when we make $z = 0$ or $x = a$. Consequently, if we set

(14) $$F(x) = (x-a)^{m'} \varphi(x),$$

we have

$$F_{m'}(a) = \varphi(a)$$

and

(15) $$A = \frac{f(a)}{\varphi(a)}.$$

In the case where the functions $f(x)$ and $F(x)$ are both real and the equation $F(x) = 0$ admits m' roots equal to $\alpha + \beta\sqrt{-1}$, the same equation also admits m' roots equal and conjugate to the first ones, and consequently represented by

$$\alpha - \beta\sqrt{-1}.$$

Under this hypothesis, if after the decomposition of the rational fraction

$$\frac{f(x)}{F(x)},$$

we combine in pairs the simple fractions which have as their denominators

$$\begin{array}{ll} \left(x-\alpha-\beta\sqrt{-1}\right)^{m'} & \text{and } \left(x-\alpha+\beta\sqrt{-1}\right)^{m'}, \\ \left(x-\alpha-\beta\sqrt{-1}\right)^{m'-1} & \text{and } \left(x-\alpha+\beta\sqrt{-1}\right)^{m'-1}, \\ \ldots\ldots\ldots\ldots\ldots, & \ldots\ldots\ldots\ldots\ldots, \end{array}$$

[319] and finally

$$x-\alpha-\beta\sqrt{-1} \quad \text{and} \quad x-\alpha+\beta\sqrt{-1},$$

the different sums obtained are the real and rational fractions which have as their respective denominators

$$\begin{array}{l} \left[(x-\alpha)^2+\beta^2\right]^{m'}, \\ \left[(x-\alpha)^2+\beta^2\right]^{m'-1}, \\ \ldots\ldots\ldots\ldots\ldots, \\ (x-\alpha)^2+\beta^2, \end{array}$$

and by which the system can be replaced by a sequence of other fractions which, with the same denominators, have as their numerators real linear functions of the variable x. Finally, it is easy to calculate directly this new sequence of fractions by beginning with those which correspond to the highest powers of $(x-\alpha)^2+\beta^2$. For example, let us seek the one which has as its denominator

$$\left[(x-\alpha)^2+\beta^2\right]^{m'} = \left(x-\alpha-\beta\sqrt{-1}\right)^{m'}\left(x-\alpha+\beta\sqrt{-1}\right)^{m'}.$$

From the principles established in section I, the fraction is

$$\frac{u}{\left[(x-\alpha)^2+\beta^2\right]^{m'}}, \tag{16}$$

provided that we make

$$\begin{cases} u = \dfrac{1}{2\beta\sqrt{-1}}\left[\dfrac{f\left(\alpha+\beta\sqrt{-1}\right)}{\varphi\left(\alpha+\beta\sqrt{-1}\right)}\left(x-\alpha+\beta\sqrt{-1}\right) \right. \\ \qquad\qquad\qquad \left. - \dfrac{f\left(\alpha-\beta\sqrt{-1}\right)}{\varphi\left(\alpha-\beta\sqrt{-1}\right)}\left(x-\alpha-\beta\sqrt{-1}\right)\right] \end{cases} \tag{17}$$

and

$$\varphi(x) = \frac{F(x)}{\left[(x-\alpha)^2+\beta^2\right]^{m'}}. \tag{18}$$

We add that if we successively set

$$x=\alpha+\beta\sqrt{-1}+z \quad\text{and}\quad x=\alpha-\beta\sqrt{-1}+z$$

in the preceding formula, [320] we conclude, taking into account the second of equations (8), that

$$\varphi\left(\alpha+\beta\sqrt{-1}+z\right) = \frac{F_{m'}\left(\alpha+\beta\sqrt{-1}\right)+zF_{m'+1}\left(\alpha+\beta\sqrt{-1}\right)+\dots}{\left(2\beta\sqrt{-1}+z\right)^{m'}},$$

$$\varphi\left(\alpha-\beta\sqrt{-1}+z\right) = \frac{F_{m'}\left(\alpha-\beta\sqrt{-1}\right)+zF_{m'+1}\left(\alpha-\beta\sqrt{-1}\right)+\dots}{\left(-2\beta\sqrt{-1}+z\right)^{m'}},$$

and consequently

$$\begin{cases} \varphi\left(\alpha+\beta\sqrt{-1}\right) = \dfrac{F_{m'}\left(\alpha+\beta\sqrt{-1}\right)}{\left(2\beta\sqrt{-1}\right)^{m'}} \quad\text{and} \\ \varphi\left(\alpha-\beta\sqrt{-1}\right) = (-1)^{m'}\dfrac{F_{m'}\left(\alpha-\beta\sqrt{-1}\right)}{\left(2\beta\sqrt{-1}\right)^{m'}}. \end{cases} \tag{19}$$

Chapter 12
On recurrent series.

12.1 General considerations on recurrent series.

[321] A series

(1) $$a_0, \quad a_1x, \quad a_2x^2, \quad \ldots, \quad a_nx^n, \quad \ldots,$$

ordered according to the ascending integer powers of the variable x, is called *recurrent* when in this series, starting after a given term, the coefficient of any power of the variable is expressed as a linear function of a fixed number of the coefficients of lesser powers, and consequently it suffices to *run back*[1] to the values of these last coefficients to deduce the one we are seeking. Thus, for example, the series

(2) $$1, \quad 2x, \quad 3x^2, \quad \ldots, \quad (n+1)x^n, \quad \ldots$$

is recurrent, considering that if we make

$$a_n = n+1,$$

we always have, for values of n greater than 1,

(3) $$a_n = 2a_{n-1} - a_{n-2}.$$

In general, series (1) is recurrent if, for all values of n greater than a certain limit, the coefficients

$$a_n, \quad a_{n-1}, \quad a_{n-2}, \quad \ldots, \quad a_{n-m}$$

of several consecutive powers of x are found related to each other [322] by an equation of the first degree. Let

(4) $$ka_{n-m} + la_{n-m+1} + \ldots + pa_{n-1} + qa_n = 0$$

[1] Cauchy uses the French verb *recourir* here. He seems to be commenting on the etymology of "recurrent" (*récurrent* in French), which has its origins in the Latin verb *currere*, "to run."

R.E. Bradley, C.E. Sandifer, *Cauchy's Cours d'analyse*, Sources and Studies in the History of Mathematics and Physical Sciences, DOI 10.1007/978-1-4419-0549-9_12,

be the equation in question, where k, l, ..., p and q denote determined constants. The sequence of these constants forms what we call the *recurrence relation*[2] of the series, the recurrence for which the constants themselves are the different *terms*.

In series (1), assumed to be recurrent, the variable x and its coefficients a_0, a_1, a_2, ..., a_n, can be either real quantities or imaginary expressions. Given this, represent the modulus of the expression a_n by ρ_n, and consequently the numerical value of this expression whenever it is real. We conclude immediately from the principles established in Chapters VI and IX that series (1) is either convergent or divergent depending on whether the modulus or the numerical value of x is less than or greater than the smallest of the limits towards which the expression $(\rho_n)^{-\frac{1}{n}}$ converges, when n grows indefinitely.

12.2 Expansion of rational fractions into recurrent series.

Any time that a rational fraction can be expanded into a convergent series ordered according to ascending integer powers of the variable, that series is recurrent, as we will see.

First consider the rational fraction

$$\frac{A}{(x-a)^m}, \tag{1}$$

in which a and A denote two constants, real or imaginary, and m an integer number. It can be put into the form

$$(-1)^m \frac{A}{a^m}\left(1-\frac{x}{a}\right)^{-m},$$

[323] and it is expandable, as well as the expression

$$\left(1-\frac{x}{a}\right)^{-m},$$

into a convergent series ordered according to the ascending integer powers of the variable x if the numerical value of the ratio $\frac{x}{a}$ in the real case, or the modulus of the same ratio in the imaginary case, is a quantity contained between the limits 0 and 1. This condition is satisfied if the modulus of the variable x, a modulus which reduces to the numerical value of the same variable when it becomes real,[3] is less than the modulus of the constant a, and we have, under this hypothesis,

[2] Cauchy uses the term *échelle de relation*, literally "scale [or ladder] of relation." (tr.)

[3] Cauchy writes *imaginaire* here in [Cauchy 1821, p. 391, Cauchy 1897, p. 323].

$$(2)\qquad \begin{cases} \left(1-\frac{x}{a}\right)^{-m} = 1+\frac{m}{1}\frac{x}{a}+\frac{m(m+1)}{1\cdot 2}\frac{x^2}{a^2}+\dots \\ \qquad = \frac{1\cdot 2\cdot 3\dots(m-1)}{1\cdot 2\cdot 3\dots(m-1)}+\frac{2\cdot 3\cdot 4\dots m}{1\cdot 2\cdot 3\dots(m-1)}\frac{x}{a} \\ \qquad\qquad +\frac{3\cdot 4\cdot 5\dots(m+1)}{1\cdot 2\cdot 3\dots(m-1)}\frac{x^2}{a^2}+\dots. \end{cases}$$

Consequently, we find

$$(3)\qquad \frac{A}{(x-a)^m} = (-1)^m\left(\frac{A}{a^m}+\frac{m}{1}\frac{Ax}{a^{m+1}}+\frac{m(m+1)}{1\cdot 2}\frac{Ax^2}{a^{m+2}}+\dots\right).$$

If for brevity we make

$$(4)\qquad \begin{cases} (-1)^m\frac{A}{a^m} = a_0, \\ (-1)^m\frac{m}{1}\frac{A}{a^{m+1}} = a_1, \\ (-1)^m\frac{m(m+1)}{1\cdot 2}\frac{A}{a^{m+2}} = a_2, \\ \dots\dots\dots\dots\dots\dots, \end{cases}$$

we obtain the equation

$$(5)\qquad \frac{A}{(x-a)^m} = a_0+a_1x+a_2x^2+\dots+a_nx^n+\dots.$$

[324] Now imagine that we multiply both sides of the preceding equation by $(a-x)^m$. We find that[4]

$$(6)\qquad \begin{cases} (-1)^m A = \left[a^m-\frac{m}{1}a^{m-1}x+\frac{m(m-1)}{1\cdot 2}a^{m-2}x^2-\dots\pm x^m\right] \\ \qquad \times\left(a_0+a_1x+a_2x^2+\dots\right) \\ \qquad = a^m\left(a_0+a_1x+a_2x^2+\dots+a_m\,x^m+a_{m+1}x^{m+1}+\dots\right) \\ \qquad\quad -\frac{m}{1}a^{m-1}\left(a_0x+a_1x^2+\dots+a_{m-1}x^m+a_m\,x^{m+1}+\dots\right) \\ \qquad\quad +\frac{m(m-1)}{1\cdot 2}a^{m-2}\left(a_0x^2+\dots+a_{m-2}x^m+a_{m-1}x^{m+1}+\dots\right) \\ \qquad\quad -\dots\dots\dots\dots\dots\dots \\ \qquad\quad \pm\left(a_0x^m+a_1\,x^{m+1}+\dots\right), \end{cases}$$

or what amounts to the same thing,

[4] Cauchy writes a + before the ellipses in the first line of this equation in [Cauchy 1821, p. 392, Cauchy 1897, p. 324].

$$(7)\quad \begin{cases} (-1)^m A = a^m a_0 + \left(a^m a_1 - \frac{m}{1} a^{m-1} a_0\right) x \\ \qquad + \left[a^m a_2 - \frac{m}{1} a^{m-1} a_1 + \frac{m(m-1)}{1\cdot 2} a^{m-2} a_0\right] x^2 \\ \qquad + \dots\dots\dots\dots\dots\dots \\ \qquad + \left[a^m a_n - \frac{m}{1} a^{m-1} a_{n-1} + \frac{m(m-1)}{1\cdot 2} a^{m-2} a_{n-2} \right. \\ \qquad\quad \left. - \dots \pm a_{n-m}\right] x^n \\ \qquad + \dots\dots\dots\dots\dots\dots \end{cases}$$

This last formula ought to remain true any time that the modulus of the variable x is less than the modulus of the constant a, and consequently any time that we attribute to x a real value slightly different from zero. We conclude, by reasoning similar to that which we have used for the proof of theorem VI of Chapter VI (§ IV), that

$$(8)\quad \begin{cases} (-1)^m A = a^m a_0, \\ a^m a_1 - \frac{m}{1} a^{m-1} a_0 = 0, \\ a^m a_2 - \frac{m}{1} a^{m-1} a_1 + \frac{m(m-1)}{1\cdot 2} a^{m-2} a_0 = 0, \\ \dots\dots\dots\dots\dots\dots, \end{cases}$$

[325] and in general,

$$(9)\qquad a^m a_n - \frac{m}{1} a^{m-1} a_{n-1} + \frac{m(m-1)}{1\cdot 2} a^{m-2} a_{n-2} - \dots \pm a_{n-m} = 0.$$

It is essential to remark that equation (9) applies only for real integer values of n greater than or equal to m, and that whenever we suppose that $n < m$, it ought to be replaced by one of the formulas (8). Moreover, because equation (9) is linear with respect to the constants

$$a_n, \quad a_{n-1}, \quad a_{n-2}, \quad \dots, \quad a_{n-m},$$

it gives the first of these constants as a linear function of all the other ones. It follows that in the series

$$(10)\qquad a_n, \quad a_1 x, \quad a_2 x^2, \quad \dots, \quad a_n x^n, \quad \dots$$

starting from the term $a_m x^m$,[5] the coefficient of any power of x is expressed as a linear function of the m coefficients of lesser powers taken consecutively. This series is thus one of those that we have named *recurrent*.

Among the various particular formulas which we can deduce from equation (3), it is good to mention those which correspond to the two suppositions $m = 1$ and $m = 2$. We find, under the first hypothesis, that

[5] In [Cauchy 1897, p. 325], this is written as $a^m x_m$. It is given correctly in [Cauchy 1821, p. 394].

$$\frac{A}{x-a} = -\left(\frac{A}{a} + \frac{A}{a^2}x + \frac{A}{a^3}x^2 + \dots\right), \tag{11}$$

and under the second hypothesis, that

$$\frac{A}{(x-a)^2} = \frac{A}{a^2} + 2\frac{A}{a^3}x + 3\frac{A}{a^4}x^2 + 4\frac{A}{a^5}x^3 + \dots. \tag{12}$$

The two preceding formulas, where the first determines the sum of a geometric progression, remain true, and thus equation (3) as well, as long as the modulus of x is less than the modulus of a. [326] When in equation (12) we make both

$$A = 1 \quad \text{and} \quad a = 1,$$

we obtain the following

$$\frac{1}{(x-1)^2} = 1 + 2x + 3x^2 + 4x^3 + \dots, \tag{13}$$

which has for its right-hand side the sum of series (2) (§ I), and supposes that the modulus of x is less than 1.

Now consider any rational fraction

$$\frac{f(x)}{F(x)}, \tag{14}$$

where $f(x)$ and $F(x)$ are two integer functions of the variable x. Represent by a, b, c, ... the various roots of the equation

$$F(x) = 0, \tag{15}$$

by m' the number of roots equal to a, by m'' the number of roots equal to b, by m''' the number of roots equal to c, ..., and by k the coefficient of the highest power of x in the polynomial $F(x)$, so that we have

$$F(x) = k(x-a)^{m'}(x-b)^{m''}(x-c)^{m'''}\dots. \tag{16}$$

For the decomposition of the rational fraction $\frac{f(x)}{F(x)}$ into simple fractions, the method explained in the preceding chapter gives an equation of the form

$$(17)\quad \begin{cases} \dfrac{f(x)}{F(x)} = R + \dfrac{A}{(x-a)^{m'}} + \dfrac{A_1}{(x-a)^{m'-1}} + \ldots + \dfrac{A_{m'-1}}{x-a} \\ \quad + \dfrac{B}{(x-b)^{m''}} + \dfrac{B_1}{(x-b)^{m''-1}} + \ldots + \dfrac{B_{m''-1}}{x-b} \\ \quad + \dfrac{C}{(x-c)^{m'''}} + \dfrac{C_1}{(x-c)^{m'''-1}} + \ldots + \dfrac{C_{m'''-1}}{x-c} \\ \quad + \ldots\ldots\ldots\ldots\ldots\ldots\ldots\ldots\ldots\ldots, \end{cases}$$

where $A, A_1, \ldots, B, B_1, \ldots, C, C_1, \ldots$, etc. denote determined constants [327] and R is an integer function of x which vanishes when the degree of the polynomial $f(x)$ is less than that of the polynomial $F(x)$. Given this, imagine that the modulus of the variable x is less than the moduli of the various roots $a, b, c, \ldots$, and consequently less than the smallest of these moduli. We can expand each of the simple fractions that make up the right-hand side of equation (17) into a convergent series ordered according to the ascending powers of the variable x. Then, by adding the expansions formed like this to the polynomial R, we obtain a new convergent series, still ordered according to the ascending powers of x and where the sum is equal to the rational fraction $\frac{f(x)}{F(x)}$. Let

$$(18)\qquad a_0, \quad a_1x, \quad a_2x^2, \quad \ldots, \quad a_nx^n, \quad \ldots$$

be the new series in question here. The formula

$$(19)\qquad \frac{f(x)}{F(x)} = a_0 + a_1x + a_2x^2 + \ldots$$

remains true any time this new series is convergent, that is to say any time the modulus of the variable x is less than the smallest of the numbers that serve as the moduli of the roots of equation (15). I add that series (18) is still a recurrent series. We will easily prove this as follows.

Denote by m the sum of the integer numbers $m', m'', m''', \ldots$, or what amounts to the same thing, the degree of the polynomial $F(x)$, and consequently make

$$(20)\qquad F(x) = kx^m + lx^{m-1} + \ldots + px + q,$$

where $k, l, \ldots, p$ and q represent constants, real or imaginary. Equation (19) becomes

$$(21)\qquad \frac{f(x)}{kx^m + lx^{m-1} + \ldots + px + q} = a_0 + a_1x + a_2x^2 + \ldots.$$

After putting it into the form

$$(22)\qquad f(x) = \left(q + px + \ldots + lx^{m-1} + kx^m\right)\left(a_0 + a_1x + a_2x^2 + \ldots\right),$$

[328] we get, by expanding the right-hand side as we did for equation (6),

$$(23)\qquad \begin{cases} f(x) = qa_0 + (qa_1 + pa_0)x + \dots \\ \qquad + (qa_m + pa_{m-1} + \dots + la_1 + ka_0)x^m + \dots \\ \qquad + (qa_n + pa_{n-1} + \dots + la_{n-m+1} + ka_{n-m})x^n \\ \qquad + \dots\dots\dots\dots\dots\dots\dots\dots\dots\dots \end{cases}$$

Because this last formula ought to remain true as long as the modulus of the variable x is less than the moduli of the constants a, b, c, ..., we can prove, by reasoning similar to that which we have used to establish theorem VI of Chapter VI (§ IV), that the coefficients of like powers of x in the two sides are necessarily equal to each other. It follows: 1° that the coefficients of the various powers of x in the different terms of the polynomial $f(x)$ are respectively equal to the coefficients of the same powers of the series, the sum of which constitutes the right-hand side of equation (23); and 2° that in this series the coefficients of the powers where the exponent surpasses the degree of the polynomial $f(x)$ reduce to zero. Moreover, if we consider a term of the series in which the exponent n of the variable x surpasses the degree of the polynomial $f(x)$, and is at the same time equal to or greater than m, the term is of the form

$$(qa_n + pa_{n-1} + \dots + la_{n-m+1} + ka_{n-m})x^n.$$

Thus, any time the value of n is greater than the degree of the polynomial $f(x)$ and is also equal to or greater than the degree m of the polynomial $F(x)$, the coefficients

$$a_n, \quad a_{n-1}, \quad \dots, \quad a_{n-m+1}, \quad a_{n-m}$$

are found to satisfy the linear equation

$$(24)\qquad qa_n + pa_{n-1} + \dots + la_{n-m+1} + ka_{n-m} = 0.$$

Consequently, for such a value of n, the coefficent a_n of the power x^n is expressed as a linear function of those coefficients of m lesser powers taken consecutively. Series (18) [329] is thus one of those that we call *recurrent*. Its recurrence relation is composed of the constants

$$k, \quad l, \quad \dots, \quad p, \quad q,$$

respectively equal to the coefficients of the various powers of x in the polynomial $F(x)$.

Among the series which represent the expansions of the fractions contained in the right-hand side of formula (17) and which are all convergent in the case where the modulus of the variable x remains less than the moduli of the various roots of equation (15), at least one would become divergent if the modulus of the variable came to surpass that of some root. Consequently, series (18), still convergent in the first case, is divergent in the second. On the other hand, if we make the integer number n increase indefinitely, and if we denote by ρ_n the modulus of the coefficient a_n in series (18), this series is convergent or divergent (see § I) depending on whether the modulus of x is less than or greater than the smallest of the limits of $(\rho_n)^{-\frac{1}{n}}$.

Because the two rules of convergence that we have just stated must necessarily agree with each other, we can conclude that *the smallest of the moduli which correspond to roots of equation* (15) *is precisely equal to the smallest of the limits of the expression* $(\rho_n)^{-\frac{1}{n}}$.

When the two functions $f(x)$ and $F(x)$ are real, the coefficient a_n is real as well, and its modulus ρ_n is no different from its numerical value. If under the same hypothesis, the equation $F(x) = 0$ has no real roots, the root that has the smallest numerical value is, from what we have just said, equal (up the sign) to the smallest of the limits of $(\rho_n)^{-\frac{1}{n}}$. Finally, if the ratio $\frac{\rho_n}{\rho_{n+1}}$ converges to a fixed limit, we can substitute it (Chap. II, § III, theorem II) for the desired limit of the expression $(\rho_n)^{-\frac{1}{n}}$.[6] This remark leads to the rule that Daniel Bernoulli has given for determining numerically [330] the smallest (ignoring the sign) of all the quantities which represent the roots, supposed to be real, of an algebraic equation.

12.3 Summation of recurrent series and the determination of their general terms.

When a series ordered according to the ascending powers of the variable x is at the same time convergent and recurrent, it always has a rational fraction as its sum. Indeed, let

(1) $$a_0, \quad a_1x, \quad a_2x^2, \quad \ldots, \quad a_nx^n, \quad \ldots$$

be such a series. Suppose that for values of n above a certain limit, the coefficient a_n of the power x^n is determined as a linear function of the n coefficients of the lesser powers by an equation of the form

(2) $$ka_{m-n} + la_{n-m+1} + \ldots + pa_{n-1} + qa_n = 0,$$

such that the constants

$$k, \quad l, \quad \ldots, \quad p \text{ and } q$$

form the recurrence relation of the series. If we multiply the sum of the series, namely

$$a_0 + a_1x + a_2x^2 + \ldots$$

by the polynomial

$$kx^m + lx^{m-1} + \ldots + px + q,$$

the product obtained is the sum of a new series in which the coefficient of x^n, calculated as in Chapter VI (§ IV, theorem V), vanishes for values of n greater than the assigned limit. In other words, the product in question is a new polynomial of a degree indicated by this limit. If we denote this new polynomial by $f(x)$, we have

[6] This was given as $(\rho)^{-\frac{1}{n}}$ in [Cauchy 1821, p. 400, Cauchy 1897, p. 329].

(3) $$f(x) = \left(kx^m + lx^{m-1} + \ldots + px + q\right)\left(a_0 + a_1 x + a_2 x^2 + \ldots\right)$$

[331] and consequently,

(4) $$a_0 + a_1 x + a_2 x^2 + \ldots = \frac{f(x)}{kx^m + lx^{m-1} + \ldots + px + q}.$$

Thus, any series which is ordered according to the ascending and integer powers of the variable x and which is both convergent and recurrent has as its sum a rational fraction for which the denominator is a polynomial in which the successive powers of x have for coefficients the different terms of the recurrence relation of the series.

When we describe a recurrent series by giving only its first terms and the recurrence relation which serves to determine from the first terms all those which follow, then with the aid of the method which we have just indicated, we easily determine the rational fraction which represents the sum of the series in the case where it remains convergent. Once this rational fraction is calculated, we can substitute a sum of simple fractions for it, possibly augmented by an integer function of the variable x. If we then seek the recurrent series which expresses the expansions of the simple fractions in question for conveniently chosen values of x, and we add the general terms of these same series, we obtain the general term of the proposed series.

NOTES.[1]

NOTE I.

ON THE THEORY OF POSITIVE AND NEGATIVE QUANTITIES.

[333] There has been much dispute about the nature of positive and negative quantities, and various theories have been given on this subject. The one we have adopted (see the Preliminaries, pages 2 and 3)[2] appears to us to be the best for clarifying all the difficulties. First we will state it in a few words. Then we will show how we deduce the rule of signs.

Just as we see the idea of number born from the measurement of magnitudes, so also we acquire the idea of quantity (positive or negative) when we consider each magnitude of a given kind as being able to serve as the increase or diminution of another fixed size of the same kind. To indicate this intention, we represent the sizes that ought to serve as increases by numbers preceded by the sign $+$, and the sizes that ought to serve as diminutions by the numbers preceded by the sign $-$. Given this, the signs $+$ or $-$ placed in front of numbers can be compared, following the remark that has been made,[3] with adjectives placed near their nouns. We designate

[1] The last portion of the *Cours d'analyse* is a collection of 9 appendices, which Cauchy calls "Notes." On p. 1 of the introduction [Cauchy 1821, p. ii, Cauchy 1897, p. ii], he describes them as "the derivations which may be useful both to professors and students of the Royal Colleges, as well as to those who wish to make a special study of analysis."

[2] See [Cauchy 1821, pp. 2–3, Cauchy 1897, pp. 18–19]. Curiously, this reference in [Cauchy 1897] was still to pages 2 and 3, even though that edition has no pages numbered 2 or 3. (tr.)

[3] Cauchy's footnote (1) reads "*Transactions philosophiques*, année 1806," a reference is to [Buée 1806]. Abbé Buée (1748–1826) in turn cited Carnot, Frend and Euler, so it seems that Buée, hence

R.E. Bradley, C.E. Sandifer, *Cauchy's Cours d'analyse*, Sources and Studies in the History of Mathematics and Physical Sciences, DOI 10.1007/978-1-4419-0549-9_BM2,

numbers preceded by the sign $+$ as *positive quantities*, and numbers preceded by the sign $-$ as *negative quantities*. Finally, we agree to include the absolute numbers which are not preceded by any sign among the class of positive quantities, and it is for this reason that we sometimes dispense with writing the sign $+$ before the numbers which ought to represent quantities of this kind.

In Arithmetic we always operate on numbers for which the particular value is known, and which are consequently given as figures, while in Algebra, where we consider the general properties of numbers, [334] we ordinarily represent these same numbers by letters. There, a quantity is expressed by a letter preceded by the sign $+$ or $-$. Moreover, nothing prevents representing the quantities by simple letters as well as by numbers. It is an artifice which augments the resources of Analysis, but when we wish to use it, it is necessary to take account of the following conventions.

Following what we have said above, in the case where the letter A represents a number, we can denote the positive quantity for which the numerical value is equal to A either by $+A$ or by A alone, while $-A$ denotes the opposite quantity, that is to say the negative quantity for which A is the numerical value. Thus, in the case where a represents a quantity, we regard the two expressions a and $+a$ as synonyms, and we denote by $-a$ the opposite quantity.

Following these conventions, if we represent either a number or any quantity by A, and if we make

$$a = +A \quad \text{and} \quad b = -A,$$

then we have

$$+a = +A, \qquad +b = -A,$$
$$-a = -A \ \text{ and } -b = +A.$$

In the last four equations, if we replace a and b with their values between parentheses, we get the formulas

$$(1) \qquad \begin{cases} +(+A) = +A, & +(-A) = -A, \\ -(+A) = -A & \text{and } -(-A) = -A. \end{cases}$$

In each of these formulas, the sign of the right-hand side is what we call the product of the two signs of the left-hand side. To *multiply* two signs by each other is to form their product. Inspection alone of equations (1) suffices to establish the *rule of signs*, contained in the theorem which I am going to state.

Theorem I. — *The product of two signs that are the same is always $+$ and the product of two opposite signs is always $-$.*

It also follows from the same equations that when one of the signs is $+$, the product of two signs is equal to the other one. Thus, if we have several signs to

Cauchy, was fully aware of the controversy raging in England at the time, regarding the nature of negative numbers. This paper by Buée is described as "perhaps the very first purely mathematical theory of time" [Windred 1933].

multiply together, we can ignore all the $+$ signs. From this remark, we easily deduce the following propositions:

Theorem II. — *If we multiply together several signs* [335] *in any order, the product is always* $+$ *whenever the number of* $-$ *signs is even, and the product is* $-$ *in the opposite case.*

Theorem III. — *The product of as many signs as we like remains the same, in any order in which we multiply them.*

An immediate consequence of the above definitions is that the multiplication of signs has no relation to the multiplication of numbers. However, we need not be surprised if we note that the idea of the product of two signs arises as one of the first steps that we make in Analysis, because in addition or subtraction of a monomial, we really multiply the sign of this monomial by the sign $+$ or $-$.

Starting from the principles which we have just established, we easily clear up all difficulties which the use of the signs $+$ and $-$ can present in the operations of Algebra and of Trigonometry. We need only distinguish carefully the operations relative to numbers from those which apply to positive or negative quantities. We especially ought to clarify precisely the goals of each kind of operation, to define their results and to describe their principal properties. This is what we are going to try to do in a few words, for the various operations which we commonly use.

Addition and subtraction.

Sums and differences of numbers. — To add the number B to the number A, or in other words, to subject the number A to an increase $+B$ is what we call an *arithmetic addition*. The result of this operation is called the *sum*. We indicate this by placing the increase $+B$ next to the number A, as follows:

$$A+B.$$

We will not prove it, but we admit as evident that *the sum of several numbers remains the same in whatever order we add them*. This is a fundamental axiom on which rest Arithmetic, Algebra and all the sciences of calculation.

Arithmetic subtraction is the inverse of addition. It consists of taking away from a first number A some second number B, that is to say of finding a third number C which added to the second number reproduces the first number. This is also what we call subjecting to the number A the diminution $-B$. The result of this operation is called the *difference*. We indicate it by placing [336] the diminution $-B$ following the number A, as follows:

$$A-B.$$

Sometimes we indicate the difference $A-B$ with the name the *excess* or the *remainder* or the *arithmetic relation* between the two numbers A and B.

Sums and differences of quantities. — We have explained in the preliminaries what it is to add two quantities together. In adding several quantities to each other, we obtain what we call their *sum*. Based on the axiom about the addition of numbers, it is easy to prove the following proposition:

Theorem IV. — *The sum of several quantities remains the same in whatever order we add them.*

We indicate the unique sum of several quantities by the simple juxtaposition either of the letters which represent their numerical values or of the quantities themselves, with each letter preceded by the sign which it must have to express the corresponding quantity. Moreover, the different letters can be arranged in any order, and it is permitted to suppress the $+$ sign before the first letter. Let us consider, for example, the quantities

$$a, \quad b, \quad c, \quad \ldots, \quad -f, \quad -g, \quad -h, \quad \ldots.$$

Their sum could be represented by the expression

$$a-f-g+b-h+c+\ldots.$$

In such an expression, each of the quantities

$$a, \quad b, \quad c, \quad \ldots, \quad -f, \quad -g, \quad -h, \quad \ldots$$

is what we call a *monomial*. The expression itself is a *polynomial*, for which the monomials in question are the different *terms*.

When a polynomial contains only two, three, four, ..., terms, it takes the name *binomial, trinomial, quadrinomial*, ...

We easily prove that two polynomials for which the terms are equal and of contrary signs represent two opposite quantities.

The *difference* between a first quantity and a second is a third quantity which, added to the second, reproduces the first. On the basis of this definition, we prove that *to subtract a second quantity b from a first quantity* [337] *a, it suffices to add the quantity opposite to b, that is to say* $-b$*, to the first quantity.* We thus conclude that the difference of two quantities a and b ought to be represented by

$$a-b.$$

Note. — Subtraction, being the inverse of addition, can always be indicated in two ways. Thus, for example, to express that the quantity c is the difference of two quantities a and b, we can write either

$$a-b=c \quad \text{or} \quad a=b+c.$$

Multiplication and division.

Products and quotients of numbers. — To *multiply* the number A by the number B is to operate on the number A precisely as we operate on one to obtain B. The result of this operation is what we call the *product* of A by B. To better understand the preceding definition of *multiplication*, it is necessary to distinguish different cases, depending on the kind of number B is. This number may be rational, that is to say integer or fractional, or it may be irrational, that is to say not rational.

To obtain B when B is an integer number, it suffices to add one to itself several times consecutively. Thus, to form the product of A by B, we must add the number A to itself the same number of times, that is to say the sum of as many numbers equal to A as there are ones in B.

When B is a fraction which has numerator m and denominator n, the operation by which we arrive at the number B consists of separating the number one into n equal parts and then repeating the result m times. Thus, we obtain the product of A by B by separating the number A into n equal parts and then repeating one of these parts m times.

When B is an irrational number, we can obtain rational numbers that approach it more and more closely. We can easily see that under the same hypothesis the product of A by the rational numbers in question approach a certain limit more and more closely. This limit is the product of A by B. If we suppose, for example, that $B = 0$, we find a zero limit, and we conclude that the product of any number by zero vanishes.

In the multiplication of A by B, we call the number A the *multiplicand* and [338] the number B the *multiplier*. These numbers are also designated together under the name the *factors* of the product.

To indicate the product of A by B, we use any one of the following three notations:

$$B \times A, \quad B \cdot A \quad \text{or} \quad BA.$$

The product of several numbers remains the same in whatever order we multiply them. This proposition, when it concerns just two or three integer factors, is derived from the axiom about the addition of numbers. We can then prove it successively: 1° for two or three rational factors; 2° for two or three irrational factors; and finally 3° for any number of factors, rational or irrational.

To *divide* the number A by the number B is to find a third number for which its product by B is equal to A. The operation by which we arrive at this is called *division* and the result of this operation is the *quotient*. Moreover, the number A takes the name of *dividend* and the number B that of *divisor*.

To indicate the quotient of A by B, we use at will one of the two following notations:

$$\frac{A}{B} \quad \text{or} \quad A : B.$$

Sometimes we indicate the quotient $A : B$ by the name *ratio* or *geometric relation* of the two numbers A and B.

The equality of two geometric ratios $A:B$ and $C:D$, or in other words the equation

$$A : B = C : D$$

is what we call a *geometric proportion.* Ordinarily, instead of the sign = we use the following ::, which has the same meaning, and we write

$$A : B :: C : D.$$

Note. — From the definition, when B is an integer number, to divide A by B is to find a number which, repeated B times, reproduces A. Thus, it is to separate the number A into as many equal parts as there are ones in B. We conclude easily from this remark that if m and n denote two integer numbers, the nth part of one ought to be represented by

$$\frac{1}{n},$$

[339] and the fraction which has numerator m and denominator n by

$$m \times \frac{1}{n}.$$

Indeed, this is the notion by which we naturally denote the fraction in question. However, because we easily prove that the product

$$m \times \frac{1}{n}$$

is equivalent to the quotient of m by n, that is to say to $\frac{m}{n}$, it follows that the same fraction can be represented more simply by the notation

$$\frac{m}{n}.$$

Products and quotients of quantities. — The *product* of a first quantity by a second is a third quantity which has for its numerical value the product of the numerical values of the two others, and for its sign the product of their signs. To *multiply* two quantities by each other is to form their product. The first of the two quantities is called the *multiplier* and the other the *multiplicand*, and the two of them are both factors of the product.

Using these definitions, we easily establish the following proposition:

Theorem V. — *The product of several quantities remains the same in whatever order we multiply them.*

To prove this proposition, it suffices to combine the similar proposition about numbers with theorem III about signs (see above).[4]

[4] [Cauchy 1821, p. 405, Cauchy 1897, p. 335].

To *divide* a first quantity by a second is to find a third quantity which, multiplied by the second, reproduces the first. The operation by which we arrive at this is called *division*. The first quantity is the *dividend*, the second is the *divisor* and the result of the operation is the *quotient*. Sometimes we indicate the quotient by the name of *ratio* or *geometric relation* of the two given quantities. On the basis of the preceding definitions, we easily prove that *the quotient of two quantities has as its numerical value the quotient of their numerical values, and as its sign the product of their signs.*

[340] Multiplication and division of quantities are indicated just like the multiplication and division of numbers.

We say that two quantities are *inverses* of each other when the product of these two quantities is one. From this definition, the quantity a has $\frac{1}{a}$ as its inverse, and reciprocally.

We have remarked above that what we call a *fraction* in Arithmetic is equal to the ratio or quotient of two integer numbers. In Algebra we also denote the ratio or quotient of any two quantities by the name *fraction*. Thus if a and b represent two quantities, their ratio $\frac{a}{b}$ is an algebraic fraction.

Again we observe that division, being an inverse operation of multiplication, can always be indicated in two ways. Thus, for example, to express that the quantity c is the quotient of two quantities a and b, we can write either

$$\frac{a}{b} = c \quad \text{or} \quad a = bc.$$

Products and quotients of numbers enjoy general properties to which we often have recourse. We have already spoken of the one whereby the product remains the same in whatever order we may multiply its factors. Other properties, no less remarkable, are found in the formulas which I am about to write.

Let

$$a, \quad b, \quad c, \quad \ldots, \quad k, \quad a', \quad b', \quad \ldots, \quad a'', \quad b'', \quad \ldots, \quad \ldots$$

be several sequences of quantities, positive or negative. For all possible values of these quantities, we have

$$(2) \qquad \begin{cases} k\,(a+b+c+\ldots) = ka+kb+kc+\ldots, \\ \dfrac{a+b+c+\ldots}{k} = \dfrac{a}{k}+\dfrac{b}{k}+\dfrac{c}{k}+\ldots, \\ \dfrac{a}{b}\times\dfrac{a'}{b'}\times\dfrac{a''}{b''}\times\ldots = \dfrac{aa'a''\ldots}{bb'b''\ldots}, \\ \dfrac{k}{\frac{a}{b}} = \dfrac{bk}{a} = \dfrac{b}{a}\times k. \end{cases}$$

The four preceding formulas give rise to a multitude of consequences [341] which would be too long to list here in detail. We conclude from the third formula, for example: 1° that the fractions

$$\frac{a}{b} \quad \text{and} \quad \frac{ka}{kb}$$

are equal to each other, where a, b and k denote any quantities; 2° that the fraction $\frac{a}{b}$ has $\frac{b}{a}$ as its inverse; and 3° that to divide one quantity k by another quantity a, it suffices to multiply k by the inverse of a, that is to say by $\frac{1}{a}$.

Elevation of powers. Extraction of roots.

Powers and roots of numbers. Positive exponents. — To *raise* the number A to the power indicated by the number B is to look for a third number which is formed from A by multiplication as B is formed from one by addition. The result of this operation made on the number A is what we call its power of *degree* B. To understand the preceding definition of the elevation to powers well, it is necessary to distinguish three cases, depending on whether the number B is integer, fractional or irrational.

When B denotes an integer number, this number is the sum of several ones. The power of A of degree B thus ought to be the product of as many factors equal to A as there are ones in B.

When B represents a fraction $\frac{m}{n}$ (m and n being two integer numbers), to represent this fraction it is necessary: 1° to find a number which, repeated n times, produces one; and 2° repeat the number in question m times. Thus, to obtain the power of A of degree $\frac{m}{n}$, it is necessary: 1° to find a number such that the multiplication of n factors equal to this number reproduces A; and 2° to form a product of m factors equal to this same number. When we suppose in particular that $m = 1$, the power of A under consideration reduces to that of degree $\frac{1}{n}$, and it is found to be determined by the single condition that the number A be equivalent to the product of n factors equal to this same power.

When B is an irrational number, we can then obtain rational numbers with values approaching it more and more closely. We easily prove that under the same hypothesis, powers of A indicated by the rational [342] numbers in question approach more and more closely towards a certain limit. This limit is the power of A of degree B.

In the elevation of the number A to the power of degree B, the number A is called the *root* and the number B, which indicates the degree of the power, is the *exponent*. To represent the power of A of degree B, we use the following notation

$$A^B.$$

From the preceding definitions, the first power of a number is nothing but the number itself. Its second power is the product of two factors equal to this number, its third power three such factors, and so on. Geometric considerations have led us to indicate the second power by the name *square* and the third power by the name *cube*. As for the power of degree zero, it is the limit towards which the power of degree B converges when the number B decreases indefinitely. It is easy to show that this limit reduces to one, from which it follows that we have, in general,

$$A^0 = 1.$$

We always suppose that the value of the number A remains finite and different from zero.

To *extract the root* indicated by the number B of a number A is to find a third number which, raised to the power of degree B produces A. The operation by which we accomplish this is called *extraction* and the result of the operation is the root of A of *degree* B. The number B which indicates the degree of the root is called the *index.* To represent it, we use the following notation:

$$\sqrt[B]{A}.$$

The roots of second and third degree are ordinarily indicated by the names *square roots* and *cube roots*. When it is a matter of a square root, we almost always dispense with writing the index 2 along with the sign $\sqrt{}$. Thus the two notations

$$\sqrt[2]{A} \quad \text{and} \quad \sqrt{A}$$

ought to be considered as equivalent.

Note. — The extraction of roots of numbers, being the inverse of their elevation to powers, can always be indicated in two ways. Thus, for example, to express that the number C is equal to the root of A of [343] degree B, we can write either

$$A = C^B \quad \text{or} \quad C = \sqrt[B]{A}.$$

We remark again that, by virtue of these definitions, if we denote any integer number by n, then $A^{\frac{1}{n}}$ is a number such that the multiplication of n factors equal to this number produces A. In other words, we have

$$\left(A^{\frac{1}{n}}\right)^n = A,$$

from which we conclude that

$$A^{\frac{1}{n}} = \sqrt[n]{A}.$$

Thus, when n is an integer number, the power of A of degree $\frac{1}{n}$, and the nth root of A are equivalent expressions. We prove easily that it is the same in the case where we replace the integer number n by any number.

Powers of numbers. Negative exponents. — To *raise* the number A to the *power* indicated by the *negative exponent* $-B$ is to divide one by A^B. The value of the expression

$$A^{-B}$$

is thus found to be determined by the equation

$$A^{-B} = \frac{1}{A^B},$$

which we can also put into the form

$$A^B A^{-B} = 1.$$

Consequently, if we raise the same number to two powers indicated by two opposite quantities, we obtain as results two positive quantities that are inverses to each other.

Powers and real roots of quantities. — In the definitions which we have given of powers and roots of numbers corresponding to exponents, either integer or fractional, if we substitute the word *quantities* in place of *numbers*, we obtain the following definitions for powers and real roots of quantities.

To *raise* the quantity a to the *real power of degree* m, where m is an integer number, [344] is to form the product of as many factors equal to a as there are ones in m.

To *raise* the quantity a to the *real power of degree* $\frac{m}{n}$, where m and n are two integer numbers and, to avoid all uncertainty, where the fraction $\frac{m}{n}$ is reduced to its simplest expression, is to form a product of m factors chosen so that the nth power of each of them is equal to the quantity a.

To *extract* the *real root of degree* m or $\frac{m}{n}$ of the quantity a is to find a new quantity which, raised to the real power of degree m or $\frac{m}{n}$ produces a. From this definition, the nth real root of a quantity is evidently the same thing as its real power of degree $\frac{1}{n}$. Moreover, we easily prove that the root of degree $\frac{n}{m}$ equals the power of degree $\frac{m}{n}$.

Finally, to *raise* the quantity a to the *real power of degree* $-m$ or $-\frac{m}{n}$ is to divide one by the same quantity a raised to the real power of degree m or $\frac{m}{n}$.

In these operations of which we have just spoken, the number or the quantity which marks the degree of a real power of a is called the *exponent* of this power, while the number which marks the degree of a real root is named the *index* of this root.

Every power of a which corresponds to an exponent for which the numerical value is an integer, that is to say to an exponent of the form $+m$ or $-m$, where m represents an integer number, admits a unique real value which we denote by the notation

$$a^m \quad \text{or} \quad a^{-m}.$$

As for the roots and powers for which the numerical value is fractional, they can admit either two real values, or but one real value, or admit none at all. The real values in question here are necessarily either positive quantities or negative quantities. However, in Algebra, in addition to these quantities we also use symbols which have no meaning by themselves, but nevertheless receive the names *powers* and *roots* because of their properties. These symbols [345] are among the algebraic expressions to which we have given the name *imaginary*, as opposed to the name *real expressions*, which only applies to numbers or quantities.

Given this, it follows from the principles established in Chapter VII that the nth root of any quantity a and its powers of degree $\frac{m}{n}$ and $-\frac{m}{n}$, where n is an integer number and $\frac{m}{n}$ is an irreducible fraction, each of which admits n distinct values, real or imaginary. Conforming to the notations adopted in the same chapter, we denote any one of these values, if it is a question of the nth root, by the notation

$$\sqrt[n]{a} = ((a))^{\frac{1}{n}},$$

and if it is a question of the power which has for its exponent $\frac{m}{n}$ or $-\frac{m}{n}$, by the notation

$$((a))^{\frac{m}{n}} \quad \text{or} \quad ((a))^{-\frac{m}{n}}.$$

We add that the expression $((a))^{\frac{1}{n}}$ is contained as a particular case of the more general expression $((a))^{\frac{m}{n}}$. By calling A the numerical value of a, we find that the real values of the two expressions

$$((a))^{\frac{m}{n}} \quad \text{and} \quad ((a))^{-\frac{m}{n}} \quad \text{are:}$$

1° If n denotes an odd number and

$$a \text{ is } +A \ldots\ldots\ldots\ldots +A^{\frac{m}{n}} \text{ and } +A^{-\frac{m}{n}},$$
$$a \text{ is } -A \ldots\ldots\ldots\ldots -A^{\frac{m}{n}} \text{ and } -A^{-\frac{m}{n}};$$

2° If n denotes an even number and

$$a \text{ is } +A \ldots\ldots\ldots\ldots \pm A^{\frac{m}{n}} \text{ and } \pm A^{-\frac{m}{n}}.$$

In the last case, when we suppose that a is negative, all the values of each of the expressions $((a))^{\frac{m}{n}}$ and $((a))^{-\frac{m}{n}}$ become imaginary.

If we make the fraction $\frac{m}{n}$ vary in such a way that it approaches indefinitely an irrational number B, the denominator n then grows beyond any assignable limit, and likewise the number of imaginary values [346] which each of the expressions

$$((a))^{\frac{m}{n}} \quad \text{and} \quad ((a))^{-\frac{m}{n}}$$

take on. Consequently, we cannot admit into calculation the notations

$$((a))^{B} \quad \text{and} \quad ((a))^{-B}$$

or the notation

$$((a))^{b},$$

when we make $b = \pm B$, unless we consider such a notation itself as representing an infinity of imaginary expressions. To avoid this inconvenience, we never employ the algebraic expression

$$((a))^{b}$$

in the case where the numerical value of b is irrational. Under this hypothesis, only when a takes a positive value $+A$ can we make use of the notation

$$a^b \quad \text{or} \quad (a)^b,$$

which we ought to consider as equivalent to

$$+A^b$$

(see Chapter VII, § IV).

Powers of numbers and quantities enjoy several remarkable properties which are easy to prove. Among others, we note those contained in the formulas which I am going to write.

Let a, a', a'', ..., b, b', b'', ... be any quantities, positive or negative,. Let A, A', A'', ... be any numbers and let m, m', m'', ... be integer numbers. We have

$$(3) \qquad \begin{cases} A^b A^{b'} A^{b''} \ldots = A^{b+b'+b''+\ldots}, \\ A^b A'^b A''^b \ldots = (AA'A''\ldots)^b, \\ \left(A^b\right)^{b'} = A^{b'b}, \end{cases}$$

and

$$(4) \qquad \begin{cases} a^{\pm m} a^{\pm m'} a^{\pm m''} \ldots = a^{\pm m \pm m' \pm m'' \pm \ldots}, \\ a^m a'^m a''^m \ldots = (aa'a''\ldots)^m, \\ a^{-m} a'^{-m} a''^{-m} \ldots = (aa'a''\ldots)^{-m}, \\ (a^m)^{m'} = (a^{-m})^{-m'} = a^{mm'} \quad \text{and} \\ (a^m)^{-m'} = (a^{-m})^{m'} = a^{-mm'}, \end{cases}$$

where each of the numbers m, m', m'', ... in the first equation (4) must be affected with the same sign on both sides of the equation. [347] Formulas (3) and (4) give rise to a multitude of consequences, among which we will content ourselves to indicate the following. We get from the second formula (3) that

$$A^b \left(\frac{1}{A}\right)^b = 1^b = 1,$$

and we then conclude that

$$\left(\frac{1}{A}\right)^b = \frac{1}{A^b}.$$

Thus, if we raise two positive quantities that are inverses to each other to the same power, the results are always two inverse quantities.

Formation of exponentials and logarithms.

When we regard the number A as fixed and the quantity x as a variable in the expression A^x, the power A^x takes the name *exponential.* Under the same hypothesis, if for a particular value of x we have

$$A^x = B,$$

then this particular value is what we call the *logarithm* of the number B in the system for which the *base* is A. We indicate this logarithm by placing before the number the initial letters ln or log, like[5]

$$\ln B \quad \text{or} \quad \log B.$$

However, as such a notation does not tell the base of the system of logarithms to which it refers, it is important to state in the discussion the value of this base. Given this, if we use the characteristic log to denote logarithms taken in the system for which the base is A, the equation

$$A^x = B$$

implies the following one

$$x = \log B.$$

Sometimes, when we must treat logarithms taken in different systems at the same time, we distinguish among them with the aid of one of several accents placed to the right of the letters log, and as a consequence we denote by these letters without accents the logarithms of a first system, by the same letters followed by a single accent logarithms of a second system, etc.

Based on the preceding definitions and on the general properties of powers of numbers, we easily recognize: 1° that one [348] has zero for its logarithm in all systems; 2° that in any system of logarithms for which the base exceeds one, every number greater than one has a positive logarithm, and every number less than one has a negative logarithm; 3° that in any system of logarithms for which the base is less than one, every number less than one has a positive logarithm and every number greater than one has a negative logarithm; and finally 4° that in two systems for which the bases are inverses to one another, the logarithms of the same number are equal and of contrary signs. Moreover, we easily prove the formulas which establish the principal properties of logarithms, among which we ought to note these which I am going to write.

If we denote by $B, B', B'', \ldots, C$ any numbers, by the characteristics log and log$'$ the logarithms taken in two different systems for which the bases are A and A', and by k any quantity, positive or negative, we have

---

[5] As mentioned in the Preface and in a footnote in the *Preliminaries*, we use the more modern notations "ln" and "log" to avoid confusion, whereas Cauchy used "*l*" and "L", respectively. If Cauchy means the natural logarithm, we always use "ln." (tr.)

$$(5)\qquad \begin{cases} \log BB'B'' \ldots = \log B + \log B' + \log B'' + \ldots, \\ \log B^k = k \log B, \\ B^{\log C} = A^{\log B \cdot \log C} = C^{\log B}, \\ \dfrac{\log C}{\log B} = \dfrac{\log' C}{\log' B}. \end{cases}$$

From the first of these formulas, we get

$$\log B + \log \frac{1}{B} = \log 1 = 0$$

and consequently

$$\log \frac{1}{B} = -\log B.$$

From this it follows that two positive quantities that are inverse to each other have equal logarithms of contrary signs. We add that the fourth formula can be deduced easily from the second. Indeed, suppose that the quantity k represents the logarithm of the number C in the system for which the base is B. We have

$$C = B^k$$

and consequently

$$\log C = k \log B \quad \text{and} \quad \log' C = k \log' B,$$

from which we conclude immediately that

$$\frac{\log C}{\log B} = \frac{\log' C}{\log' B} = k.$$

[349] We can also remark that if we take $B = A$, then because $\log A = 1$, we get from the fourth formula that

$$\log' C = \log' A \cdot \log C,$$

or, taking for brevity $\log' A = \mu$,

$$\log' C = \mu \log C.$$

Thus, to pass from a system of logarithms for which the base is A to one for which the base is A', it suffices to multiply the logarithms taken in the first system by a certain coefficient μ equal to the logarithm of A taken in the second system.

The logarithms of which we have just spoken are those which we call *real logarithms* because they always reduce to positive or negative quantities. However, other than these quantities, there exist imaginary expressions which, because of their properties, also bear the name of *logarithms*. We return to this subject in Chapter IX, in which we reveal the theory of imaginary logarithms.

Formation of trigonometric lines and arcs of a circle.

We have remarked in the Preliminaries that a length measured on a curved or straight line can sometimes be represented by a number, sometimes by a quantity, depending on whether we simply regard it as the measure of this length, or if we consider it as being moved along the given line in one sense or another, relative to a fixed point which we call the *origin,* to serve as the growth or diminution of another constant length ending at this point. We have added that in a circle for which the plane is taken to be vertical, we ordinarily fix the origin of the arcs as the endpoint of the radius taken horizontally from left to right, and that, with respect to this origin, the arcs are counted as positive or negative depending on whether, to describe them, we begin by going up from there or by going down. Finally, we have indicated the origins of several trigonometric lines which correspond to these same arcs in the case where the radius of the circle is reduced to one. We will return to this topic shortly and complete the ideas which pertain to it.

First, we easily establish with regard to lengths measured on the same line or curve relative to a given origin the following propositions:

[350] **Theorem VI.** — *Let a, b, c, ... be any quantities, positive or negative. To obtain on a line, straight or curved, the extremity of the length*

$$a+b+c+\ldots$$

measured with respect to a given origin and in the direction determined by the sign of the quantity

$$a+b+c+\ldots,$$

it suffices to move along this line: 1° *the length a starting from the origin in the direction determined by the sign of a;* 2° *the length b starting from the extremity of a in the direction determined by the sign of b; and* 3° *the length c starting from the extremity of b in the direction determined by the sign of c, and so on.*

Theorem VII. — *Let a and b be any two quantities. Suppose also that we move along a straight line or curve starting from a given origin:* 1° *a length equal to the numerical value of a in the direction determined by the sign of a; and* 2° *a length equal to the numerical value of b in the direction determined by the sign of b. To pass from the extremity of the first length to that of the second, or reciprocally, along the line under consideration, it suffices to move a third length equal to the numerical value of the difference* $a-b$.

Theorem VIII. — *Supposing the same things as in the preceding theorem, the extremity of the length represented by*

$$\frac{a+b}{2}$$

is situated on the given line at a point at equal distances from the extremities of the lengths a and b (where the distances are measured along the line itself).

Now we apply these theorems to arcs measured on the circumference of a circle for which the plane is vertical and for which the radius equals one, the origin of the arcs being fixed at the extremity of the radius drawn horizontally from left to right. If we denote the ratio of the circumference to its diameter by π, following common usage, because the diameter is equal to 2, the entire circumference is found to be expressed by the number 2π, half of the circumference by the number π, and the quarter by $\frac{\pi}{2}$. Moreover, if we denote by a any arc, [351] positive or negative, we conclude from theorem VI that, to obtain the extremity of the arc

$$a+2m\pi \quad \text{or} \quad a-2m\pi,$$

(where m is an integer number), it is necessary to move along the circumference, starting with the extremity of the arc a, either in the direction of the positive arcs or in the direction of the negative arcs, a length equal to $2m\pi$, that is to say to travel m times about the entire circumference in one direction or the other, which necessarily returns to the point from which we started. It follows that the extremities of the arcs

$$a \quad \text{and} \quad a \pm 2m\pi$$

coincide.

Likewise we conclude from theorems VI or VII: 1° that the extremities of the arcs

$$a \quad \text{and} \quad a \pm \pi$$

contain between themselves an arc equal to π and as a consequence they consist of the extremities of the same diameter; and 2° that the extremities of the arcs

$$a \quad \text{and} \quad a \pm \frac{\pi}{2}$$

contain between themselves a quarter of the circumference, and so they coincide with the endpoints of two radii perpendicular to each other.

Finally, we conclude from theorem VIII: 1° that the extremities of the arcs

$$a \quad \text{and} \quad \pi - a$$

are located at equal distances from the extrtemity of the arc

$$\frac{\pi}{2},$$

and as a consequence are placed symmetrically about the vertical diameter; and 2° that the extremities of the arcs

$$a \quad \text{and} \quad \frac{\pi}{2} - a$$

are situated at equal distances from the extremity of the arc

$$\frac{\pi}{4}.$$

[352] The arcs

$$\pi - a \quad \text{and} \quad \frac{\pi}{2} - a$$

in question here are, respectively, called the *supplement* and the *complement* of the arc a. In other words, two arcs represented by two quantities a and b are *supplements* or *complements* of each other depending on whether we have

$$a+b=\pi \quad \text{or} \quad a+b=\frac{\pi}{2}.$$

Because angles at the center which have for a common side the radius taken as the origin of the arcs grow or diminish proportionally with the arcs which they serve to measure, and because these angles themselves can be considered as the increases or decreases of one of these taken at will, nothing prevents us from denoting angles by the same quantities as arcs. This is a convention which has been effectively adopted. We also say that two angles are complements or supplements of each other when the corresponding arcs are themselves complements or supplements of each other.

Now we move on to the study of trigonometric lines, and towards this end we consider a single arc represented by the quantity a. If we project it successively: 1° on the vertical diameter; and 2° on the horizontal diameter, the two projections are what we call the *sine* and the *versed sine* of the arc a.[6] We can observe that the first of these is at the same time the projection on the vertical diameter of the radius which passes through the extremity of the arc. If we prolong this same radius until it intersects the tangent of the circle taken from the origin of the arcs, the part of this tangent contained between the origin and the point of intersection is what we call the trigonometric *tangent* of the arc a. Finally, the length measured on the radius extended between the center and the point of intersection is the *secant* of this same arc.

The *cosine* and *versed cosine* of an arc, its *cotangent* and its *cosecant* are nothing but the sine, versed sine, tangent and secant of its complement, and they constitute, along with the sine, the versed sine, the tangent and the secant of the same arc, the complete system of *trigonometric lines.*

From what has been said above, the sine of an arc is measured on the vertical diameter, the versed sine on the horizontal diameter, the tangent on the line which touches the circle at the origin of the arcs, and the secant on the moving diameter which passes through the extremity of the given arc. Moreover, the sine and the secant have for their common origin the center of the circle, while the origin [353] of the tangents and versed sines correspond to that of the arcs. Finally, we generally agree to represent by positive quantities the trigonometric lines of the arc a in the

[6] Some readers may have expected the projection onto the horizontal axis to be the cosine, but the cosine is the projection of the *radius*, not the *arc*. For a more complete account of the versed sine and other topics in the history of trigonometry, see [Van Brummelen 2009, Ch. 3–4].

case where the arc is positive and less than a quarter of the circumference, from which it follows that we ought to measure the sine and the tangent positively from the base upwards, the versed sine from right to left, and the secant in the direction of the radius towards the extremity of the arc a.

On the basis of the principles which we have just adopted, we immediately recognize that the versed sine, and consequently the versed cosine, are always positive, and moreover, we determine without trouble the signs which ought to affect the other trigonometric lines of an arc for which the endpoint is given. To make this determination easier, we imagine the circle divided into four equal parts by two diameters perpendicular to each other, one horizontal and the other vertical, and these four parts of the circle are, respectively, designated as the first, second, third and fourth quarters of the circle. The first two quarters of the circle are situated above the horizontal diameter, namely the first on the right and the second on the left. The last two are situated below the same diameter, namely the third on the left and the fourth on the right. Given this, because the extremities of two arcs that are complements of each other are equally distant from the extremity of the arc $\frac{\pi}{4}$, we conclude that they are placed symmetrically on either side of the diameter which divides the first and the third quarters of the circle into two equal parts. If we then look for what signs ought to be attributed to the various trigonometric lines of an arc other than the versed sign and the versed cosine, according to whether the extremity of this arc falls in one quarter of the circle or in another, we find that the signs are, respectively,

	In the 1st quarter of the circle	In the 2nd quarter of the circle	In the 3rd quarter of the circle	In the 4th quarter of the circle
For sine and cosecant	$+$	$+$	$-$	$-$
For cosine and secant	$+$	$-$	$-$	$+$
For tangent and cotangent	$+$	$-$	$+$	$-$

On this subject, we can remark that the sign of the tangent is always the product of the sign of the sine by the sign of the cosine.

The preceding considerations now lead us to recognize that the cosine of an arc corresponds with the projection of the radius which passes through the extremity of this arc onto the horizontal diameter, and that on this diameter it ought to be measured positively from left to right, starting from the center taken as the origin. The versed cosine can be measured on the vertical diameter [354] from the highest point on the circumference taken as the origin to the endpoint of the sine. The cotangent, measured positively from left to right along the horizontal tangent to the circle at the origin of the versed cosines, reduces to the length contained between this origin and the extension of the moving diameter the half of which is the radius taken to the extremity of the arc. Finally the cosecant, measured along the moving diameter, is measured positively in the direction of the radius in question and starting from the center taken as the origin to the extremity of the cotangent.

In the preliminaries we have sufficiently developed the system of notations used to represent the various trigonometric lines and the arcs to which they correspond. We shall not return to this subject, and we will content ourselves to observe that the trigonometric lines of an arc are at the same time supposed to belong to the angle at the center of the circle which it measures and which we designate by the same quantity. Thus, for example, if a, b, ... represent any quantities, we can say that the notations

$$\sin a, \quad \cos b, \quad \dots$$

express equally the sine of the arc or of the angle a, the cosine of the arc or of the angle b, ...

We end this note by recalling some remarkable properties of trigonometric lines.

First, if we denote by a any quantity, we find that the sine and the cosine of the angle a are always related to each other by the equation

$$\sin^2 a + \cos^2 a = 1, \tag{6}$$

and that the other trigonometric lines can be expressed by means of these first two as follows:

$$\left\{\begin{aligned} \operatorname{siv} a &= 1 - \cos a, \ \tan a = \tfrac{\sin a}{\cos a}, \ \sec a = \tfrac{1}{\cos a}, \\ \operatorname{cosiv} a &= 1 - \sin a, \ \cot a = \tfrac{\cos a}{\sin a}, \ \csc a = \tfrac{1}{\sin a}. \end{aligned}\right. \tag{7}$$

From formulas (6) and (7) we easily deduce several other equations, for example

$$\cot a = \frac{1}{\tan a}, \quad \sec^2 a = 1 + \tan^2 a, \quad \csc^2 a = 1 + \cot^2 a, \quad \dots. \tag{8}$$

It is also easy to see that if the positive quantity R represents the length [355] of a straight line between two points and α represents the angle, acute or obtuse, formed by this straight line with a fixed axis, the projection of the given length on the fixed axis is measured by the numerical value of the product

$$R \cos \alpha,$$

and the projection of the same length on a perpendicular axis is measured by the numerical value of the product

$$R \sin \alpha.$$

Finally, we recognize without trouble that if by starting from a point taken at random on the circumference of a circle of radius one, we move along this circumference in one direction or the other a length equal to the numerical value of any quantity c, the smallest arc contained between the endpoints of this length is less than or greater than $\frac{\pi}{2}$, depending on whether $\cos c$ is positive or negative.

Admitting these principles, imagine that on the circumference of which we speak we determine: 1° the extremities A and B of the arcs represented by any two quantities a and b; and 2° the extremity N of a third arc represented by $\frac{a+b}{2}$.[7] In addition, let M be the midpoint of the chord which joins the points A and B, and suppose that the point M projects onto the horizontal diameter of the circle to a certain point P. If the lengths measured on the diameter starting from the center taken for the origin are counted positively from left to right, like cosines, the distance from the center to the point P ought to be represented (by virtue of theorem VIII) by the quantity

$$\frac{\cos a + \cos b}{2}.$$

Moreover, because (by virtue of the same theorem) the point N is situated at equal distances from the points A and B, the diameter which passes through the point N contains the midpoint M of the chord $\overline{AB}$ and the distance from this midpoint M to the center of the circle is equal (ignoring the sign) to the cosine of each of the arcs $\overline{NA}$ and $\overline{NB}$, or what amounts to the same thing, to

$$\cos\left(\frac{a+b}{2} - a\right) = \cos\left(\frac{a+b}{2} - b\right) = \cos\frac{a-b}{2}.$$

To obtain the horizontal projection of this distance, it suffices to multiply it by the cosine of the acute angle contained between the radius taken horizontally [356] from left to right and the diameter which contains the point N, that is to say by a factor equal (up to sign) by $\cos\frac{a+b}{2}$. In other words, the distance from the center to the point P has for its measure the numerical value of the product

$$\cos\frac{a-b}{2}\cos\frac{a+b}{2}.$$

I add that this product is positive or negative according to whether the point M is situated to the right or to the left of the vertical diameter. Indeed, $\cos\frac{a+b}{2}$ is positive or negative according to whether the point N is situated to the right side or the left side with respect to this diameter. Also, $\cos\frac{a-b}{2}$ is positive or negative – and consequently the product

$$\cos\frac{a-b}{2}\cos\frac{a+b}{2}$$

is of the same sign as $\cos\frac{a+b}{2}$ or of the opposite sign – according to whether each of the arcs $\overline{NA}$ and $\overline{NB}$ is less than or greater than $\frac{\pi}{2}$, which in turn follows whether the point M is situated on the same side as the point N or on the opposite side. Moreover, because the vertical line which passes through the point M also contains the point P, it follows from the preceding remark that the distance from the center to the point P, even in the case where we pay attention to the signs, can be represented by the

[7] At this point Cauchy is embarking on a delicate argument which, following the example of Lagrange, he is determined to carry out without the aid of diagrams. The reader who wishes to follow this argument carefully should note that the arcs a and b uniquely determine the point N, although their extremities A and B do not. However, Cauchy will soon show that the additional information of the sign of $\cos(\frac{a-b}{2})$ suffices to determine N.

product

$$\cos\frac{a-b}{2}\cos\frac{a+b}{2}.$$

Thus, this product and the quantity $\frac{\cos a+\cos b}{2}$ have the same sign as well as the same numerical value, and we have, as a consequence, for all possible values of the quantities a and b,

$$\cos a+\cos b=2\cos\frac{a-b}{2}\cos\frac{a+b}{2}. \tag{9}$$

If we replace b by $b+\pi$ in equation (9), we get

$$\cos a-\cos b=2\sin\frac{b-a}{2}\sin\frac{a+b}{2}. \tag{10}$$

Moreover, if in equations (9) and (10) we substitute for the angles a and b their [357]
complements $\frac{\pi}{2}-a$ and $\frac{\pi}{2}-b$, we obtain the following:

$$\left\{\begin{aligned} \sin a+\sin b &= 2\cos\tfrac{a-b}{2}\sin\tfrac{a+b}{2},\\ \sin a-\sin b &= 2\sin\tfrac{a-b}{2}\cos\tfrac{a+b}{2}. \end{aligned}\right. \tag{11}$$

Once formulas (9), (10) and (11) are established, we then easily deduce a great number of others. We find, for example,

$$\left\{\begin{aligned} &\frac{\sin a-\sin b}{\sin a+\sin b}=\frac{\tan\frac{1}{2}(a-b)}{\tan\frac{1}{2}(a+b)},\\ &\frac{\cos b-\cos a}{\cos b+\cos a}=\tan\tfrac{1}{2}(a-b)\tan\tfrac{1}{2}(a+b), \end{aligned}\right. \tag{12}$$

$$\left\{\begin{aligned} &\cos(a-b)+\cos(a+b)=2\cos a\cos b,\\ &\cos(a-b)-\cos(a+b)=2\sin a\sin b, \end{aligned}\right. \tag{13}$$

$$\left\{\begin{aligned} &\sin(a+b)+\sin(a-b)=2\sin a\cos b,\\ &\sin(a+b)-\sin(a-b)=2\sin b\cos a, \end{aligned}\right. \tag{14}$$

$$\left\{\begin{aligned} &\cos(a\pm b)=\cos a\cos b\mp\sin a\sin b,\\ &\sin(a\pm b)=\sin a\cos b\pm\sin b\cos a, \end{aligned}\right. \tag{15}$$

$$\tan(a\pm b)=\frac{\tan a\pm\tan b}{1\mp\tan a\tan b}, \tag{16}$$

$$\left\{\begin{aligned} &\cos 2a=\cos^2 a-\sin^2 a=2\cos^2 a-1=1-2\sin^2 a,\\ &\sin 2a=2\sin a\cos a. \end{aligned}\right. \tag{17}$$

Now let a, b and c be any three angles. From the first formula (13), we get

$$(18)\qquad \left\{\begin{array}{l}\cos(a+b+c)+\cos(b+c-a)+\cos(c+a-b)\\ \quad+\cos(a+b-c)=4\cos a\cos b\cos c.\end{array}\right.$$

In the preceding formula, if we write $\frac{1}{2}a$, $\frac{1}{2}b$ and $\frac{1}{2}c$, instead of a, b and c and then we suppose that

$$(19)\qquad a+b+c=\pi,$$

we find

$$(20)\qquad \sin a+\sin b+\sin c=4\cos\frac{a}{2}\cos\frac{b}{2}\cos\frac{c}{2}.$$

[358] Under the same hypothesis, formula (16) gives

$$(21)\qquad \tan a+\tan b+\tan c=\tan a\tan b\tan c.$$

Equation (20) ought to remain true, along with equation (19), when we replace two of the angles a, b and c with their supplements, and then change the sign of the third one. Then we conclude

$$(22)\qquad \left\{\begin{array}{l}\sin b+\sin c-\sin a=4\cos\frac{a}{2}\sin\frac{b}{2}\sin\frac{c}{2},\\ \sin c+\sin a-\sin b=4\sin\frac{a}{2}\cos\frac{b}{2}\sin\frac{c}{2},\\ \sin a+\sin b-\sin c=4\sin\frac{a}{2}\sin\frac{b}{2}\cos\frac{c}{2}.\end{array}\right.$$

Combining these last formulas with equation (20), we deduce the following:

$$(23)\qquad \left\{\begin{array}{l}\cos^2\frac{1}{2}a=\dfrac{(\sin a+\sin b+\sin c)(\sin b+\sin c-\sin a)}{4\sin b\sin c},\\ \sin^2\frac{1}{2}a=\dfrac{(\sin c+\sin a-\sin b)(\sin a+\sin b-\sin c)}{4\sin b\sin c}.\end{array}\right.$$

Finally, if we imagine that a, b and c denote the three angles of a triangle and that their opposite sides are, respectively, A, B and C, the six products, equal in pairs, namely

$$B\sin c=C\sin b,\quad C\sin a=A\sin c\quad \text{and}\quad A\sin b=B\sin A,$$

represent the perpendiculars dropped from the vertices to the three sides. It follows that we have

$$(24)\qquad \frac{\sin a}{A}=\frac{\sin b}{B}=\frac{\sin c}{C},$$

and equations (23) become

$$(25)\qquad \left\{\begin{array}{l}\cos^2\frac{1}{2}a=\dfrac{(A+B+C)(B+C-A)}{4BC},\\ \sin^2\frac{1}{2}a=\dfrac{(C+A-B)(A+B-C)}{4BC}.\end{array}\right.$$

Moreover, by taking into consideration formulas (19) and (24), we get from the first
[359] equation (12)

$$\tan\tfrac{1}{2}(a-b) = \frac{A-B}{A+B}\cot\tfrac{1}{2}c. \tag{26}$$

Formulas (19), (24), (25) and (26) suffice to determine three of the six elements of a rectilinear triangle when the other three elements are known and when this determination is possible. We can also remark that the values of $\cos a$ and $\sin a$, deduced from equations (25) with the aid of formulas (17), are, respectively,[8]

$$\begin{cases} \cos a = \dfrac{B^2+C^2-A^2}{2BC} \quad \text{and} \\ \sin a = \dfrac{\sqrt{(A+B+C)(B+C-A)(C+A-B)(A+B-C)}}{2BC}. \end{cases} \tag{27}$$

The first of these values can be drawn directly from a known theorem of Geometry. As for the second, it gives a means of expressing the area of a triangle as a function of its three sides. Indeed, this area, equal to the product of the base C by half the height corresponding to $B\sin a$ is[9]

$$\tfrac{1}{2}BC\sin a = \tfrac{1}{4}\sqrt{(A+B+C)(B+C-A)(C+A-B)(A+B-C)}. \tag{28}$$

[8] The first formula in (27) is the Law of Cosines.

[9] Formula (28) is known as Heron's formula. The area is often given as $\sqrt{S(S-A)(S-B)(S-C)}$ where $S = (A+B+C)/2$ is the semiperimeter of the triangle.

Note II – On formulas that result from the use of the signs > or <, and on the averages among several quantities.

[360] Let a and b be two unequal quantities. The two formulas

$$a > b \quad \text{and} \quad a < b$$

serve equally to express that the first quantity, a, surpasses the second, b, that is to say that the difference

$$a - b$$

is positive. On the basis of this principle, we easily establish the propositions that I am going to state:

Theorem I. — *If a, a', a'', ..., b, b', b'', ... represent quantities subject to the conditions*

$$\begin{gathered} a > b, \\ a' > b', \\ a'' > b'', \\ \ldots\ldots, \end{gathered}$$

then we also have

$$a + a' + a'' + \ldots > b + b' + b'' + \ldots.$$

Proof. — Indeed, when the quantities

$$a - b, \quad a' - b', \quad a'' - b'', \quad \ldots$$

are positive, we can be sure that their sum

$$a + a' + a'' + \ldots - (b + b' + b'' + \ldots)$$

is positive as well.

Theorem II. —*If A, A', A'', ..., B, B', B'', ... represent numbers* [361] *subject to the conditions*

$$A > B,$$
$$A' > B',$$
$$A'' > B'',$$
$$\dots\dots\dots,$$

then we also have

$$AA'A''\dots > BB'B''\dots.$$

Proof. — Indeed, because each of the differences is positive by hypothesis,

$$A-B, \quad A'-B', \quad A''-B'', \quad \dots,$$

each of the products

$$\begin{aligned}(A-B)A'A''\dots &= AA'A''\dots - BA'A''\dots,\\ B(A'-B')A''\dots &= BA'A''\dots - BB'A''\dots,\\ BB'(A''-B'')\dots &= BB'A''\dots - BB'B''\dots,\\ \dots\dots\dots\dots&\dots\dots\dots\dots\dots,\end{aligned}$$

is positive as well, and consequently, so is their sum

$$AA'A''\dots - BB'B''\dots.$$

Theorem III. — *Let a, b and r be any three quantities and suppose that*

$$a > b.$$

We then conclude that if r is positive, then

$$ra > rb,$$

and if r is negative, then

$$ra < rb.$$

Proof. — Indeed, the product

$$r(a-b) = ra - rb$$

is positive in the first case and negative in the second.

Corollary. – Suppose that a and b are positive. If we successively take

$$r = \frac{1}{a} \quad \text{and} \quad r = \frac{1}{b},$$

we then conclude that

$$1 > \frac{b}{a} \quad \text{and} \quad \frac{a}{b} > 1.$$

We are thus brought back to the proposition, obvious by itself, [362] that a fraction is less than or greater than 1 according to whether the larger of its two terms is its denominator or its numerator.

Theorem IV. — *Let A and A′ be two numbers that satisfy the condition*

$$A > A',$$

and let b be any quantity. If b is positive we have

$$A^b > A'^b,$$

and if b is negative,

$$A^b < A'^b.$$

Proof. — Indeed, because the quotient $\frac{A}{A'}$ is > 1, the fraction

$$\frac{A^b}{A'^b} = \left(\frac{A}{A'}\right)^b$$

is evidently greater than or less than 1 according to whether the quantity b is positive or negative.

Theorem V. — *Denote any number by A and let b and b′ be two quantities subject to the condition*

$$b > b'.$$

We then conclude that if A is greater than 1, *then*

$$A^b > A^{b'},$$

and if A is less than 1, *then*

$$A^b < A^{b'}.$$

Proof. — Indeed, because the quantity $b - b'$ is positive by hypothesis, the fraction

$$\frac{A^b}{A^{b'}} = A^{b-b'}$$

is evidently greater than or less than 1 according to whether $A > 1$ or $A < 1$.

Theorem VI. — *Let* log *be the characteristic of logarithms taken in the system for which the base is A, and denote by B and B′ two numbers subject to the condition*

$$B > B'.$$

If A is greater than 1, *we have*

$$\log B > \log B',$$

[363] *and if A is less than* 1, *we have*

$$\log B < \log B'.$$

Proof. — Indeed, the logarithm

$$\log \frac{B}{B'} = \log B - \log B'$$

is positive in the first case and negative in the second.

Corollary. — If we use the symbol ln to indicate the Napierian logarithms taken in the system for which the base is

$$e = 2.7182818\ldots, \tag{1}$$

[Chapter VI, § I, equation (5)], then the condition

$$B > B'$$

always entails the formula

$$\ln B > \ln B'.$$

To the preceding theorems we add the following, from which we can deduce several important consequences.

Theorem VII. — *Let x be any quantity. We have*

$$1 + x < e^x, \tag{2}$$

where the letter e denotes, as usual, the base of the Napierian logarithms.

Proof. — Because the right-hand side of formula (2) still remains positive, the stated theorem is evident by itself if the quantity $1 + x$ is negative. Thus it suffices to examine the case where we suppose that

$$1 + x > 0. \tag{3}$$

Now, for all possible real values of x, equation (23) of Chapter VI (§ IV) gives

$$\begin{cases} e^x = 1 + \dfrac{x}{1} + \dfrac{x^2}{1\cdot 2} + \dfrac{x^3}{1\cdot 2\cdot 3} + \dfrac{x^4}{1\cdot 2\cdot 3\cdot 4} + \dfrac{x^5}{1\cdot 2\cdot 3\cdot 4\cdot 5} + \ldots \\ \quad = 1 + x + \dfrac{x^2}{2}\left(1 + \dfrac{x}{3}\right) + \dfrac{x^4}{2\cdot 3\cdot 4}\left(1 + \dfrac{x}{5}\right) + \ldots. \end{cases} \tag{4}$$

[364] Because the products[1]

[1] The first term is given as $\frac{x^2}{3}(1 + \frac{x}{3})$ in [Cauchy 1897, p. 364]. It is given correctly in [Cauchy 1821, p. 442]. (tr.)

$$\frac{x^2}{2}\left(1+\frac{x}{3}\right), \quad \frac{x^4}{2\cdot 3\cdot 4}\left(1+\frac{x}{5}\right), \quad \ldots$$

are positive not only when the quantity x is positive but also when x is negative but it has a numerical value less than 1, we get from equation (4) that whenever condition (3) is satisfied,

$$e^x > 1+x.$$

Corollary I. — In the case where $1+x$ is positive, if we take the Napierian logarithms of both sides of formula (2), we obtain the following:

$$\ln(1+x) < x \tag{5}$$

(see the corollary of theorem VI). This last formula remains true whenever its left-hand side is real.

Corollary II. — Let x, y, z, ... be several quantities subject to the conditions

$$1+x>0, \quad 1+y>0, \quad 1+z>0, \quad \ldots. \tag{6}$$

By virtue of formula (2), we have

$$1+x<e^x, \quad 1+y<e^y, \quad 1+z<e^z, \quad \ldots,$$

and so we conclude (theorem II) that

$$(1+x)(1+y)(1+z)\ldots < e^{x+y+z\ldots}. \tag{7}$$

This last formula remains true whenever its left-hand side contains only positive factors.

Corollary III. — In the preceding corollary, if we suppose that

$$x=a\alpha, \quad y=a'\alpha', \quad z=a''\alpha'', \quad \ldots,$$

where α, α', α'', ... denote positive quantities and a, a', a'', ... denote other quantities, respectively, greater than

$$-\frac{1}{\alpha}, \quad -\frac{1}{\alpha'}, \quad -\frac{1}{\alpha''}, \quad \ldots,$$

[365] then formula (7) becomes

$$(1+a\alpha)\left(1+a'\alpha'\right)\left(1+a''\alpha''\right)\ldots < e^{a\alpha+a'\alpha'+a''\alpha''+\ldots}.$$

Moreover, if the quantities a, a', a'', ... are all less than a certain limit A, then we have (by virtue of theorems I and III) that

$$a\alpha+a'\alpha'+a''\alpha''+\ldots < A\left(\alpha+\alpha'+\alpha''+\ldots\right),$$

and consequently we finally have

$$(1+a\alpha)\left(1+a'\alpha'\right)\left(1+a''\alpha''\right)\ldots < e^{A\left(\alpha+\alpha'+\alpha''+\ldots\right)}. \tag{8}$$

Formula (8) can be used to good advantage in the approximate solution of differential equations.

Now we move on to theorems on averages. As we have already said (*Preliminaries*),[2] we call an *average* among several given quantities a new quantity contained between the smallest and the largest of those under consideration. From this definition, the quantity h is an average between two quantities g and k, or among several quantities among which one of these values is the largest and the other is the smallest, if the two differences

$$g-h \quad \text{and} \quad h-k$$

are of the same sign. Given this, if we use the notation

$$M\left(a,a',a'',\ldots\right)$$

for denoting an average among the quantities a, a', a'', ..., as we did in the *Preliminaries*, we establish the following propositions without trouble:

Theorem VIII. — *Let a, a', a'', ... and h be several quantities subject to the condition*

$$h=M\left(a,a',a'',\ldots\right), \tag{9}$$

and let r be an entirely arbitrary quantity. Then we always have

$$rh=M\left(ra,ra',ra'',\ldots\right). \tag{10}$$

Proof. — Indeed, let g denote by the largest and k denote the smallest of the quantities a, a', a'', ... The two differences

$$g-h \quad \text{and} \quad h-k$$

[366] are positive, and consequently the products

$$r(g-h) \quad \text{and} \quad r(h-k)$$

or in other words, the two differences

$$rg-rh \quad \text{and} \quad rh-rk$$

are of the same sign. Thus we have

$$rh=M(rg,rk)$$

and *a fortiori*,

[2] See [Cauchy 1821, p. 14, Cauchy 1897, p. 27].

$$rh = M\left(ra, ra', ra'', \ldots\right),$$

given that rg and rk are necessarily two of the products

$$ra, \quad ra', \quad ra'', \quad \ldots.$$

Theorem IX. — *Let A, A', A'', $\ldots$ and H be several numbers which satisfy the condition*

(11) $$H = M\left(A, A', A'', \ldots\right),$$

and let b be any quantity. Then we have

(12) $$H^b = M\left(A^b, A'^b, A''^b, \ldots\right).$$

Proof. — Indeed, let G and K be the largest and the smallest of the numbers A, A', A'', $\ldots$. Because the differences

$$G - H \quad \text{and} \quad H - K$$

are positive, we conclude from theorem IV that the following

$$G^b - H^b \quad \text{and} \quad H^b - K^b$$

are of the same sign. Thus we have

$$H^b = M\left(G^b, K^b\right),$$

and *a fortiori*,

$$H^b = M\left(A^b, A'^b, A''^b, \ldots\right).$$

Corollary. — In particular, if we make $b = \frac{1}{2}$, we find

$$\sqrt{H} = M\left(\sqrt{A}, \sqrt{A'}, \sqrt{A''}, \ldots\right).$$

[367] **Theorem X.** — *Let A denote any number and let b, b', b'', $\ldots$ and h be several quantities subject to the condition*

(13) $$h = M\left(b, b', b'', \ldots\right).$$

Then we have

(14) $$A^h = M\left(A^b, A^{b'}, A^{b''}, \ldots\right).$$

Proof. — Denote by g the greatest and k the smallest of the quantities b, b', b'', $\ldots$. Because the two differences

$$g - h \quad \text{and} \quad h - k$$

are positive, we conclude from theorem V that the quantities

$$A^g - A^h \quad \text{and} \quad A^h - A^k$$

are of the same sign. Thus we have

$$A^h = M\left(A^g, A^k\right) = M\left(A^b, A^{b'}, A^{b''}, \ldots\right).$$

Theorem XI. — *Let* log *be the characteristic of logarithms in the system for which the base is A and denote by B, B′, B″, … and H several numbers subject to the condition*

(15) $$H = M\left(B, B', B'', \ldots\right).$$

Whatever A may be, we have

(16) $$\log H = M\left(\log B, \log B', \log B'', \ldots\right).$$

Proof. — Indeed, suppose that we represent by G the largest and by K the smallest of the numbers B, B', B'', …. Then because the two fractions

$$\frac{G}{H} \quad \text{and} \quad \frac{H}{K}$$

are greater than 1, the logarithms

$$\log\frac{G}{H} \quad \text{and} \quad \log\frac{H}{K},$$

or in other words, the differences

$$\log B - \log H \quad \text{and} \quad \log H - \log K$$

are of the same sign. Thus we have

$$\log H = M\left(\log G, \log K\right) = M\left(\log B, \log B', \log B'', \ldots\right).$$

[368] **Theorem XII.** — *Let b, b′, b″, … be several quantities of the same sign, n in number, and let a, a′, a″, … be any quantities, also n in number. Then we have*

(17) $$\frac{a + a' + a'' + \ldots}{b + b' + b'' + \ldots} = M\left(\frac{a}{b}, \frac{a'}{b'}, \frac{a''}{b''}, \ldots\right).$$

Proof. — Let g be the largest and k the smallest of the quantities

$$\frac{a}{b}, \quad \frac{a'}{b'}, \quad \frac{a''}{b''}, \quad \ldots.$$

Then the differences

$$g-\frac{a}{b} \quad \text{and} \quad \frac{a}{b}-k,$$
$$g-\frac{a'}{b'} \quad \text{and} \quad \frac{a'}{b'}-k,$$
$$g-\frac{a''}{b''} \quad \text{and} \quad \frac{a''}{b''}-k,$$
$$\cdots\cdots\cdots\cdots\cdots\cdots$$

are all positive. By multiplying the first two by b, the following two by b', etc., we obtain the products

$$gb-a \quad \text{and} \quad a-kb,$$
$$gb'-a' \quad \text{and} \quad a'-kb',$$
$$gb''-a'' \quad \text{and} \quad a''-kb'',$$
$$\cdots\cdots\cdots\cdots\cdots\cdots$$

which are all of the same sign as the quantities b, b', b'', ... Consequently, the sums of these two kinds of products, namely

$$g\,(b+b'+b''+\ldots)-(a+a'+a''+\ldots) \quad \text{and}$$
$$a+a'+a''+\ldots-k\,(b+b'+b''+\ldots),$$

and the quotients of these sums by $b+b'+b''+\ldots$, namely

$$g-\frac{a+a'+a''+\ldots}{b+b'+b''+\ldots} \quad \text{and} \quad \frac{a+a'+a''+\ldots}{b+b'+b''+\ldots}-k,$$

are also quantities of the same sign. From this we conclude that

$$\frac{a+a'+a''+\ldots}{b+b'+b''+\ldots}=M\,(g,k)=M\left(\frac{a}{b},\frac{a'}{b'},\frac{a''}{b''},\ldots\right)$$

[see in the *Preliminaries* theorem I and formula (6)].

[369] *Corollary I.* — Suppose that the quantities b, b', b'', ... reduce to 1. We find that

(18) $$\frac{a+a'+a''+\ldots}{n}=M(a,a',a'',\ldots).$$

The left-hand side of the preceding formula is what we call the *arithmetic mean* of the quantities a, a', a'', ...

Corollary II. — Because the average among several equal quantities is equal to each of them, if the fractions $\frac{a}{b}$, $\frac{a'}{b'}$, $\frac{a''}{b''}$, ... are equal, we have

(19) $$\frac{a+a'+a''+\ldots}{b+b'+b''+\ldots}=\frac{a}{b}=\frac{a'}{b'}=\frac{a''}{b''}=\ldots,$$

and this is easy to prove directly.

Corollary III. — If we denote by α, α', α'', ... new quantities which have the same sign, then by virtue of equation (17) we have

$$(20)\quad \begin{cases} \dfrac{\alpha a+\alpha' a'+\alpha'' a''+\dots}{\alpha b+\alpha' b'+\alpha'' b''+\dots} = M\left(\dfrac{\alpha a}{\alpha b},\dfrac{\alpha' a'}{\alpha' b'},\dfrac{\alpha'' a''}{\alpha'' b''},\dots\right) \\ \qquad\qquad = M\left(\dfrac{a}{b},\ \dfrac{a'}{b'},\ \dfrac{a''}{b''},\ \dots\right). \end{cases}$$

This last formula suffices to establish theorem III of the *Preliminaries.*

Theorem XIII. — *Let A, A′, A″, ..., B, B′, B″, ... be two sequences of numbers taken at will, each of which we suppose has the same number of terms, n. With these two sequences form the roots*

$$\sqrt[B]{A},\quad \sqrt[B']{A'},\quad \sqrt[B'']{A''},\quad \dots.$$

Then we have

$$(21)\qquad \sqrt[B+B'+B''+\dots]{AA'A''\dots} = M\left(\sqrt[B]{A},\ \sqrt[B']{A'},\ \sqrt[B'']{A''},\ \dots\right).$$

Proof. — The logarithms of the quantities

$$\sqrt[B+B'+B''+\dots]{AA'A''\dots},\quad \sqrt[B]{A},\quad \sqrt[B']{A'},\quad \sqrt[B'']{A''},\quad \dots,$$

indicated by the characteristic ln are, respectively,

$$\frac{\ln A+\ln A'+\ln A''+\dots}{B+B'+B''+\dots},\quad \frac{\ln A}{B},\quad \frac{\ln A'}{B'},\quad \frac{\ln A''}{B''},\quad \dots,$$

[370] and equation (17) gives the following relation among these logarithms:

$$\frac{\ln A+\ln A'+\ln A''+\dots}{B+B'+B''+\dots} = M\left(\frac{\ln A}{B},\ \frac{\ln A'}{B'},\ \frac{\ln A''}{B''},\ \dots\right).$$

Now if we return from logarithms to numbers, as is permitted by virtue of theorem X, we again find formula (21).

Corollary I. — By supposing that the numbers B, B', B'', ... reduce to 1, we have simply

$$(22)\qquad \sqrt[n]{AA'A''\dots} = M\left(A,\ A',\ A'',\ \dots\right).$$

The left-hand side of the preceding formula is what we call the *geometric mean* of the numbers A, A', A'', ...

Corollary II. — If all the roots

$$\sqrt[B]{A},\quad \sqrt[B']{A'},\quad \sqrt[B'']{A''},\quad \dots$$

are equal, then their average is equal to each of them. Thus we have

$$(23) \qquad \sqrt[B+B'+B''+\dots]{AA'A''\dots} = \sqrt[B]{A} = \sqrt[B']{A'} = \sqrt[B'']{A''} = \dots,$$

which would be easy to prove directly.

The numerical value of an average among several given quantities is not always an average among their numerical values. Thus, for example, -1 is an average between -2 and $+3$; however, 1 is not an average value between 2 and 3. Among the various ways of obtaining an average among numerical values of n quantities

$$a, \quad a', \quad a'', \quad \dots,$$

one of the simplest consists of first forming the arithmetic mean among the squares,

$$a^2, \quad a'^2, \quad a''^2, \quad \dots,$$

and then extracting the square root of the result. In operating in this way, we first find

$$\frac{a^2 + a'^2 + a''^2 + \dots}{n} = M\left(a^2, a'^2, a''^2, \dots\right),$$

[371] and then, taking into account the corollary of theorem IX, we find

$$(24) \qquad \frac{\sqrt{a^2 + a'^2 + a''^2 + \dots}}{\sqrt{n}} = M\left(\sqrt{a^2}, \sqrt{a'^2}, \sqrt{a''^2}, \dots\right).$$

Now because the positive quantities

$$\sqrt{a^2}, \quad \sqrt{a'^2}, \quad \sqrt{a''^2}, \quad \dots$$

represent precisely the numerical values of the given quantities

$$a, \quad a', \quad a'', \quad \dots,$$

it follows from formula (24) that we obtain an average among the values if we divide the very simple expression

$$\sqrt{a^2 + a'^2 + a''^2 + \dots}$$

by $\sqrt{n}$. This expression, which is greater than the largest of the numerical values in question, is what we could call the *modulus* of the system of quantities $a, a', a'', \dots$. The modulus of a system of two quantities a and b is nothing other than the modulus itself of the imaginary expression $a + b\sqrt{-1}$ (see Chapter VII, § II). In any case, real expressions of the form

$$\sqrt{a^2 + a'^2 + a''^2 + \dots}$$

enjoy some very remarkable properties. In Geometry, they serve to determine the measured lengths of a straight line and the areas of plane surfaces by means of their orthogonal projections. In Algebra, they are the subject of several important theorems, among which I will content myself to state those which follow.

Theorem XIV. — *If the fractions*

$$\frac{a}{b}, \quad \frac{a'}{b'}, \quad \frac{a''}{b''}, \quad \ldots$$

are equal, then the numerical value of each of them is expressed by the ratio

$$\frac{\sqrt{a^2+a'^2+a''^2+\ldots}}{\sqrt{b^2+b'^2+b''^2+\ldots}},$$

so that we have

$$\frac{a}{b}=\frac{a'}{b'}=\frac{a''}{b''}=\ldots=\pm\frac{\sqrt{a^2+a'^2+a''^2+\ldots}}{\sqrt{b^2+b'^2+b''^2+\ldots}}, \tag{25}$$

[372] *where the sign* $+$ *or the sign* $-$ *is adopted according to whether the given fractions are positive or negative.*

Proof. — Indeed, under the given hypothesis, the fractions

$$\frac{a^2}{b^2}, \quad \frac{a'^2}{b'^2}, \quad \frac{a''^2}{b''^2}, \quad \ldots$$

are equal, and as a consequence we have

$$\frac{a^2}{b^2}=\frac{a'^2}{b'^2}=\frac{a''^2}{b''^2}=\ldots=\frac{a^2+a'^2+a''^2+\ldots}{b^2+b'^2+b''^2+\ldots}.$$

By extracting the square roots, we recover formula (25).

Theorem XV. — *Let* $a, a', a'', \ldots$ *be any* n *real quantities. If these quantities are not equal to each other, then the numerical value of the sum*

$$a+a'+a''+\ldots$$

is less than the product

$$\sqrt{n}\sqrt{a^2+a'^2+a''^2+\ldots},$$

so that we have

$$\text{val.num.}\left(a+a'+a''+\ldots\right)<\sqrt{n}\sqrt{a^2+a'^2+a''^2+\ldots}. \tag{26}$$

Proof. — Indeed, if to the square of the sum

$$a+a'+a''+\ldots$$

we add the squares of the differences among the quantities a, a', a'', ... combined in pairs in every possible manner, namely

$$(a-a')^2, \quad (a-a'')^2, \quad \ldots, \quad (a'-a'')^2, \quad \ldots,$$

we find

$$(27)\quad \begin{cases} (a+a'+a''+\ldots)^2+(a-a')^2+(a-a'')^2+\ldots+(a'-a'')^2+\ldots \\ \quad =n\left(a^2+a'^2+a''^2+\ldots\right), \end{cases}$$

and we conclude that

$$\left(a+a'+a''+\ldots\right)^2<n\left(a^2+a'^2+a''^2+\ldots\right).$$

By taking the positive square roots of both sides of this last formula, we obtain precisely formula (26).

[373] *Corollary.* — If we divide both sides of formula (26) by n, we find

$$(28)\qquad \text{val.num.}\frac{a+a'+a''+\ldots}{n}<\frac{\sqrt{a^2+a'^2+a''^2+\ldots}}{\sqrt{n}}.$$

Thus the numerical value of the arithmetic mean among several quantities a, a', a'', ... is less than the ratio

$$\frac{\sqrt{a^2+a'^2+a''^2+\ldots}}{\sqrt{n}},$$

which represents an average among the numerical values of these same quantities, as we have remarked above.

Scholium I. — When the quantities a, a', a'', ... become equal, we evidently have

$$\text{val.num.}\left(a+a'+a''+\ldots\right)=\sqrt{n}\sqrt{a^2+a'^2+a''^2+\ldots}=na.$$

Scholium II. — If we successively set $n=2$, $n=3$, ... in equation (27), we conclude that

$$(29)\begin{cases} (a+a')^2+(a-a')^2=2\left(a^2+a'^2\right), \\ (a+a'+a'')^2+(a-a')^2+(a-a'')^2+(a'-a'')^2=3\left(a^2+a'^2+a''^2\right), \\ \ldots\ldots\ldots\ldots\ldots\ldots\ldots\ldots\ldots\ldots \end{cases}$$

Theorem XVI.[3] — *Let a, a', a'', ..., α, α', α'', ... be two sequences of quantities and suppose that each of these sequences contains n terms. If the ratios*

[3] This is now known as the Cauchy–Schwarz Inequality.

$$\frac{a}{\alpha}, \quad \frac{a'}{\alpha'}, \quad \frac{a''}{\alpha''}, \quad \ldots$$

are not all equal to each other, then the sum

$$a\alpha + a'\alpha' + a''\alpha'' + \ldots$$

is less than the product

$$\sqrt{a^2 + a'^2 + a''^2 + \ldots}\sqrt{\alpha^2 + \alpha'^2 + \alpha''^2 + \ldots},$$

so that we have

$$(30) \qquad \left\{ \begin{array}{l} \text{val.num.}\,(a\alpha + a'\alpha' + a''\alpha'' + \ldots) \\ \quad < \sqrt{a^2 + a'^2 + a''^2 + \ldots}\sqrt{\alpha^2 + \alpha'^2 + \alpha''^2 + \ldots}. \end{array} \right.$$

[374] *Proof.* — Indeed, if to the square of the sum

$$a\alpha + a'\alpha' + a''\alpha'' + \ldots$$

we add the numerators of the fractions which represent the squares of the differences between the ratios

$$\frac{a}{\alpha}, \quad \frac{a'}{\alpha'}, \quad \frac{a''}{\alpha''}, \quad \ldots$$

combined with each other in every possible way, namely

$$(a\alpha' - a'\alpha)^2, \quad (a\alpha'' - a''\alpha)^2, \quad \ldots, \quad (a'\alpha'' - a''\alpha')^2, \quad \ldots,$$

we find

$$(31) \qquad \left\{ \begin{array}{l} (a\alpha + a'\alpha' + a''\alpha'' + \ldots)^2 + (a\alpha' - a'a)^2 \\ \qquad + (a\alpha'' - a''a)^2 + \ldots + (a'\alpha'' - a''\alpha')^2 + \ldots \\ \quad = \left(a^2 + a'^2 + a''^2 + \ldots\right)\left(\alpha^2 + \alpha'^2 + \alpha''^2 + \ldots\right), \end{array} \right.$$

and we conclude that

$$(a\alpha + a'\alpha' + a''\alpha'' + \ldots)^2 < \left(a^2 + a'^2 + a''^2 + \ldots\right)\left(\alpha^2 + \alpha'^2 + \alpha''^2 + \ldots\right).$$

By extracting the square roots of both sides of this last formula, we obtain precisely formula (30).

Corollary. — If we divide both sides of formula (30) by n, we find

$$(32) \qquad \left\{ \begin{array}{l} \text{val.num.}\,\dfrac{a\alpha + a'\alpha' + a''\alpha'' + \ldots}{n} \\ \quad < \dfrac{\sqrt{a^2 + a'^2 + a''^2 + \ldots}}{\sqrt{n}} \dfrac{\sqrt{\alpha^2 + \alpha'^2 + \alpha''^2 + \ldots}}{\sqrt{n}}. \end{array} \right.$$

Thus the arithmetic mean among the products

$$a\alpha, \quad a'\alpha', \quad a''\alpha'', \quad \ldots$$

has a numerical value less than the product of the two ratios that represent the averages among the numerical values of the two kinds of quantities contained in the two sequences

$$a,\ a',\ a'',\ \ldots \text{ and}$$
$$\alpha,\ \alpha',\ \alpha'',\ \ldots.$$

[375] *Scholium I.* — When the ratios

$$\frac{a}{\alpha}, \quad \frac{a'}{\alpha'}, \quad \frac{a''}{\alpha''}, \quad \ldots$$

become equal, we get from formula (31) that

$$(a\alpha + a'\alpha' + a''\alpha'' + \ldots)^2 = (a^2 + a'^2 + a''^2 + \ldots)\left(\alpha^2 + \alpha'^2 + \alpha''^2 + \ldots\right),$$

and consequently

$$\begin{aligned} &\text{val.num.}\,(a\alpha + a'\alpha' + a''\alpha'' + \ldots) \\ &\quad = \sqrt{a^2 + a'^2 + a''^2 + \ldots}\sqrt{\alpha^2 + \alpha'^2 + \alpha''^2 + \ldots}. \end{aligned}$$

It is easy to arrive directly at the same result.

Scholium II. — If we successively set

$$n = 2, \quad n = 3, \quad \ldots,$$

in formula (31), then we conclude that

$$(33) \quad \left\{ \begin{aligned} &(a\alpha + a'\alpha')^2 + (a\alpha' - a'\alpha)^2 = (a^2 + a'^2)\left(\alpha^2 + \alpha'^2\right), \\ &(a\alpha + a'\alpha' + a''\alpha'')^2 + (a\alpha' - a'\alpha)^2 \\ &\qquad + (a\alpha'' - a''\alpha)^2 + (a'\alpha'' - a''\alpha')^2 \\ &\quad = (a^2 + a'^2 + a''^2)\left(\alpha^2 + \alpha'^2 + \alpha''^2\right), \\ &\ldots\ldots\ldots\ldots\ldots\ldots\ldots\ldots\ldots\ldots\ldots\ldots \end{aligned} \right.$$

The first of the preceding equations agrees with equation (8) of Chapter VII (§ I). The second can be written as follows

$$(34) \quad \left\{ \begin{aligned} &(a\alpha' - a'\alpha)^2 + (a\alpha'' - a''\alpha)^2 + (a'\alpha'' - a''\alpha')^2 \\ &\quad = (a^2 + a'^2 + a''^2)\left(\alpha^2 + \alpha'^2 + \alpha''^2\right) - (a\alpha + a'\alpha' + a''\alpha'')^2, \end{aligned} \right.$$

and in this form it can be used with good advantage in the theory of radii of curvature of curves traced on any surfaces, thus in several questions of Mechanics.

We end this note with the proof of a theorem worthy of remark, which leads to comparing the geometric mean of several numbers with their arithmetic mean. It consists of the following:

Theorem XVII.[4] — *The geometric mean of several numbers A, B, C, D, ... is always less than their arithmetic mean.*

[376] *Proof.* — Let n be the number of the letters $A, B, C, D, \ldots$. It suffices to prove in general that

$$\sqrt[n]{ABCD\ldots} < \frac{A+B+C+D+\ldots}{n}, \tag{35}$$

or what amounts to the same thing,

$$ABCD\ldots < \left(\frac{A+B+C+D+\ldots}{n}\right)^n. \tag{36}$$

Now in the first place, it is evident, for $n = 2$, that

$$AB = \left(\frac{A+B}{2}\right)^2 - \left(\frac{A-B}{2}\right)^2 < \left(\frac{A+B}{2}\right)^2,$$

and by taking successively $n = 4$, $n = 8$, ..., and finally $n = 2^m$, we conclude that

$$\begin{aligned} ABCD \quad &< \left(\frac{A+B}{2}\right)^2 \left(\frac{C+D}{2}\right)^2 < \left(\frac{A+B+C+D}{2}\right)^2, \\ ABCDEFGH &< \left(\frac{A+B+C+D}{4}\right)^4 \left(\frac{E+F+G+H}{4}\right)^4 \\ &< \left(\frac{A+B+C+D+E+F+G+H}{8}\right)^8, \\ &\ldots\ldots\ldots\ldots\ldots\ldots\ldots\ldots\ldots\ldots \end{aligned}$$

and

$$ABCD\ldots < \left(\frac{A+B+C+D+\ldots}{2^m}\right)^{2^m}. \tag{37}$$

In the second place, if n is not a term of the geometric progression

$$2, \quad 4, \quad 8, \quad 16, \quad \ldots,$$

we denote by 2^m a term of this progression greater than n, and we make

$$K = \frac{A+B+C+D+\ldots}{n},$$

[4] This is the Arithmetic–Geometric Mean Theorem, originally due to Gauss.

then, returning to formula (37) and supposing that in the left-hand side of this formula, the last $2^m - n$ factors are equal to K, we find

$$ABCD\ldots K^{2^m - n} < \left[\frac{A+B+C+D+\ldots+(2^m-n)K}{2^m}\right]^{2^m},$$

or in other words,

$$ABCD\ldots K^{2^m-n} < K^{2^m}.$$

[377] Thus we have as a consequence

$$ABCD\ldots < K^n = \left(\frac{A+B+C+D+\ldots}{n}\right)^n,$$

which is what we set out to prove.

Corollary. — We conclude generally from formula (36)

(38) $$A+B+C+D+\ldots > n\sqrt[n]{ABCD\ldots},$$

whatever the number of the letters A, B, C, D, … may be. Thus, for example,

(39) $$\begin{cases} A+B > 2\sqrt{AB}, \\ A+B+C > 3\sqrt[3]{ABC}, \\ \ldots\ldots\ldots\ldots\ldots\ldots\ldots\ldots \end{cases}$$

Note III – On the numerical solution of equations.

[378] To solve *numerically* one or several equations is to find the values in numbers of the unknowns which they contain. This evidently requires that the constants contained in the equations be themselves constrained to numbers. We will concern ourselves here only with equations that contain one unknown, and we will begin by establishing, in this connection, the following theorems.

Theorem I.[1] — *Let $f(x)$ be a real function of the variable x, which remains continuous with respect to this variable between the limits $x = x_0$ and $x = X$. If the two quantities $f(x_0)$ and $f(X)$ have opposite signs, we can satisfy the equation*

$$f(x) = 0 \tag{1}$$

with one or several real values of x contained between x_0 and X.

Proof. — Let x_0 be the smaller of the two quantities x_0 and X. Let

$$X - x_0 = h,$$

and denote by m any integer number larger than 1. Because one of the two quantities $f(x_0)$ and $f(X)$ is positive and the other negative, if we form the sequence

$$f(x_0), \quad f\left(x_0 + \frac{h}{m}\right), \quad f\left(x_0 + 2\frac{h}{m}\right), \quad \ldots, \quad f\left(X - \frac{h}{m}\right), \quad f(X),$$

and if we suppose that, in this sequence, we successively compare the first term with the second, the second with the third, the third with the fourth, etc., eventually we must find one or more times that two consecutive terms have opposite signs. Let

[1] This is a special case of the Intermediate Value Theorem. See Chapter II, § II, theorem IV, p. 32 [Cauchy 1821, p. 43, Cauchy 1897, p. 50]. Cauchy's proof there was quite intuitive. The version of the theorem in Chapter II is given here as corollary II [Cauchy 1821 p. 463, Cauchy 1897, p. 381]. See also [Grabiner 2005, pp. 69–75].

$$f(x_1) \quad \text{and} \quad f(X')$$

[379] be two such terms, x_1 being the smaller of the two corresponding values of x. We evidently have[2]

$$x_0 < x_1 < X' < X$$

and

$$X' - x_1 = \frac{h}{m} = \frac{1}{m}(X - x_0).$$

Having determined x_1 and X' as we have just said, we can likewise locate two other values x_2 and X'' between x_1 and X', which give results of opposite signs when substituted into $f(x)$, and which satisfy the conditions

$$x_1 < x_2 < X'' < X'$$

and

$$X'' - x_2 = \frac{1}{m}(X' - x_1) = \frac{1}{m^2}(X - x_0).$$

In continuing like this, we obtain: 1° an increasing series of values of x, namely

(2) $$x_0,\ x_1,\ x_2,\ \ldots;$$

and 2° a series of decreasing values

(3) $$X,\ X',\ X'',\ \ldots,$$

which exceed the corresponding values of the first series by quantities, respectively, equal to the products

$$1 \times (X - x_0),\ \frac{1}{m} \times (X - x_0),\ \frac{1}{m^2} \times (X - x_0),\ \ldots,$$

and they eventually differ from the terms of the first series by as little as we might wish. We must conclude that the general terms of series (2) and (3) converge towards a common limit. Let a be that limit. Because the function $f(x)$ is continuous from $x = x_0$ to $x = X$, the general terms of the following series

$$f(x_0),\ f(x_1),\ f(x_2),\ \ldots,$$
$$f(X),\ f(X'),\ f(X''),\ \ldots$$

converge likewise towards the common limit $f(a)$. As they approach that limit they always have opposite signs, so it is clear [380] that the quantity $f(a)$, being necessarily finite, cannot differ from zero. As a consequence, it satisfies the equation

(1) $$f(x) = 0,$$

2 Here, Cauchy does not make the distinction between "less than" and "less than or equal to."

by assigning to the variable x the particular value a, contained between x_0 and X. In other words,

$$x = a \tag{4}$$

is a *root* of equation (1).

Scholium I. — Suppose we have extended series (2) and (3) to the terms

$$x_n \quad \text{and} \quad X^{(n)},$$

(where n denotes any integer number). If we take the half-sum of these terms as the value approximating the root a, the error made is less than their half-difference, namely

$$\frac{1}{2}\frac{X - x_0}{m^n}.$$

Because this last expression decreases indefinitely as n increases, it follows that, by calculating a sufficient number of terms of the two series, we eventually obtain values as close to the root a as we wish.

Scholium II. — If there exist several real roots of equation (1) between the limits x_0 and X, the preceding method locates some, and sometimes all of them. Then, we find for x_1 and X' or for x_2 and X'', ... several systems of values which enjoy the same properties. [3]

Scholium III. — If the function $f(x)$ is constantly increasing or constantly decreasing from $x = x_0$ to $x = X$, then between these limits there exists but a single value of x that satisfies equation (1).

Corollary I. — If equation (1) has no real roots between the limits x_0 and X, then the two quantities

$$f(x_0) \quad \text{and} \quad f(X)$$

have the same sign.

[381] *Corollary II.* — If in the statement of theorem I, we replace the function $f(x)$ by

$$f(x) - b$$

(where b denotes a constant quantity), then we obtain precisely theorem IV of Chapter II (§ II).[4] Under the same hypothesis, and following the method indicated above, we can determine numerically the roots of the equation

$$f(x) = b \tag{5}$$

[3] Cauchy is saying in other words that if $m > 2$, then we may see more than one sign change in $f(x)$ in the interval at any given stage.

[4] See p. 32 [Cauchy 1821, p. 43, Cauchy 1897, p. 50].

contained between x_0 and X.

Note. — When equation (1) has several roots contained between x_0 and X, in calculating series (2) and (3), we are not always assured of finding the smallest or the largest of the roots in the interval. However, we could do this following another method that Mr. Legendre has used in his *Supplément à la Théorie des nombres*.[5] This second method follows immediately from the two theorems I am about to state.

Theorem II. — *As in theorem I, suppose that the function $f(x)$ remains continuous from $x = x_0$ to $x = X$ (where X is greater than x_0), and denote by $\varphi(x)$ and $\chi(x)$ two auxiliary functions also continuous on the given interval, but also subject to:* 1° *that they both increase constantly*[6] *with x on this interval; and* 2° *that they give for the difference*

$$\varphi(x) - \chi(x),$$

an expression which initially is negative when we give x the particular value x_0, and which always remains equal (up to sign) to $f(x)$. If the equation

$$f(x) = 0 \tag{1}$$

has one or several real roots between x_0 and X, then the values of x given by

$$x_0,\ x_1,\ x_2,\ x_3,\ \ldots, \tag{6}$$

and derived from one another by means of the formulas

$$\varphi(x_1) = \chi(x_0),\ \varphi(x_2) = \chi(x_1),\ \varphi(x_3) = \chi(x_2),\ \ldots \tag{7}$$

make an increasing series of quantities, for which the general term converges towards the smallest of these roots. On the other hand, if equation (1) *does not have*
[382] *real roots contained between x_0 and X, then the general term of series* (6) *eventually exceeds X.*[7]

Proof. — Let us suppose in the first place that the equation $f(x) = 0$ has one or several real roots between x_0 and X, and denote by a the smallest of these roots. It satisfies the equation in question, or what amounts to the same thing, the following:

$$\varphi(x) - \chi(x) = 0. \tag{1}$$

Taking $x = a$, we have as a consequence

$$\varphi(a) = \chi(a). \tag{8}$$

[5] See [Legendre 1816, p. 43].

[6] Legendre used the word *omale* to describe these functions that we now call "monotonic." Galois also used the term. It seems that Cauchy did not adopt the word, though he must have known of Legendre's use of it, and its use seems to have died out.

[7] See [Galuzzi 2001] for more on Legendre's method.

Moreover, because the function $\chi(x)$ is constantly increasing with x from $x = x_0$ to $x = X$ and a is greater than x_0, we have

$$\chi(a) > \chi(x_0).$$

By combining these last two formulas with the first of equations (7), namely

$$\chi(x_0) = \varphi(x_1),$$

we conclude that

$$\varphi(a) > \varphi(x_1)$$

and consequently

$$a > x_1. \tag{9}$$

In the same way, by combining the three formulas

$$\varphi(a) = \chi(a), \quad \chi(a) > \chi(x_1) \quad \text{and} \quad \chi(x_1) = \varphi(x_2),$$

where the second follows immediately from formula (9), we find

$$\varphi(a) > \varphi(x_2),$$

and consequently

$$a > x_2. \tag{10}$$

By continuing like this, we are assured that all the terms of series (6) are less than the root a. I will add that these various terms form an increasing sequence of quantities, and indeed, because the difference

$$\varphi(x) - \chi(x)$$

[383] is negative by hypothesis for $x = x_0$, we have

$$\varphi(x_0) < \chi(x_0).$$

But $\chi(x_0) = \varphi(x_1)$, so

$$\varphi(x_0) < \varphi(x_1)$$

and

$$x_0 < x_1. \tag{11}$$

Moreover, because x_1 is contained between x_0 and a, no real root of the equation

$$\varphi(x) - \chi(x) = 0$$

is found contained between the limits x_0 and x_1, and consequently (see theorem I, corollary I)

$$\varphi(x_0) - \chi(x_0) \quad \text{and} \quad \varphi(x_1) - \chi(x_1)$$

are quantities of the same sign, that is to say, both of them are negative. So we have

$$\varphi(x_1) < \chi(x_1),$$

and consequently, because $\chi(x_1) = \varphi(x_2)$,

$$\varphi(x_1) < \varphi(x_2),$$

and so

(12) $$x_1 < x_2,$$

etc. Thus, the quantities

$$x_0, x_1, x_2, \ldots$$

form a series for which the general term x_n increases constantly with n without ever surpassing the root a, and it necessarily converges to a root equal to or less than this root. Let us call this limit l. Because, by virtue of equations (7), we have, for every n,

$$\varphi(x_{n+1}) = \chi(x_n),$$

we conclude, by letting n increase indefinitely and passing to the limits,

(13) $$\varphi(l) = \chi(l).$$

Thus, the quantity l is itself a root of equation (1), and because this quantity is greater than x_0 without being greater than the root a, we evidently have

(14) $$l = a.$$

[384] In the second place, let us suppose that equation (1) has no real roots between x_0 and X. We will now prove under this hypothesis that the general term x_n of series (6) grows constantly with x, at least as long as this term remains less than X. Indeed, as long as this condition is satisfied, the difference

$$\varphi(x_n) - \chi(x_n)$$

has (theorem I, corollary I) the same sign as

$$\varphi(x_0) - \chi(x_0),$$

that is to say negative, and consequently, we establish formulas (11), (12), ..., as above. Moreover, x_n cannot converge towards a fixed limit l less than X, because the existence of this limit would evidently involve equation (13), and consequently the existence of a real root contained between x_0 and X. Thus necessarily, under the given hypothesis, the value of x_n eventually surpasses the limit X.

Corollary I. — The conditions to which the auxiliary functions $\varphi(x)$ and $\chi(x)$ are subject in the statement of theorem II can be satisfied in infinitely many ways. However, among the infinitely many values we could give to the function $\varphi(x)$, it is important to choose one that permits the easy solution of equations (7), that is to say in general, any equation of the form

$$\varphi(x) = \text{const.}$$

After choosing the value of $\varphi(x)$ as we just said, we calculate without difficulty the various terms of series (6), and it suffices to find the limit towards which they converge to obtain the smallest of the roots of equation (1) contained between x_0 and X. If these same terms eventually surpass X, then equation (1) does not have a real root in the interval from x_0 to X.

Corollary II. — If we take

$$x_0 = 0,$$

and if also equation (1) has positive roots, then the quantities $x_1, x_2, \ldots$ are all less than the smallest root of this kind, and they give its value more and more closely.

Theorem III. — *As in theorem I, suppose that the function $f(x)$ remains continuous from $x = x_0$ to $x = X$ (where X is greater than x_0), and* [385] *denote by $\varphi(x)$ and $\chi(x)$ two auxiliary functions also continuous on the given interval, but also subject to:* 1° *that they both increase constantly with x on this interval; and* 2° *that they give for the difference*

$$\varphi(x) - \chi(x)$$

an expression which becomes positive when we give x the particular value X, and which always remains equal (up to sign) to $f(x)$. If the equation

$$f(x) = 0 \tag{1}$$

has one or several real roots between x_0 and X, then the values of x given by

$$X, \quad X', \quad X'', \quad X''', \quad \ldots \tag{15}$$

and derived from one another by means of the formulas

$$\varphi(X') = \chi(X), \quad \varphi(X'') = \chi(X'), \quad \varphi(X''') = \chi(X''), \quad \ldots \tag{16}$$

make a decreasing series of quantities for which the general term converges towards the largest of these roots. On the other hand, if equation (1) *does not have real roots contained between x_0 and X, then the general term of series* (15) *eventually descends below x_0.*

The proof of this third theorem is so similar to that of the second that for brevity we will dispense with recounting it here.

Corollary I. — Among the infinitely many values we could give to the function $\varphi(x)$ in a way that satisfies the given conditions, it is important to choose one that permits the easy solution of equations (16), that is to say in general, any equation of the form

$$\varphi(x) = \text{const.}$$

After choosing the value of $\varphi(x)$ as we just said, we calculate without difficulty the various terms of series (15), and it suffices to find the limit towards which they converge to obtain the largest of the roots of equation (1) contained between x_0 and X. If these same terms eventually fall below x_0, then equation (1) does not have a real root in the interval from x_0 to X.

[386] *Corollary II.* — If equation (1) has positive roots and if X surpasses the largest root of this kind, the quantities $X', X'', \ldots$ always remain larger than this root and they give its value more and more closely.

Scholium I. — If equation (1) has but one real root a, contained between x_0 and X, the general terms of series (6) and (15), where the first is increasing and the second is decreasing, converge towards a common limit equal to this root. Then, if we extend these series up to the terms

$$x_n \quad \text{and} \quad X^{(n)},$$

and then if we take the half-sum of these two terms as a value close to the root a, the resulting error is less than

$$\frac{X^{(n)} - x_n}{2}.$$

Scholium II. — To show an application of the principles that we have just established, consider in particular the equation

$$x^m - A_1 x^{m-1} - A_2 x^{m-2} - \ldots - A_{m-1} x - A_m = 0, \tag{17}$$

where m denotes any integer number and where

$$A_1, \quad A_2, \quad \ldots, \quad A_{m-1}, \quad A_m$$

denote quantities, positive or zero. Because the left-hand side of this equation is negative for $x = 0$ and positive for very large values of x, it follows that it has at least one root that is positive and finite. Moreover, this same equation is not different from the following

$$\frac{A_1}{x} + \frac{A_2}{x^2} + \ldots + \frac{A_{m-1}}{x^{m-1}} + \frac{A_m}{x^m} = 1,$$

where the right-hand side remains invariable, while the left-hand side decreases constantly for positive and increasing values of x, so it evidently admits but a single real positive root. Let a be that root, and let A be the largest of the numbers

$$A_1, \quad A_2, \quad \ldots, \quad A_{m-1}, \quad A_m.$$

Finally, denote as usual an average of these numbers by the notation

$$M(A_1, A_2, \ldots, A_{m-1}, A_m).$$

[387] By making $x = a$ and taking into account formula (11) of the *Preliminaries*, we get from equation (17)

$$\begin{aligned} a^m &= A_1 a^{m-1} + A_2 a^{m-2} + \ldots + A_{m-1} a + A_m \\ &= \left(a^{m-1} + a^{m-2} + \ldots + a + 1\right) M(A_1, A_2, \ldots, A_{m-1}, A_m) \\ &= \frac{a^m - 1}{a - 1} M(A_1, A_2, \ldots, A_{m-1}, A_m) < A \frac{a^m - 1}{a - 1}, \end{aligned}$$

and consequently

$$a - 1 < A \frac{a^m - 1}{a^m} < A$$

and

(18) $$a < A + 1.$$

As a consequence, the positive root of equation (17) is contained between the limits 0 and $A+1$. On the other hand, by letting

$$A_r a^{m-r} \quad \text{and} \quad A_s a^{m-s}$$

denote the smallest and the largest of the terms contained in the polynomial

$$A_1 a^{m-1} + A_2 a^{m-2} + \ldots + A_{m-1} a + A_m,$$

and denoting the number of these that are different from zero by[8] $n \leq m$ we evidently have

$$\begin{aligned} a^m &> n A_r a^{m-r} \quad \text{and} \\ a^m &< n A_s a^{m-s}, \end{aligned}$$

and consequently

$$\begin{aligned} a &> (n A_r)^{\frac{1}{r}} \quad \text{and} \\ a &< (n A_s)^{\frac{1}{s}}. \end{aligned}$$

It is clear that the root a is contained between the smallest and the largest of the numbers

(19) $$n A_1, \quad (n A_2)^{\frac{1}{2}}, \quad (n A_3)^{\frac{1}{3}}, \quad \ldots, \quad (n A_m)^{\frac{1}{m}}.$$

Finally, because by virtue of theorem I (corollary I), the left-hand side of equation (17) remains negative from $x = 0$ to $x = a$ and positive from $x = a$ to $x = \infty$, it

---

[8] Cauchy is simultaneously defining n and noting that it has the property that it is less than or equal to m. He used the notation "$n =$ or $< m$" in [Cauchy 1821, p. 472, 474]. In [Cauchy 1897, p. 387, 389], the editors used a symbol resembling $n \overline{\overline{<}} m$. There is a similar situation in Note VI with the symbol $\geq$ in [Cauchy 1821, pp. 533–534, Cauchy 1897, pp. 437–438].

follows that we can also choose as a lower bound[9] of the root a the largest of the integer numbers which make negative the expression

$$x^m - A_1 x^{m-1} - A_2 x^{m-2} - \ldots - A_{m-1} x - A_m, \tag{20}$$

[388] and as an upper bound the smallest of these that make it positive. Now let

$$x_0 \quad \text{and} \quad X$$

be the lower and upper bounds found following the rules that we have just given. Moreover, if we make

$$\begin{cases} \varphi(x) = x^m \quad \text{and} \\ \chi(x) = A_1 x^{m-1} + A_2 x^{m-2} + \ldots + A_{m-1} x + A_m, \end{cases} \tag{21}$$

then theorems II and III are applicable to equation (17), and because under this hypothesis, each of equations (7) and (16) can be reduced to the form

$$x^m = \text{const.},$$

it becomes easy to calculate the quantities contained in the two series

$$X,\ X',\ X'',\ X''',\ \ldots \quad \text{and}$$
$$x_0,\ x_1,\ x_2,\ x_3,\ \ldots,$$

where the general terms are values approaching the root a from above and from below.

Scholium III. — Consider again the equation

$$x^m + A_1 x^{m-1} + A_2 x^{m-2} + \ldots + A_{m-1} x - A_m = 0, \tag{22}$$

where m still denotes an integer number and where

$$A_1, \quad A_2, \quad \ldots, \quad A_{m-1}, \quad A_m$$

denote quantities, positive or zero, the largest of which is equal to A. By taking $\frac{1}{x}$ as the unknown,[10] we can rewrite this equation in the following form:

$$\left(\frac{1}{x}\right)^m - \frac{A_{m-1}}{A_m}\left(\frac{1}{x}\right)^{m-1} - \frac{A_{m-2}}{A_m}\left(\frac{1}{x}\right)^{m-2} - \ldots - \frac{A_1}{A_m}\frac{1}{x} - \frac{1}{A_m} = 0, \tag{23}$$

[9] Here we translate Cauchy's words *limite inférieure* as "lower bound." He does not seem to mean "limit inferior" in its modern sense. Likewise we translate *limite supérieure* as "upper bound" in the next phrase. (tr.)

[10] Note that equation (23) is algebraically equivalent to equation (22). Cauchy has not substituted $\frac{1}{x}$ in place of x in equation (22).

which is similar to that of equation (17). We thus conclude that equation (22) admits but one positive root less than the quotient

$$\text{(24)} \qquad \frac{1}{\frac{A}{A_m}+1},$$

[389] and this root is contained, not only between the smallest and the largest of the quantities

$$\text{(25)} \quad \frac{A_m}{nA_{m-1}}, \quad \left(\frac{A_m}{nA_{m-2}}\right)^{\frac{1}{2}}, \quad \left(\frac{A_m}{nA_{m-3}}\right)^{\frac{1}{3}}, \quad \ldots, \left(\frac{A_m}{nA_1}\right)^{\frac{1}{m-1}}, \quad \left(\frac{A_m}{n}\right)^{\frac{1}{m}},$$

where $n \leq m$ represents the number of variable terms contained in the left-hand side of equation (22), but the root a is also contained between the largest of the integer numbers that make the following expression negative

$$\text{(26)} \qquad x^m + A_1 x^{m-1} + A_2 x^{m-2} + \ldots + A_{m-1} x - A_m$$

and the smallest of those that make it positive. Following these remarks, after having determined two limits, one greater than and one less than the root in question, in order to approach it more closely, it suffices to apply theorems II and III to equation (23) and to consider $\frac{1}{x}$ as the unknown that we are trying to find.

Scholium IV. — If equation (1) had two real roots contained between x_0 and X, but extremely close to each other, the general terms of series (6) and (15) would appear at first to converge towards the same limit, and we would not be able to extend the series very long before we would perceive the difference between the limits towards which they are effectively converging. The same remark applies to series (2) and (3). Consequently, the solution methods based only on theorem I or else on theorems II and III are not always useful in all cases to find the number of real roots of a numerical equation. However, they always give the value of any single root which is found contained between any two given limits as accurately as we might wish.

In the particular case where the numerical equation that we are considering has for its left-hand side a real integer function of the variable x, we can determine the number of real roots all at once, as M. Lagrange has shown, and calculate their approximate values. To do this easily, it is best to start by reducing the given equation so that it has only unequal roots, and then proceeding as follows.

Let

$$\text{(27)} \qquad F(x) = 0$$

be the given equation. Denote by a, b, c, ... the various roots, real or [390] imaginary, and let m be the degree of the left-hand side, for which we suppose that the coefficient of this highest power of x is reduced to 1. Finally, let m' be the number of these roots equal to a, m'' the number equal to b, m''' the number equal to c, ... Then we have

$$m' + m'' + m''' + \ldots = m \tag{28}$$

and

$$F(x) = (x-a)^{m'} (x-b)^{m''} (x-c)^{m'''} \ldots. \tag{29}$$

Let z be a new variable. We conclude that

$$\frac{F(x+z)}{F(x)} = \left(1+\frac{z}{x-a}\right)^{m'} \left(1+\frac{z}{x-b}\right)^{m''} \left(1+\frac{z}{x-c}\right)^{m'''} \ldots. \tag{30}$$

Now if we make

$$F(x+z) = F(x) + zF_1(x) + z^2 F_2(x) + \ldots, \tag{31}$$

and if we expand the expressions

$$\left(1+\frac{z}{x-a}\right)^{m'}, \quad \left(1+\frac{z}{x-b}\right)^{m''}, \quad \left(1+\frac{z}{x-c}\right)^{m'''}, \quad \ldots$$

according to increasing powers of z, equation (30) becomes

$$\begin{aligned} &1 + z\frac{F_1(x)}{F(x)} + z^2\frac{F_2(x)}{F(x)} + \ldots \\ &= \left(1+\tfrac{m'}{x-a}z+\ldots\right)\left(1+\tfrac{m''}{x-b}z+\ldots\right)\left(1+\tfrac{m'''}{x-c}z+\ldots\right)\ldots \\ &= 1+\left(\tfrac{m'}{x-a}+\tfrac{m''}{x-b}+\tfrac{m'''}{x-c}+\ldots\right)z+\ldots. \end{aligned}$$

Then, by equating corresponding coefficients of the first power of z on the two sides, we find

$$\left\{\begin{aligned} \tfrac{F_1(x)}{F(x)} &= \tfrac{m'}{x-a}+\tfrac{m''}{x-b}+\tfrac{m'''}{x-c}+\ldots \\ &= \tfrac{m'(x-b)(x-c)\ldots+m''(x-a)(x-c)\ldots+m'''(x-a)(x-b)\ldots+\ldots}{(x-a)(x-b)(x-c)\ldots}. \end{aligned}\right. \tag{32}$$

Because the preceding formula has for its right-hand side an algebraic fraction that is evidently irreducible, it follows that it is enough to divide the [391] left-hand side $F(x)$ of equation (27) by the greatest common divisor of the two polynomials $F(x)$ and $F_1(x)$ to reduce this equation to the following

$$(x-a)(x-b)(x-c)\ldots = 0, \tag{33}$$

which has only unequal roots.

We will not stop here to show how we could use these same principles to deduce various equations whose distinct roots would be equal, either to the simple roots, or to the double roots, or to the triple roots, etc., of the given equation. Here we will only add some remarks relative to the case where we begin by supposing that all the roots of equation (27) are distinct from one another. Each of these numbers m', m'', m''', ... then reduces to 1, and we get from formula (32)

$$(34)\quad F_1(x) = (x-b)(x-c)\ldots + (x-a)(x-c)\ldots + (x-a)(x-b)\ldots + \ldots,$$

and consequently

$$(35)\quad \begin{cases} F_1(a) = (a-b)(a-c)\ldots, \\ F_1(b) = (b-a)(b-c)\ldots, \\ F_1(c) = (c-a)(c-b)\ldots, \\ \ldots\ldots\ldots\ldots\ldots\ldots\ldots\ldots, \end{cases}$$

and

$$(36)\quad F_1(a)F_1(b)F_1(c)\ldots = (-1)^{\frac{m(m-1)}{2}}(a-b)^2(a-c)^2\ldots(b-c)^2\ldots.$$

Thus, under the given hypothesis, the product of squares of the differences between the roots of equation (27) is equivalent, ignoring the sign, to the product

$$F_1(a)F_1(b)F_1(c)\ldots,$$

and consequently to the last term of the equation in z given by the elimination of x between the two following

$$(37)\quad F(x) = 0 \quad \text{and} \quad z - F_1(x) = 0.$$

Then, by calling the numerical value of the last term H, we have

$$(38)\quad (a-b)^2(a-c)^2\ldots(b-c)^2\ldots = \pm H.$$

Under the same hypothesis, because the values of $F_1(a)$, $F_1(b)$, ... given by formulas (35) are never zero, if we denote a real root of equation (27) by a, it suffices to give very small values of the number α for [392] the two quantities

$$\begin{aligned} F(a+\alpha) &= \alpha F_1(a) + \alpha^2 F_2(a) + \ldots \quad \text{and} \\ F(a-\alpha) &= -\alpha F_1(a) + \alpha^2 F_2(a) - \ldots \end{aligned}$$

to have opposite signs. Moreover, if we represent by x_0 and X a lower bound and an upper bound where a is the only real root contained between them, then by virtue of theorem I (corollary I), $F(X)$ has the same sign as $F(a+\alpha)$ and $F(x_0)$ the same sign as $F(a-\alpha)$, and consequently the two quantities

$$F(x_0) \quad \text{and} \quad F(X)$$

have opposite signs.

When equation (27) does not have equal roots, or it has been disencumbered of those that it did have, then for this equation it becomes easy to determine not only two limits between which all the real roots are contained, but also a sequence of quantities which, taken two by two, serve, respectively, as limits of the various roots of this kind, and finally values as close to these same roots as we might want. We will establish this by solving, one after another, the three following problems.

Problem I. — *To determine two limits between which all of the real roots of the equation*

(27) $$F(x) = 0$$

are contained.

Solution. — By hypothesis, $F(x)$ is a real polynomial of degree m with respect to x, and in which the highest power of x has 1 for its coefficient. If we denote the successive coefficients of the lesser powers by

$$a_1, \quad a_2, \quad \ldots, \quad a_{m-1}, \quad a_m,$$

and the numerical values of these same coefficients by

$$A_1, \quad A_2, \quad \ldots, \quad A_{m-1}, \quad A_m,$$

then we have identically

(39) $$\begin{cases} F(x) = x^m + a_1 x^{m-1} + a_2 x^{m-2} + \ldots + a_{m-1} x + a_m \\ \qquad\; = x^m \pm A_1 x^{m-1} \pm A_2 x^{m-2} \pm \ldots \pm A_{m-1} x \pm A_m. \end{cases}$$

[393] Now let k be a number greater than the unique positive root of equation (17) (theorem III, scholium II). Polynomial (20) is positive whenever we suppose that $x \geq k$. Consequently, it suffices to give x a numerical value greater than the number k for the sum of the numerical values of the terms

$$A_1 x^{m-1}, \quad A_2 x^{m-2}, \quad \ldots, \quad A_{m-1} x, \quad A_m$$

to become less than the numerical value of x^m. As a result, the left-hand side of equation (7) can never vanish while the value of x is located inside the limits

$$-k \quad \text{and} \quad +k.$$

Thus all the roots, positive and negative, of equation (27) are contained between these same limits.

Scholium I. — If the number k is subject to the sole condition of surpassing the positive root of equation (17), then we can suppose it to be equal either to the largest of the expressions (19), or to the smallest of the integer numbers which, when substituted in place of x in polynomial (20), give a positive result.

Scholium II. — We could easily assure that the number k, determined as we have just said, is greater not only than all the numerical values of the real roots of equation (27), but also than the moduli of all the imaginary roots. Indeed, let

$$x = r\left(\cos t + \sqrt{-1}\sin t\right)$$

be such a root. At the same time we have two real equations

$$(40)\quad \begin{cases} r^m \cos mt \pm A_1 r^{m-1} \cos(m-1)t \\ \quad \pm A_2 r^{m-2} \cos(m-2)t \pm \ldots \pm A_{m-1} r \cos t \pm A_m = 0 \end{cases}$$

and

$$(41)\quad \begin{cases} r^m \sin mt \pm A_1 r^{m-1} \sin(m-1)t \\ \quad \pm A_2 r^{m-2} \sin(m-2)t \pm \ldots \pm A_{m-1} r \sin t = 0. \end{cases}$$

By adding the first equation multiplied by $\cos mt$ to the second one multiplied by $\sin mt$, we conclude

$$(42)\quad \begin{cases} r^m \pm A_1 r^{m-1} \cos t \pm A_2 r^{m-2} \cos 2t \pm \ldots \\ \quad \pm A_{m-1} r \cos(m-1)t \pm A_m \cos mt = 0. \end{cases}$$

Now it is clear that we would not satisfy this last equation by supposing [394] $r > k$, because under this hypothesis the numerical value of r^m exceeds the sum of the numerical values of the terms

$$A_1 r^{m-1}, \quad A_2 r^{m-2}, \quad \ldots, \quad A_{m-1} r, \quad A_m,$$

and *a fortiori* it exceeds the sum of the numerical values that these same terms acquire when they are multiplied by the cosines.

Scholium III. — By comparing the polynomial (26) with the left-hand sides of equations (27) and (40), we easily prove that if g denotes a number less than the unique positive root of equation (22), then g is a lower limit not only for the numerical values of all the real roots of equation (27), but also of the moduli of all the imaginary roots. This is what happens, for example, if we take g to be the smallest of expressions (25), or the largest of the integer numbers which, substituted in place of x in polynomial (26), give a negative result. The number g being determined as we have just said, all the positive roots of equation (27) are contained between the limits

$$+g \quad \text{and} \quad +k$$

and the negative roots of the same equation are between the limits

$$-k \quad \text{and} \quad -g.$$

Scholium IV. — When we only propose to obtain a lower limit on the smallest of the positive roots or an upper limit on the largest of them, we can sometimes do this by using the corollary to theorem XVII (preceding Note). Indeed, suppose that all the terms of the polynomial $F(x)$ except one have the same sign. Then equation (27) takes the following form:

$$(43)\qquad \begin{cases} x^m + A_1 x^{m-1} + \ldots + a_{s-1} x^{m-s+1} \\ \quad + A_{s+1} x^{m-s-1} + \ldots + A_{m-1} x + A_m = A_s x^{m-s}. \end{cases}$$

Now let n be the number of terms in the left-hand side of equation (43) which do not reduce to zero, and

$$Bx^\mu$$

the geometric mean of these terms, where B denotes the geometric mean of their coefficients. By virtue of the corollary to theorem XVII (Note II), every real and positive value of x that satisfies the given equation, [395] or what amounts to the same thing, serves as its root, necessarily satisfies the condition

$$A_s x^{m-s} > nBx^\mu$$

and consequently, one of the two following

$$(44)\qquad x > \left(\frac{nB}{A_s}\right)^{\frac{1}{m-s-\mu}},$$

or

$$(45)\qquad x < \left(\frac{A_s}{nB}\right)^{\frac{1}{\mu-m+s}},$$

namely, the first, if $m-s$ is greater than μ, and the second in the contrary case. It is worth observing that if the number s vanishes, then A_s reduces to the coefficient of x^m, that is to say, to 1.

Scholium V. — It is again easy to obtain two limits, one less than and the other greater than the positive roots of equation (27), by the method I am going to indicate. We observe initially that any equation for which the left-hand side gives but one change of sign, that is any equation that presents itself in the form

$$A_0 x^m + A_1 x^{m-1} + \ldots - A_n x^{m-n} - A_{n+1} x^{m-n-1} - \ldots = 0$$

or in the following

$$-A_0 x^m - A_1 x^{m-1} - \ldots + A_n x^{m-n} + A_{n+1} x^{m-n-1} + \ldots = 0,$$

where $A_0, A_1, \ldots, A_n, A_{n+1}, \ldots$ denote any numbers, admits but one positive root, evidently equal to the only positive value of x for which the fraction

$$\frac{A_0 x^n + A_1 x^{n-1} + A_2 x^{n-2} + \ldots}{A_n + A_{n+1}\left(\frac{1}{x}\right) + A_{n+2}\left(\frac{1}{x}\right)^2 + \ldots},$$

which increases without stopping from $x=0$ to $x=\infty$, can be reduced to 1. Consequently the left-hand side of such an equation has the same sign as its first or its last terms, depending on whether the value of x is larger than the root in question, or it is contained between zero and this same root. Given this, imagine that in polynomial (39), $-A_s x^s$ is the first negative term after x^m, that $+A_n x^n$ is the first positive term after $-A_s x^s$, that $-A_v x^v$ is the first negative term after $A_n x^n$, that $+A_w x^w$ is the first [396] positive term after $-A_v x^v$, ..., so that equation (27) becomes

$$\begin{aligned} x^m &+ A_1 x^{m-1} + \dots \\ &- A_s x^{m-s} - A_{s+1} x^{m-s-1} - \dots \\ &+ A_u x^{m-u} + A_{u+1} x^{m-u-1} + \dots \\ &- A_v x^{m-v} - A_{v+1} x^{m-v-1} - \dots \\ &+ A_w x^{m-w} + A_{w+1} x^{m-w-1} + \dots \pm A_m = 0. \end{aligned}$$

We conclude from the preceding remarks that every positive value of x that satisfies equation (27) ought to be: 1° less than the largest of the positive roots of the equations

$$\begin{array}{llll} x^m & +A_1 x^{m-1} & +\dots - A_s x^{m-s} - A_{s+1} x^{m-s-1} & - \dots = 0, \\ A_n x^{m-n} & +A_{n+1} x^{m-n-1} & +\dots - A_v x^{m-v} - A_{v+1} x^{m-v-1} & - \dots = 0, \\ \dots\dots & \dots\dots & \dots\dots\dots\dots & \dots\dots ; \end{array}$$

and 2° greater than the smallest of the same roots, when A_m is preceded by the sign $-$, and in the contrary case, to the smallest positive roots of equations of the form

$$\begin{array}{l} -A_s x^{m-s} - A_{s+1} x^{m-s-1} - \dots + A_n x^{m-n} + A_{n+1} x^{m-n-1} + \dots = 0, \\ -A_v x^{m-v} - A_{v+1} x^{m-v+1} - \dots + A_w x^{m-w} + A_{w+1} x^{m-w-1} + \dots = 0, \\ \dots\dots\dots\dots\dots\dots\dots\dots\dots\dots\dots\dots \end{array}$$

Sometimes the two conditions which we have just stated are mutually exclusive, and then we can be assured that equation (27) has no positive roots.

Problem II. — *To find the number of real roots of equation* (27) *with a sequence of quantities which, taken in pairs, serve as limits to these same roots.*

Solution. — Suppose that equation (27) is reduced so that it has only unequal roots. Then, if we denote by k (see the preceding problem) an upper limit of all of the real roots, and by h a number less than the smallest difference between these roots, and finally by $k_1, k_2, \dots, k_n$ some other numbers chosen so that in the sequence

$$(46) \qquad -k,\ -k_1,\ -k_2,\ \dots,\ -k_n,\ 0,\ k_n,\ \dots,\ k_2,\ k_1,\ k,$$

the difference between any term and the term which precedes it is always a positive quantity less than or equal to h,[11] it is clear that two consecutive terms of the sequence (46) never contain more than one real root between them. Moreover, when

[11] From here to equation (49), Cauchy will be working to get a lower bound on the size of h.

we substitute in the place of x in the polynomial $F(x)$, two quantities [397] between which there is at most a single real root, the results obtained are of the same sign or of contrary signs. Put another way, the comparison of these two results gives either no change in sign or a variation in sign depending on whether there does or does not exist a real root between the two quantities in question. Consequently, if we take the terms of sequence (46) for the successive values of the variable x and then form the sequence of corresponding values of the polynomial $F(x)$, this new sequence gives as many variations in sign as equation (27) has real roots, and each of these roots is contained between two consecutive values of x which, substituted into $F(x)$, give results with contrary signs. Thus, all the difficulty consists in finding a suitable value for the number h. We do that in the following way.

Denote by H the numerical value of the last term in the equation in z given by eliminating x in formulas (37). The number H, as we have already remarked, is equivalent (ignoring the sign) to the product of the squares of the differences between the roots, real and imaginary, of equation (27). Consequently, $H^{\frac{1}{2}}$ is equivalent to the product of the moduli of these differences (where the modulus of each real difference is just its numerical value). Given this, let a and b be two distinct roots of equation (27). If these two roots are real, each of them has a numerical value less than k, and the numerical value of their difference, that is to say the difference or the sum of their numerical values, never surpasses $2k$. On the other hand, if one or both of these roots becomes imaginary, denoting their moduli by r_1 and r_2 and two real arcs by t_1 and t_2, we can suppose that

$$a = r_1\left(\cos t_1 + \sqrt{-1}\sin t_1\right) \quad \text{and}$$
$$b = r_2\left(\cos t_2 + \sqrt{-1}\sin t_2\right).$$

From this we deduce that

$$a-b = r_1\cos t_1 - r_2\cos t_2 + (r_1\sin t_1 - r_2\sin t_2)\sqrt{-1},$$
$$\begin{aligned}\operatorname{mod.}(a-b) &= \left[(r_1\cos t_1 - r_2\cos t_2)^2 + (r_1\sin t_1 - r_2\sin t_2)^2\right]^{\frac{1}{2}}\\ &= \left[r_1^2 - 2r_1r_2\cos(t_1-t_2) + r_2^2\right]^{\frac{1}{2}} < \left(r_1^2 + 2r_1r_2 + r_2^2\right)^{\frac{1}{2}}.\end{aligned}$$

Thus we have

$$\operatorname{mod.}(a-b) < r_1 + r_2,$$

and consequently

(47) $$\operatorname{mod.}(a-b) < 2k,$$

[398] provided that the number k has been chosen, as in the first problem, in a way that surpasses not only the numerical values of all the real roots, but also the moduli of all the imaginary roots. In the same way, we prove that each of the differences

$$a-c, \quad \ldots, \quad b-c, \quad \ldots$$

has for its modulus a number less than $2k$. If we form all the moduli of this kind, numbering $\frac{m(m-1)}{2}$, and if we set aside one of them, for example, the modulus of the difference $a-b$, then we conclude that the product of all the others is a number less than the expression

$$(2k)^{\frac{m(m-1)}{2}-1}.$$

Then, if we multiply this expression by the modulus of the difference $a-b$, we find a result greater than the product of the moduli of all the differences, that is to say a result greater than $H^{\frac{1}{2}}$. In other words, we have

$$(2k)^{\frac{m(m-1)}{2}-1} \times \operatorname{mod.}(a-b) > H^{\frac{1}{2}},$$

or what amounts to the same thing,

$$\operatorname{mod.}(a-b) > \frac{H^{\frac{1}{2}}}{(2k)^{\frac{m(m-1)}{2}-1}}. \tag{48}$$

When the roots a and b are real, the modulus of the difference $a-b$ reduces to its numerical value. Consequently, if we set

$$h = \frac{H^{\frac{1}{2}}}{(2k)^{\frac{m(m-1)}{2}-1}}, \tag{49}$$

then we obtain a number h less than the smallest difference between the real roots of equation (27).

Scholium I. — It would be easy to prove that if each of the numbers $A_1, A_2, \ldots, A_m$ (problem I) is integer, then the number H is integer as well. Consequently, under this hypothesis, the number H, which cannot vanish as long as the roots of equation (27) are unequal to each other, has a value equal to or greater than 1. Given this, formula (48) gives

$$\operatorname{mod.}(a-b) > \frac{1}{(2k)^{\frac{m(m-1)}{2}-1}}. \tag{50}$$

[399] We conclude that, to obtain a number h less than the smallest difference between the roots, it suffices to take

$$h = \frac{1}{(2k)^{\frac{m(m-1)}{2}-1}}. \tag{51}$$

Scholium II. — Let

$$Z = 0 \tag{52}$$

be the equation in z given by the eliminating x from formulas (37). If, by the method indicated above (problem I, scholium III), we determine a limit G less than the moduli of all of the roots, real or imaginary, of equation (52), and if we still denote the roots of equation (27) $a, b, c, \ldots$, then we have

$$\bmod . F_1(a) > G,$$

or what amounts to the same thing [see equations (35)],

$$\bmod . (a-b)(a-c)\ldots > G.$$

We conclude that

$$\bmod . (a-b) > \frac{G}{\bmod . (a-c)\ldots}$$

and consequently,

$$\bmod . (a-b) > \frac{G}{(2k)^{m-2}}, \tag{53}$$

because the differences

$$a-b, \quad a-c, \quad \ldots,$$

which involve the root a combined successively with each of the others, are $m-1$ in number, or if we set aside the difference $a-b$, they are $m-2$ in number. Given this, it is clear that the number h still satisfies the required conditions if we take

$$h = \frac{G}{(2k)^{m-2}}. \tag{54}$$

Scholium III. — Having determined h by one of the preceding methods, we are able to choose for the sequence of numbers

$$k_1, \quad k_2, \quad \ldots, \quad k_n$$

a decreasing arithmetic progression for which the difference is equal to or less than h, in which the terms must always [400] lie between the limits 0 and k. Moreover, if we denote by g (see problem I, scholium III), a limit less than the numerical values of all of the real roots of equation (27), we are evidently able to remove from series (46) all the terms, positive and negative, whose numerical values are smaller than g and write in their place just the two terms

$$-g \quad \text{and} \quad +g.$$

If after modifying sequence (46) as we have just said, we substitute successively into the polynomial $F(x)$: 1° the negative terms of this series from $-k$ to $-g$; and 2° the positive terms from $+g$ to $+k$, and every time that two consecutive terms of the first or the second kind give results of contrary signs, we are certain that a real root, negative in the first case, positive in the second, is contained between these two terms.

Scholium IV. — When, by whatever means, we have determined for equation (27) a value close to the real root a, either above or below, then in a great number of cases we can obtain a value close to this same root in the contrary direction, and thus fix two limits, one greater than the real roots less than a, and the other less than the real roots greater than a, by applying the proposition I am about to state.

As usual, let

$$F_1(x), \quad F_2(x), \quad F_3(x), \quad \ldots$$

represent the coefficients of the first, second, third, … powers of z in the expansion of $F(x+z)$, a, b, c, … the various roots of equation (27) *and k a number greater than their moduli. In addition, suppose that the quantity ξ has a value close to the real root a, where the difference $a-\xi$ and the quantity α determined by the equation*

$$\alpha = -\frac{F(\xi)}{F_1(\xi)} \tag{55}$$

are so small, ignoring the signs, that in the polynomial

$$F_1(\xi)+2(2\alpha)F_2(\xi)+3(2\alpha)^2F_3(\xi)+4(2\alpha)^3F_4(\xi)+\ldots, \tag{56}$$

the numerical value of the first term surpasses the sum of the numerical values of all the others. Finally, denote by G a number less than [401] *the excess of the first numerical value of the given sum. We are certain:* 1° *that the real root a is contained between the limits*

$$\xi \quad \text{and} \quad \xi+2\alpha;$$

and 2° *that the difference $a-b$ or $b-a$ between the root a and a new real root b does not surpass*

$$\frac{G}{(2k)^{m-2}}. \tag{57}$$

To prove the preceding proposition, we first observe that under the given hypothesis, because polynomial (56) has the same sign as its first term, we can say as much *a fortiori* about the two polynomials

$$\begin{cases} -3F_1(\xi)+(2\alpha)F_2(\xi)-(2\alpha)^2F_3(\xi)+(2\alpha)^3F_4(\xi)-\ldots, \\ \quad F_1(\xi)+(2\alpha)F_2(\xi)+(2\alpha)^2F_3(\xi)+(2\alpha)^3F_4(\xi)+\ldots, \end{cases} \tag{58}$$

which we obtain by expanding the fractions

$$\frac{F(\xi-2\alpha)}{\alpha} \quad \text{and} \quad \frac{F(\xi+2\alpha)}{\alpha}$$

according to the ascending powers of α and using equation (55). Consequently, because the first terms of the two polynomials are of opposite signs, the same is true for the two fractions and for their numerators

$$F(\xi-2\alpha) \quad \text{and} \quad F(\xi+2\alpha).$$

So there is at least one real root of equation (27) between the limits

$$\xi - 2\alpha \quad \text{and} \quad \xi + 2\alpha.$$

I add that there is only one of them, and indeed, it is easy to see that if several real roots were contained between these limits and if a and b denote two such roots taken one after the other from the sequence, we would find for the values of the expressions

$$F_1(a) = (a-b)(a-c)\ldots, \quad \text{and}$$
$$F_1(b) = (b-c)(b-a)\ldots,$$

two quantities with contrary signs. Consequently the equation

(59) $$F_1(x) = 0$$

[402] would have a real root contained between a and b, of the form

$$\xi + z,$$

where the quantity z is contained between the limits -2α and $+2\alpha$. Now this cannot be accepted, because if we replace z in formula (31) by $y+z$ and if we expand the left-hand side of this formula, modified according to the ascending powers of y, we get

$$F(x+z) + yF_1(x+z) + \ldots = F(x) + (y+z)F_1(x) + (y+z)^2 F_2(x) + \ldots,$$

then, in equating the coefficients of the first power of y on both sides,

(60) $$F_1(x+z) = F_1(x) + 2zF_2(x) + 3z^2F_3(x) + 4z^3F_4(x) + \ldots.$$

Consequently, the expansion of

(61) $$F_1(\xi + z)$$

becomes

(62) $$F_1(\xi) + 2zF_2(\xi) + 3z^2F_3(\xi) + 4z^3F_4(\xi) + \ldots.$$

Because in polynomial (56) the numerical value of the first term exceeds the sum of the numerical values of all the others, the same is true *a fortiori* for polynomial (62), as long as the numerical value of z is supposed to be less than that of 2α. It follows that under this hypothesis expression (61) does not vanish. Thus equation (59) does not have real roots contained between the limits $\xi - 2\alpha$ and $\xi + 2\alpha$, and equation (27) has only one root between these limits. The root in question is necessarily the one closest to the quantity ξ, and which we have denoted by a. On the other hand, because the fraction

$$\frac{F(\xi + 2\alpha)}{\alpha}$$

is equivalent to the second of the two polynomials (58) and has the same sign as the first term of this polynomial, namely

$$F_1(\xi) = -\frac{F(\xi)}{\alpha},$$

we ought to conclude that

$$F(\xi) \quad \text{and} \quad F(\xi + 2\alpha)$$

[403] are two quantities of contrary signs and that the root a is found bound between the two limits

$$\xi \quad \text{and} \quad \xi + 2\alpha.$$

As for the second part of the proposition stated above, it is an immediate consequence of scholium II, because the quantity G evidently lies below, ignoring the sign, polynomial (62), that is to say of the expansion of $F_1(\xi + z)$, as long as the numerical value of z does not exceed that of 2α, and consequently is less than the quantity $F_1(a)$, which we deduce from $F_1(\xi + z)$ by setting

$$z = a - \xi.$$

Thus it follows from this second part that the real roots greater than a are all greater than the limit

(63) $$a + \frac{G}{(2k)^{m-2}}$$

and the roots smaller than a are less than the limit

(64) $$a - \frac{G}{(2k)^{m-2}}.$$

Problem III. — *To find values as close as we might wish to the real roots of equation* (27).

Solution. — We begin by determining, with the aid of the preceding problem, two limits, one greater than and one less than each real positive root. Suppose in particular that the root a is of this kind, and denote by x_0 and X the two limits, below and above this root. If we form two different sums, the first with the positive terms of the polynomial $F(x)$, the second with the negative terms taken with the contrary sign, then the one which is smaller for $x = x_0$ becomes the larger for $x = X$. Represent this sum by $\varphi(x)$ and the other by $\chi(x)$. The two integer functions $\varphi(x)$ and $\chi(x)$ enjoy the properties stated in theorems II and III, and consequently, if the function $\varphi(x)$ is such that we can easily solve equations of the form

$$\varphi(x) = \text{const.},$$

then formulas (7) and (16) immediately give values closer and [404] closer to the root a. This is what happens, for example, whenever the function is given in the form

$$B(x + C)^n + D,$$

where B, C, D are any three integer numbers and n an integer number equal to or less than m, because then we obtain the successive terms of series (6) and (15) by the extraction of roots of degree n. If the function $\varphi(x)$ is not of the form that we have just indicated, we can easily put it into that form by adding an integer polynomial $\psi(x)$, in which all the terms are positive, to both sides of the equation

$$\varphi(x) = \chi(x).$$

Indeed, it is clear that the values of $\varphi(x)$ and of $\chi(x)$, modified by the addition of such a polynomial, preserve all of the same properties. Moreover, we can assign an infinity of different values to the polynomial $\psi(x)$. Suppose, for example, that

$$\varphi(x) = x^3 + 3x^2 + 8.$$

The value of $\varphi(x)$ modified by the addition of the polynomial $\psi(x)$ becomes

$$(x+1)^3 + 7$$

if we suppose that

$$\psi(x) = 3x,$$

but it becomes

$$(x+2)^3$$

if we suppose that

$$\psi(x) = 3x^2 + 12x,$$

etc. On this matter, it is worth remarking: 1° that we can always choose the integer function $\psi(x)$ so that the number B is 1; and 2° that in many cases, one of the numbers C or D can be reduced to zero.

After using the preceding method to determine the real positive roots of equation (27), it evidently suffices to obtain the negative roots as well by using the same method to seek the positive roots of the equation

$$F(-x) = 0. \tag{65}$$

Scholium. — There exist several methods of approximation other than the one we have just described, among which we must mention that of Newton. It supposes that we already know a value ξ close to the [405] root that we are seeking, and it consists of taking as a correction to this value the quantity α determined by the equation

$$\alpha = -\frac{F(\xi)}{F_1(\xi)}. \tag{55}$$

However, because this last method is not always applicable, it is important to examine the cases in which it can be used. On this subject, we are going to establish the following propositions:

Theorem IV. — *Suppose that a denotes any one of the real roots, positive or negative, of equation* (27) *and that* ξ *is a value close to this root. Suppose that we determine* α *by means of equation* (55). *If* α *is small enough, ignoring the sign, that the numerical value of the first term of polynomial* (56) *exceeds the sum of the numerical values of all the others, then of the two quantities*

$$\xi \quad \text{and} \quad \xi+\alpha,$$

the second is closer to a than the first.

Proof. — We have already seen (problem II, scholium IV) that under the given hypotheses, the root a is the only root between the limits

$$\xi \quad \text{and} \quad \xi+2\alpha.$$

Given this, if we take

(66) $$a=\xi+z,$$

then z is a quantity contained between the limits 0 and 2α and satisfies the equation

$$F(\xi+z)=0,$$

or what amounts to the same thing,

(67) $$F(\xi)+zF_1(\xi)+z^2F_2(\xi)+\ldots=0.$$

If, for convenience, we make

(68) $$q=-\frac{F_2(\xi)+zF_3(\xi)}{F_1(\xi)}$$

and consider formula (55), then equation (67) becomes

(69) $$z=\alpha+qz^2.$$

[406] Consequently, we have

(70) $$a=\xi+z=\xi+\alpha+qz^2,$$

from which it follows that by taking $\xi+\alpha$ in place of ξ for the value close to a, we commit an error that is equal, no longer to the numerical value of z, but rather to that of qz^2. Moreover, because polynomial (56) has the same sign as its first term $F_1(\xi)$, the two polynomials

(71) $$\begin{cases} F_1(\xi)+2(2\alpha)F_2(\xi)+2(2\alpha)zF_3(\xi)+\ldots=(1-4\alpha q)F_1(\xi) \quad \text{and} \\ F_1(\xi)-2(2\alpha)F_2(\xi)-2(2\alpha)zF_3(\xi)-\ldots=(1+4\alpha q)F_1(\xi) \end{cases}$$

evidently enjoy the same property, which requires that the numerical value of $2\alpha q$ and *a fortiori* that of qz be less than $\frac{1}{2}$. We conclude immediately that the numerical value of qz^2 is less than that of $\frac{1}{2}z$. Thus, of the two errors which we make by taking

$$\xi \quad \text{and} \quad \xi+\alpha$$

as values close to a, the second error is smaller than half of the first.

Scholium I. — Because we get

$$z = \frac{\alpha}{1-qz}$$

from equation (69) and because the numerical value of qz is less than $\frac{1}{2}$, we are certain that the value of z always remains between the limits

$$\tfrac{2}{3}\alpha \quad \text{and} \quad 2\alpha.$$

Scholium II. — By solving equation (69) as if the value of q were known, we find

$$z = \frac{1\pm\sqrt{1-4\alpha q}}{2q} = \frac{2\alpha}{1\mp\sqrt{1-4\alpha q}}.$$

Here the radical $\sqrt{1-4\alpha q}$ is given a double sign. However, because the value of z ought to be smaller than that of 2α, it is clear that we ought to prefer the inferior sign. Thus we have

$$z = \frac{2\alpha}{1+\sqrt{1-4\alpha q}}. \tag{72}$$

[407] Given this, if we call q_0 and Q two limits, the first less than and the second greater than the quantity q determined by formula (68), we conclude from equation (72) that the exact value of z is contained between the two expressions

$$\frac{2\alpha}{1+\sqrt{1-4\alpha q_0}} \quad \text{and} \quad \frac{2\alpha}{1+\sqrt{1-4\alpha Q}}. \tag{73}$$

Consequently, this value contains all the decimal digits common to the two expressions expressed as numbers.

Scholium III. — Suppose that of the two quantities q_0 and Q, the second has the larger numerical value and that this numerical value is less than 1. Then, if the difference $a-\xi = z$ is, ignoring the sign, smaller than the unit decimal of order n, that is to say if we have

$$\text{val.num.}\, z < \left(\frac{1}{10}\right)^n, \tag{74}$$

then the difference

$$a-(\xi+\alpha) = qz^2$$

is smaller, ignoring the sign, than a unit decimal of order $2n$. Indeed, we find that

(75) $$\text{val.num.}\, qz^2 < \left(\frac{1}{10}\right)^{2n}.$$

Thus, by taking $\xi + \alpha$ in place of ξ for the value close the root a, we double the number of exact decimal places.

If we suppose that the numerical value of Q is not only less than 1 but also less than 0.1, we conclude from formula (74) that

$$\text{val.num.}\, qz^2 < \left(\frac{1}{10}\right)^{2n+1}.$$

More generally, if we suppose that the numerical value of Q is less than $\left(\frac{1}{10}\right)^r$, where r denotes any integer number, then formula (74) implies the following

(76) $$\text{val.num.}\, qz^2 < \left(\frac{1}{10}\right)^{2n+r}.$$

Finally, if the value of Q is greater than 1 but less than $(10)^r$, we [408] find

(77) $$\text{val.num.}\, qz^2 < \left(\frac{1}{10}\right)^{2n-r}.$$

Scholium IV. — The error that we make by taking $\xi + \alpha$ as a value close to a, or the numerical value of the product qz^2, can itself be calculated by approximation. Indeed, if we consider equation (69), we find

$$qz^2 = q\left(\alpha + qz^2\right)^2 = q\alpha^2 + (2\alpha)\, q^2 z^2 + q^3 z^4.$$

Now suppose that the numerical value of 2α, and consequently that of z, is less than $\left(\frac{1}{10}\right)^n$ and that the numerical value of Q, and consequently that of q, is less than $(10)^{\mp r}$, where n and r denote two integer numbers. We evidently have

$$\text{val.num. } (2\alpha)\, q^2 z^2 < \left(\frac{1}{10}\right)^{3n\pm 2r}$$

and

$$\text{val.num.}\, q^3 z^4 < \left(\frac{1}{10}\right)^{4n\pm 3r}.$$

Moreover, if the numerical value of the fraction

(78) $$\frac{F_3(\xi) + zF_4(\xi) + \ldots}{F_1(\xi)}$$

is known to be less than $(10)^{\mp s}$, where s again denotes an integer number, we can take

$$-\alpha^2 \frac{F_2(\xi)}{F_1(\xi)}$$

for a value close to the term $q\alpha^2$ without fearing an error more considerable than

$$\left(\frac{1}{10}\right)^{3n\pm s}.$$

Consequently, if we choose $\xi+\alpha-\alpha^2\frac{F_2(\xi)}{F_1(\xi)}$ in place of $\xi+\alpha$ for a value close to the root a, that is to say if we set

$$a=\xi+\alpha-\alpha^2\frac{F_2(\xi)}{F_1(\xi)}, \tag{79}$$

then the error made affects only the decimal units of [409] the order given by the largest of the three numbers

$$3n\pm s, \quad 3n\pm 2r \quad \text{and} \quad 4n\pm 3r.$$

In the particular case where the numerical value of Q is less than $\left(\frac{1}{10}\right)^r$, and that of fraction (78) is less than $\left(\frac{1}{10}\right)^s$, the new error becomes less than

$$\left(\frac{1}{10}\right)^{3n}.$$

Thus it suffices to substitute the right-hand side of equation (79) for the quantity ξ to triple the number of exact decimal digits in the value close to a. This is also what happend, more or less, when the number n becomes very considerable. These results agree with those which M. Nicholson has obtained in a work recently published in London, which has for its title *Essay on involution and evolution, etc.*[12]

Theorem V. — *Under the same hypotheses as the preceding theorem, imagine that the first term of polynomial* (56), *that is to say the polynomial that represents the expansion of* $F_1(\xi+2\alpha)$, *has a numerical value not only greater than the sum of the numerical values of all the other terms, but also greater than the double of this sum. Then if we denote a quantity contained between the limits*

$$\xi \quad \textit{and} \quad \xi+2\alpha$$

by ξ_1, *then the second of the two quantities*

$$\xi_1 \quad \textit{and} \quad \xi_1-\frac{F(\xi_1)}{F_1(\xi_1)}$$

[12] Peter Nicholson (1765–1844); see [Nicholson 1820].

is always closer to a than the first.

Proof. — To establish the proposition that we have just stated, it suffices to observe that the numerical value of the difference

$$a-\xi_1$$

is greater than that of

$$a-\left[\xi_1-\frac{F(\xi_1)}{F_1(\xi_1)}\right]=(a-\xi_1)-\frac{F(a)-F(\xi_1)}{F_1(\xi_1)},$$

or what amounts to the same thing, that the fraction

$$\frac{F_1(\xi_1)-\frac{F(a)-F(\xi_1)}{a-\xi_1}}{F_1(\xi_1)}$$

[410] has a numerical value less than 1. Represent this fraction by $\frac{u}{v}$. It is enough to prove that

$$v-u \quad \text{and} \quad v+u,$$

in other words,

(80) $$\frac{F(a)-F(\xi_1)}{a-\xi_1} \quad \text{and} \quad 2F_1(\xi_1)-\frac{F(a)-F(\xi_1)}{a-\xi_1}$$

are two expressions with the same sign. Now if we make

(81) $$a=\xi+z \quad \text{and} \quad \xi_1=\xi+\beta,$$

where z and β are two quantities of the same sign contained between the limits 0 and 2α, and if we expand the functions

$$F(\xi_1+z), \quad F(\xi_1+\beta) \quad \text{and} \quad F_1(\xi+\beta),$$

then expressions (80) become, respectively,

$$F_1(\xi)+(\beta+z)F_2(\xi)+\left(\beta^2+\beta z+z^2\right)F_3(\xi)+\ldots$$

and

$$F_1(\xi)-(\beta+z-4\beta)F_2(\xi)-\left(\beta^2+\beta z+z^2-6\beta^2\right)F_3(\xi)+\ldots.$$

Because in each of these last polynomials the coefficient of $F_n(\xi)$ has a numerical value evidently less than 1 of the quantities

$$nz^{n-1} \quad \text{and} \quad 2n\beta^{n-1},$$

and consequently less than the double of the numerical value of the product

$$n\,(2\alpha)^{n-1},$$

it is clear that if the condition stated in theorem V is satisfied, then they both have the same sign as $F_1(\xi)$. Thus, etc.

Scholium I. — The errors committed when we take successively

$$\xi_1 \quad \text{and} \quad \xi_1 - \frac{F(\xi_1)}{F_1(\xi_1)}$$

for values close to the root a are, respectively, equal to the numerical values of the two quantities

$$a - \xi_1 \quad \text{and} \quad a - \xi_1 + \frac{F(\xi_1)}{F_1(\xi_1)}.$$

[411] Considering formulas (81), we also find

$$a - \xi_1 = z - \beta \tag{82}$$

and

$$\begin{aligned} a - \xi_1 + \frac{F(\xi_1)}{F_1(\xi_1)} &= a - \xi_1 - \frac{F(a) - F(\xi_1)}{F_1(\xi_1)} \\ &= z - \beta - \frac{F(\xi + z) - F(\xi + \beta)}{F_1(\xi + \beta)}. \end{aligned}$$

Then, by expanding the functions $F(\xi + z)$, $F(\xi + \beta)$ and $F_1(\xi + \beta)$, we find

$$\left\{ \begin{aligned} & a - \xi_1 + \tfrac{F(\xi_1)}{F_1(\xi_1)} \\ & \quad = -(z-\beta)^2 \frac{F_2(\xi) + (z + 2\beta) F_3(\xi) + (z^2 + 2\beta z + 3\beta^2) F_4(\xi) + \dots}{F_1(\xi) + 2\beta F_2(\xi) + 3\beta^2 F_3(\xi) + \dots}. \end{aligned} \right. \tag{83}$$

Given this, imagine that for all values of β and z contained between 0 and 2α the numerical value of the polynomial

$$F_2(\xi) + (z + 2\beta) F_3(\xi) + (z^2 + 2\beta z + 3\beta^2) F_4(\xi) + \dots \tag{84}$$

is always less than the limit M, and that of the polynomial

$$F_1(\xi) + 2\beta F_2(\xi) + 3\beta^2 F_3(\xi) + \dots \tag{85}$$

is always greater than the limit N. If we have

$$\text{val.num.}\,(z - \beta) < \left(\frac{1}{10}\right)^n \tag{86}$$

and

$$\frac{M}{N} < (10)^{\mp r}, \tag{87}$$

where n and r denote any two integer numbers, we conclude from equation (83) that

$$\text{(88)} \qquad \text{val.num.}\left[a-\xi_1+\frac{F(\xi_1)}{F_1(\xi_1)}\right]<\left(\frac{1}{10}\right)^{2n\mp r}.$$

It is essential to remark that to obtain suitable values of M and N, it suffices: 1° to replace z and β by 2α in polynomial (84), then to calculate the sum of the numerical values of all the terms; and 2° to replace β by 2α in polynomial (85) and then to look for the difference between the numerical value of the first term and the sum of the numerical values of all the others.

[412] *Scholium II.* — Under the same hypotheses as in theorem V, if we successively make

$$\text{(89)} \qquad \xi_1=\xi-\frac{F(\xi)}{F_1(\xi)}, \quad \xi_2=\xi_1-\frac{F(\xi_1)}{F_1(\xi_1)}, \quad \xi_3=\xi_2-\frac{F(\xi_2)}{F_1(\xi_2)}, \quad \ldots$$

then the quantities $\xi_1, \xi_2, \xi_3, \ldots$ are values that are closer and closer to the root a. Moreover, if we give M and N the same values as in scholium I, then by supposing that

$$\text{val.num.}(a-\xi)<\left(\frac{1}{10}\right)^n,$$

we conclude that

$$\begin{aligned}
&\text{val.num.}(a-\xi_1)<\left(\tfrac{1}{10}\right)^{2n\pm r},\\
&\text{val.num.}(a-\xi_2)<\left(\tfrac{1}{10}\right)^{4n\pm 3r},\\
&\text{val.num.}(a-\xi_3)<\left(\tfrac{1}{10}\right)^{8n\pm 7r},\\
&\ldots\ldots\ldots\ldots\ldots\ldots\ldots\ldots\ldots
\end{aligned}$$

These last formulas contain the proposition stated by M. Fourier in the *Bulletin de la Société philomathique*[13] (read in May 1818), relative to the number of exact decimal places given by each new operation of Newton's method.

Any time the fraction $\frac{M}{N}$ is less than 1, we can take $r=0$ and consequently the successive differences between the root a and its nearby values

$$\xi, \quad \xi_1, \quad \xi_2, \quad \xi_3, \quad \ldots,$$

are, respectively, less than the numbers

$$\left(\frac{1}{10}\right)^n, \quad \left(\frac{1}{10}\right)^{2n}, \quad \left(\frac{1}{10}\right)^{4n}, \quad \left(\frac{1}{10}\right)^{8n}, \quad \ldots.$$

Thus we find that the number of exact decimal places at least doubles for each new operation.

[13] See [Fourier 1818].

The preceding researches give several methods for solving numerical equations. In order to give a better sense of the advantages that these methods present, I will apply them to the two equations

$$x^2 - 2x - 5 = 0 \tag{90}$$

and

$$x^3 - 7x + 7 = 0, \tag{91}$$

[413] which Lagrange chose as his examples (*Résolution des equations numeriques*, Chap. IV)[14] and the first of which was earlier treated by Newton.

If we first consider equation (90), we find (theorem III, scholium II) that it has a single positive root contained between the two limits

$$\sqrt{2 \cdot 2} = 2 \quad \text{and} \quad \sqrt[3]{2 \cdot 5} = 2.15\ldots.$$

Moreover, the positive value of x that satisfies the equation

$$2x + 5 = x^2$$

satisfies (problem I, scholium IV) the condition

$$2\sqrt{5 \cdot 2x} < x^2,$$

or what amounts to the same thing, the following:

$$x > (40)^{\frac{1}{5}} = 2.09\ldots.$$

The root in question is thus contained between the numbers $2.09\ldots$ and $2.15\ldots$, so that its value, to less than a tenth part, is 2.1. To obtain a more exact value, we observe that in the present case,

$$F(x) = x^3 - 2x - 5, \quad F_1(x) = 3x^2 - 2, \quad F_2(x) = 3x \quad \text{and} \quad F_3(x) = 1,$$

and that, if we take

$$\xi = 2.1,$$

then the condition stated in theorem IV is satisfied. Given this, because from equation (55) we get

$$\alpha = \frac{5 + \frac{4}{3}\xi}{3\xi^2 - 2} - \frac{\xi}{3} = -0.005431878\ldots,$$

we find for the new values close to the unknown x

$$\xi + \alpha = 2.094568121\ldots$$

and

[14] See [Lagrange 1769].

$$\xi+\alpha-\alpha^2\frac{F_2(\xi)}{F_1(\xi)}=\xi+\alpha-\alpha^2\frac{3\xi}{3\xi^2-2}=2.0945515\ldots.$$

Finally, because the exact value of x is presented in the form $x=\xi+z$, [414] z is a quantity contained between the limits 0 and 2α and that consequently we evidently have

$$-q=\frac{F_2(\xi)+zF_3(\xi)}{F_1(\xi)}=\frac{6.3+z}{11.23}<0.6<1,$$
$$\frac{F_3(\xi)}{F_1(\xi)}=\frac{1}{11.23}<0.1,$$

and

$$\text{val.num.}\,z=\text{val.num.}\,\left(\alpha+qz^2\right)<\text{val.num.}\,\alpha+(2\alpha)^2\,\text{val.num.}\,q<0.01.$$

Then we conclude (theorem IV, scholia III and IV) that in taking

$$x=2.0945681,$$

we commit an error smaller than 0.0001, and in taking

$$x=2.0945515$$

an error smaller than 0.000001.

Instead of using the general formulas, we could carry out the calculation in the following manner. After finding 2.1 as the value close to x, we make in equation (90)

$$x=2.1+z,$$

and we find, by dividing all the terms by the coefficient of z, that

$$0.005431878\ldots+z+0.560997328\ldots z^2+0.089047195\ldots z^3=0, \tag{92}$$

or what amounts to the same thing,

$$z=-0.005431878\ldots+qz^2, \tag{93}$$

where the value of q is determined by the formula

$$q=-0.560997328\ldots-0.089047195\ldots z. \tag{94}$$

The double of the first term of equation (92) is almost 0.01, and because the left-hand side of this equation gives two results of contrary signs when we successively make

$$z=0 \quad\text{and}\quad z=-0.01,$$

we can be sure it has a real root contained between the limits 0 and -0.01. To prove that this root is unique, it suffices to observe that by virtue of formula (60) the equation

$$F_1(2.1+z)=0$$

[415] reduces to

$$1+2\times 0.560997328\ldots z+3\times 0.089047195\ldots z^2=0,$$

and this last equation is not satisfied by any value of z between the limits in question. Moreover, it is clear that for such a value of z, the quantity q determined by formula (94) remains contained between -0.560 and -0.561. Because we draw from equation (93)

$$\text{(95)}\qquad \begin{cases} z=-0.005431878\ldots-\ 0.000029505\ldots(-q) \\ \qquad\qquad\qquad\qquad\quad -\ 0.000000320\ldots(-q)^2-\ldots, \end{cases}$$

we conclude: 1° supposing that $-q=0.560$,

$$z=-0.00544850\ldots;$$

and 2° supposing that $-q=0.561$,

$$z=-0.00544853\ldots.$$

Consequently, the real positive value of x that satisfies equation (90) is contained between the limits

$$2.1-0.00544850=2.09455150$$

and

$$2.1-0.00544854=2.09455146.$$

Thus this equation has a unique positive real root very nearly equal to

$$2.0945515.$$

Moreover, it is easy to assure ourselves that it does not have any negative roots, for if it had one, we would be able to satisfy with a positive value of x the formula

$$\text{(96)}\qquad x^3-2x+5=0,$$

and this value of x (see scholium V of problem I) is at the same time less than the positive root of the equation

$$x^3-2x=0,$$

that is to say, than

$$\sqrt{2}=1.414\ldots,$$

and greater than the root of the equation

$$5-2x=0,$$

[416] that is to say than

$$\frac{5}{2} = 2.5,$$

which is absurd.

We now move on to equation (91) and begin by looking for its positive roots. To have a limit greater than the roots of this kind, it suffices to observe that the equation under consideration can be put into the form

$$x^3 + 7 = 7x,$$

and we find (problem I, scholium IV) by supposing x is positive, that

$$2\sqrt{7x^3} < 7x$$

and consequently

$$x < \frac{7}{4}.$$

Thus we can take $\frac{7}{4}$ as a value close the largest positive root. Given this, if we make

$$x = \frac{7}{4} + z$$

in equation (91), we find

$$0.5 + z + 2.40z^2 + \frac{32}{70}z^3 = 0, \tag{97}$$

or what amounts to the same thing,

$$z = -0.05 + qz^2, \tag{98}$$

where the value of q is determined by the formula

$$q = -2.40 - \frac{32}{70}z. \tag{99}$$

The double of the first term of equation (97) is 0.1. Because the left-hand side changes sign as it passes from $z = 0$ to $z = -0.1$, while the polynomial

$$1 + 2 \times 2.40z + 3 \times \frac{32}{70}z^2$$

remains always positive in this interval, it follows that it has a real root, but only one of them, contained between the limits 0 and -0.1. The [417] corresponding value of q is evidently contained between the two quantities

$$-2.354\ldots \quad \text{and} \quad -2.40.$$

From equation (98), we find that

$$(100)\begin{cases} z = -\dfrac{0.1}{1+\sqrt{1+0.2q}} \\ \quad = -0.05 - 0.0025\,(-q) - 0.00025\,(-q)^2 - 0.00003125\,(-q)^3 - \ldots. \end{cases}$$

In this last equation, if we successively make

$$q = -2.354 \quad \text{and} \quad q = -2.40,$$

we find the corresponding values of z,

$$z = -0.05788\ldots \quad \text{and} \quad z = -0.05810\ldots,$$

and we conclude that the largest positive root of the given equation is contained between the limits

$$\frac{7}{4} - 0.05788\ldots = 1.69211\ldots$$

and

$$\frac{7}{4} - 0.05810\ldots = 1.69189\ldots.$$

Thus, if we call this largest root a, its value approximated to eleven one hundred thousandths is given by the formula

$$(101) \qquad a = 1.6920.$$

Starting with this first approximate value, in just one operation we could obtain a second value in which the error would not be more than decimals of the twelfth order.

In addition to the root a which we have just considered, equation (91) evidently admits a negative root equal, up to sign, to the unique positive solution of the equation

$$(102) \qquad x^3 - 7x - 7 = 0,$$

and consequently (theorem III, scholium II) contained between the limits

$$-\sqrt{14} = -3.7416\ldots \quad \text{and} \quad -\sqrt[3]{14} = -2.41\ldots.$$

[418] Call this negative root c. The third real root b of equation (91) is evidently real and positive because the product abc of the three roots is equal to the negative of the third term, that is to say, equal to -7. Let us now determine this third root. To find it, we first look for a number G equal to or less than the numerical value of $F_1(a)$. Because in this case[15]

$$F(x) = x^3 - 7x + 7 \quad \text{and}$$
$$F_1(x) = 3x^2 - 7,$$

we conclude

[15] The first term of $F(x)$ is written as x^2 in [Cauchy 1897, p. 418]. It is correctly written as x^3 in [Cauchy 1821, p. 512]. (tr.)

$$F_1(a) = 3a^2 - 7.$$

Thus we can take

$$G = 3(1.69189)^2 - 7 = 1.5874....$$

Moreover, by virtue of what has come before, we also have

$$a < 1.6922 \quad \text{and} \quad -c < 3.7417,$$

and consequently

$$a - c < 5.4339.$$

Given this, we find (problem II, scholium II)

$$a - b > \frac{G}{a-c} > \frac{1.5874}{5.4339} = 0.29212...$$

and as a consequence we have

$$b < 1.69211... - 0.29214... < 1.40.$$

After having recognized, as we have just done, that the root b is less than the limit 1.40, suppose that

$$x = 1.40 + z.$$

Under this hypothesis, equation (91) gives

(103) $$0.05 + z - 3.75z^2 - \frac{25}{28}z^3 = 0,$$

or what amounts to the same thing,

(98) $$z = -0.05 + qz^2,$$

where the value of q is determined by the formula

(104) $$q = 3.75 + \frac{25}{28}z.$$

[419] The double of the first term of equation (103) is 0.1, and because the left-hand side of this equation changes sign when it passes from $z = 0$ to $z = -0.1$, while the polynomial

$$1 - 2 \times 3.75z - 3 \times \frac{25}{28}z^2$$

always remains positive in the interval, it results that the polynomial has a single real root contained between the limits 0 and -0.1. The corresponding value of q is evidently contained between the two quantities

$$3.66 \quad \text{and} \quad 3.75.$$

By successively substituting these two quantities in place of the letter q in equation (100), we obtain two new limits for the unknown z, namely

$$-\frac{0.1}{1+\sqrt{1.732}} = -0.04317\ldots$$

and

$$-\frac{0.1}{1+\sqrt{1.750}} = -0.04305\ldots,$$

so we conclude that the positive root is contained between

$$1.40 - 0.04317\ldots = 1.35682\ldots$$

and

$$1.40 - 0.04305\ldots = 1.35694\ldots.$$

Thus we obtain the value approximating this root to a ten thousandth part if we take

(105) $$b = 1.3569.$$

As for the negative root c of equation (91), we already know that it is contained between the limits

$$-3.7416\ldots \quad \text{and} \quad -2.41\ldots.$$

Thus we have its value approximated to within one unit if we suppose it is equal to -3. Given this, take in equation (91)

$$x = -3 + z.$$

We find

(106) $$0.05 + z - 0.45z^2 + 0.05z^3 = 0,$$

[420] or what amounts to the same thing,

(98) $$z = -0.05 + qz^2,$$

where the value of q is determined by the formula

(107) $$q = 0.45 - 0.05z.$$

Moreover, we easily recognize: 1° that equation (106) has a real root, but only one of them, contained between the limits 0 and -0.1; 2° that the corresponding value of q is contained between the two numbers

$$0.45 \quad \text{and} \quad 0.455;$$

and 3° that these two numbers substituted in place of the letter q in equation (100) give two new values close to z, namely

$$-\frac{0.1}{1+\sqrt{1.09}} = -0.048922\ldots$$

and

$$-\frac{0.1}{1+\sqrt{1.091}} = -0.048911\ldots.$$

Consequently, the value approximating c to a hundred thousandth part is

$$c = -3.04892. \tag{108}$$

Finally, we could have immediately deduced the value approximating c from formulas (101) and (105). Indeed, because in equation (91) the coefficient of x^2 is reduced to zero, we conclude

$$a+b+c=0,$$
$$c=-a-b$$

and consequently, to within a very small margin,

$$c = -(1.6920 - 1.3569) = -3.0489.$$

To end this note, we will present here two theorems, the second of which contains the rule stated by Descartes relative to the determination of the number of positive or negative roots that pertain to a polynomial of any degree. With this plan, we will begin by examining the number of variations and permanences in signs exhibited by [421] a sequence of quantities, when we suppose that the different terms of the sequence are compared to each other in the order in which they appear.

Let

$$a_0, \quad a_1, \quad a_2, \quad \ldots, \quad a_{m-1}, \quad a_m \tag{109}$$

be the sequence we are considering, composed of $m+1$ terms. If none of these terms reduce to zero, then the number of variations in sign which we obtain in comparing consecutive terms two by two is completely determined. However, if some of the terms are zero, then because in this case we can arbitrarily decide the sign of each of these terms, the number of variations in sign depends on the decisions themselves, but it depends in such a way that the number cannot fall below a certain minimum, nor rise above a certain maximum. A similar remark can be made about the number of permanences in signs. We add that to obtain the maximum number of variations in sign, it suffices to consider each term that vanishes as having the opposite sign of the term that precedes it. Imagine, for example, that sequence (109) consists of the four terms

$$+1, \quad 0, \quad 0 \quad \text{and} \quad -1.$$

Because the first of these terms is positive, we obtain the maximum number of variations in sign by considering the second term as negative and the third as positive, or what amounts to the same thing, by writing

$$+1, \quad -0, \quad +0 \quad \text{and} \quad -1.$$

Consequently, in this particular case the maximum number in question is equal to 3. On the other hand, we obtain the minimum number of variations in sign, equal to 1, by assigning each 0 term a sign the same as that of the preceding term, that is to say, by writing[16]

$$+1, \quad +0, \quad +0 \quad \text{and} \quad -1.$$

If we adopt these principles, we establish the following propositions without difficulty:

Theorem VI. — *Suppose that the constant h is real and positive and that we multiply the polynomial*

$$a_0x^m + a_1x^{m-1} + a_2x^{m-2} + \ldots + a_{m-1}x + a_m \tag{110}$$

by the linear factor $x+h$. This multiplication does not increase [422] *the maximium number of variations in sign of the successive coefficients of the descending powers of the variable x.*

Proof. — By multiplying polynomial (110) by $x+h$, we obtain a new polynomial in which the descending powers of the variable have for their respective coefficients the quantities

$$a_0, \quad a_1+ha_0, \quad a_2+ha_1, \quad \ldots, \quad a_m+ha_{m-1}, \quad ha_m. \tag{111}$$

Thus it suffices to prove that the number of variations in sign does not increase in the passage from sequence (109) to sequence (111), when we have carried this number to its maximum in both sequences by giving each term that vanishes the sign opposite to that of the preceding term. Now, I say that initially fixing the signs according to this rule, each term of sequence (111), represented by a binomial of the form

$$a_n + ha_{n-1},$$

takes the same sign as one of the terms a_n or a_{n-1} of sequence (109). This assertion is equally evident in the two cases that can present themselves, namely: 1° when the terms a_{n-1} and a_n are originally, or by virtue of the rule adopted, given opposite signs, for example, when a_n vanishes; or 2° when a_n has a value different from zero and a_{n-1} has the same sign as a_n. Consequently, if we give the quantities

$$ha_0, \quad ha_1, \quad ha_2, \quad \ldots, \quad ha_{m-1}, \quad ha_m \tag{112}$$

the same signs as the corresponding terms of sequence (109), we can, without altering in any way the succession of signs in sequence (111), replace each binomial of the form

$$a_n + ha_{n-1}$$

[16] The third term in this sequence is written as -0 in [Cauchy 1897, p. 421]. It is correctly given as $+0$ in [Cauchy 1821, p. 516]. (tr.)

by one of the two monomials a_n or ha_{n-1}. In operating like this, we obtain a new sequence in which each term of the form a_n finds itself followed by another term equal either to the monomial a_{n+1} or to the monomial ha_n, which is the second part of the binomial $a_{n+1} + ha_n$, while each term of the form ha_n is followed either by the monomial ha_{n+1} or the monomial a_{n+2}, which is the first part of the binomial $a_{n+2} + ha_{n+1}$. Given this, imagine that in the new sequence we distinguish: 1° each term of the form a_n which [423] is followed by another term of the form ha_n; and 2° each term of the form ha_n which is followed by another term of the form a_{n+2}. Let, respectively,

$$a_s, \quad ha_n, \quad a_v, \quad ha_w, \quad \ldots$$

be the different terms of these two kinds arranged following the increasing size of the indices which describe the letter a. The new sequence, composed of the monomials

$$(113) \qquad \begin{cases} a_0,\, a_1,\, \ldots,\, a_s,\, ha_s,\, ha_{s+1},\, \ldots,\, ha_u,\, a_{u+2},\, \ldots,\, a_v, \\ \qquad ha_v,\, ha_{v+1},\, \ldots,\, ha_w,\, a_{w+2},\, a_{w+3},\, \ldots,\, ha_m, \end{cases}$$

evidently exhibits only those changes of sign like those of sequence (109) along with those which can arise in the passage from ha_n to a_{n+2}, from ha_w to a_{w+2}, ... Moreover, it is easy to see that if the two quantities

$$ha_n \quad \text{and} \quad a_{n+2},$$

or what amounts to the same thing,

$$a_n \quad \text{and} \quad a_{n+2},$$

have contrary signs, then the variation in sign does nothing but replace another variation of sign in sequence (109), namely the one that had been between term a_{n+1} and one of the two terms a_n or a_{n+2}. A very similar remark applies in the case that the monomials ha_w and a_{w+2} have contrary signs, etc. We can thus conclude that the maximum number of variations in sign does not increase when we pass from sequence (109) to sequence (113), and consequently to the sequence (111). This is what we set out to prove.

Corollary. — If we multiply polynomial (110) by several linear factors of the form

$$x+h, \quad x+h', \quad x+h'', \quad \ldots,$$

where h, h', h'', ... denote positive quantities, we do not increase the maximum number of variations in sign among the coefficients of successive descending powers of the variable x.

Theorem VII. — *In the polynomial*

$$(110) \qquad F(x) = a_0x^m + a_1x^{m-1} + \ldots + a_{m-1}x + a_m,$$

[424] *let m' be the minimum number of permanences in signs, and m'' the minimum number of variations in sign among the successive coefficients of descending powers of x. Then in the equation*

(114) $$F(x) = 0,$$

the number of negative roots is equal to or less than m', the number of positive roots is less than or equal to m'', and the number of imaginary roots is equal to or greater than to the difference

$$m - \left(m' + m''\right).$$

Proof. — To establish the first part of the theorem, I observe that if we call h, h', h'', ... the negative roots of equation (114), then the polynomial $F(x)$ is divisible by the product

$$(x+h)\left(x+h'\right)\left(x+h''\right)\ldots.$$

Denote the quotient by Q. From the corollary to the preceding theorem, the maximum number of variations in sign in the polynomial $F(x)$ is less than or equal to the maximum of these variations in the polynomial Q, and as a consequence less than or equal to the degree of that polynomial. It follows that the minimum number of permanences in sign of the polynomial $F(x)$ is equal to or greater than the difference between the number m and the degree of the polynomial Q, that is to say, the number of real negative roots of the equation

(114) $$F(x) = 0.$$

To prove the second part of theorem VII, it suffices to remark that by writing $-x$ in place of x in equation (114), we interchange all at once the positive and the negative roots and the variations and permanences in sign.

Finally, because this equation is of degree m, it ought to have m roots, real and imaginary, so it is clear that the third part of the theorem is an immediate consequence of the two others.

Corollary. — To show an application of the preceding theorem, consider the particular equation

(115) $$x^m + 1 = 0.$$

[425] We find: 1° if we suppose that m is even, that

$$m' = 0 \quad \text{and} \quad m'' = 0;$$

and 2° if we suppose that m is odd,

$$m' = 1 \quad \text{and} \quad m'' = 0.$$

Consequently, in the first case equation (115) does not have any real roots, and in the second case it has only one real root, namely a negative root.

Note IV – On the expansion of the alternating function $(y-x) \times (z-x)(z-y) \times \ldots \times (v-x)(v-y)(v-z)\ldots(v-u)$.

[426] We denote the function in question[1] by φ. As we have already remarked (Chap. III, § II), each term of its expansion is equivalent, ignoring the sign, to the product of the various variables arranged in a certain order and, respectively, raised to powers indicated by the numbers

$$0, \quad 1, \quad 2, \quad 3, \quad \ldots, \quad n-1.$$

Moreover, it is easy to see that each of the products of this kind can be derived from each other with the aid of one or several exchanges operated between the variables taken two by two. Thus, for example, we derive the product

$$x^0 y^1 z^2 \ldots u^{n-2} v^{n-1}$$

from any of the products of the same form by making the letter x pass to the first position by successive exchanges, then the letter y to the second position, then the letter z to the third, etc. Moreover, because the function φ changes sign but takes the same value, up to sign, every time we exchange two variables with each other, we ought to conclude: 1° that the expansion of this function contains all the products mentioned above, some taken with the sign $+$ and the others with the sign $-$; and 2° that in the same expansion, two products, taken at random, are affected with the same sign or with contrary signs depending on whether one can derive one from the other by an even number or an odd number or exchanges. As a consequence of these remarks, we establish without difficulty the following proposition:

Theorem I. — *Add to the product*

$$x^0 y^1 z^2 \ldots u^{n-2} v^{n-1}$$

[1] In the chapter heading of this note in [Cauchy 1897, p. 426], the formula is given in a slightly different form. This version of φ is what appears on [Cauchy 1821, p. 521] as well as in the tables of contents of both editions. (tr.)

all those products which we can derive with the aid of one or several exchanges [427] *successively performed among the variables*

$$x,\quad y,\quad z,\quad \ldots,\quad u,\quad v$$

taken two by two. The number of products which we obtain is

$$1\cdot 2\cdot 3\ldots(n-1)\,n,$$

and they divide themselves into two distinct classes in such a way that we can always derive two products of the same class from each other by an even number of exchanges, and two products of different classes by an odd number of exchanges. Given this, if we attach the sign $+$ *to all the products of one class and the sign* $-$ *to all the products of the other class, then we find that the sum is the expansion of* $+\varphi$ *or the expansion of* $-\varphi$*, depending on the class to which we give the sign* $+$.

It evidently suffices to consider the preceding proposition in constructing the expansion of the alternating function $+\varphi$. However, we also ought to mention another theorem, with the aid of which we can decide immediately if two products taken at random in the expansion in question are found to have the same sign or contrary signs. We will content ourselves with stating here this second theorem without giving its proof, which we can deduce without trouble from the principles which we have already explained.

Theorem II. — *To decide if, in the expansion of the alternating function* $\pm\varphi$*, two products of the form*

$$x^0y^1z^2\ldots u^{n-2}v^{n-1}$$

are affected with the same sign or with contrary signs, we distribute the variables

$$x,\quad y,\quad z,\quad \ldots,\quad u,\quad v$$

into several groups, taking care to put two variables into the same group whenever they carry the same exponent in the two products that we are considering, and forming an isolated group for each variable that does not have a change in exponent in the passage from the first product to the second. Given this, the two products are affected with the same sign if the difference between the total number of variables and the total number of groups is an even number, and they are affected with contrary signs if this difference is an odd number.

[428] The use of the preceding theorem is facilitated by writing the two products one over the other and arranging the variables in each of these according to the size of the exponents that they carry.

To apply these two theorems stated above to an example, consider in particular five variables

$$x,\quad y,\quad z,\quad u\quad \text{and}\quad v.$$

The product of their differences, or if we wish, the alternating function

$$(y-x)(z-x)(z-y)(u-x)(u-y)(u-z)$$
$$\times (v-x)(v-y)(v-z)(v-u)$$

gives an expansion composed of 120, respectively, equal to the 120 products where 60 are preceded by the sign + and 60 by the sign −. One of the products affected by the sign + is the one which has for its factors the first letters of the binomials

$$y-x, \quad z-x, \quad z-y, \quad \ldots, \quad v-u,$$

namely

$$x^0 y^1 z^2 u^3 v^4.$$

To judge whether another product such as

$$x^0 z^1 v^2 u^3 y^4$$

ought to be taken with the sign + or with the sign −, it suffices to observe that if we compare the two products in question here with respect to the changes which take place among the given variables when we pass from one to the other, we are led to divide these variables into three groups, the first containing the single variable x, a second containing the three variables y, z and v, and the third the single variable u. If from the number of variables, five, we subtract the number of groups, three, we find there remains two, that is to say, an even number. Consequently, the two products ought to be affected with the same sign, and because the first is preceded by the sign +, the second ought to be as well.

Note V – On Lagrange's interpolation formula.

[429] When we wish to determine an integer function of x of degree $n-1$, given a certain number of particular values assumed to be known, it suffices to use formula (1) of Chapter IV (§ I). This formula, first given by Lagrange, may easily be deduced from the principles outlined in section I of Chapter III. Indeed, let us denote the function we seek by

$$u = a + bx + cx^2 + \ldots + hx^{n-1} \tag{1}$$

and denote its particular values by

$$u_0, u_1, u_2, \ldots, u_{n-1},$$

corresponding to the values

$$x_0, x_1, x_2, \ldots, x_{n-1}$$

of the variable x. The unknowns in this problem are the coefficients a, b, c, …, h of the various powers of x in the polynomial u. In order to determine these unknowns, we have the equations of condition

$$\left\{\begin{array}{ll} u_0 & = a + bx_0 + cx_0^2 + \ldots + hx_0^{n-1}, \\ u_1 & = a + bx_1 + cx_1^2 + \ldots + hx_1^{n-1}, \\ u_2 & = a + bx_2 + cx_2^2 + \ldots + hx_2^{n-1}, \\ \multicolumn{2}{l}{\ldots\ldots\ldots\ldots\ldots\ldots\ldots\ldots\ldots\ldots,} \\ u_{n-1} & = a + bx_{n-1} + cx_{n-1}^2 + \ldots + hx_{n-1}^{n-1}. \end{array}\right. \tag{2}$$

Given this, in order to obtain the explicit value of the function u, we need only eliminate the coefficients a, b, c, …, h from among the formulas (1) and (2). We do this by adding equation (1) to equations (2), after [430] multiplying these latter ones by quantities chosen in such a way as to make the sum of the right-hand sides disappear. Let

$$-X_0, -X_1, -X_2, \ldots, -X_{n-1}$$

be the quantities in question. We have

$$
\begin{aligned}
& u - X_0u_0 - X_1u_1 - X_2u_2 - \ldots - X_{n-1}u_{n-1} \\
&= (1 - X_0 - X_1 - X_2 - \ldots - X_{n-1})\, a \\
&\quad + (x - x_0X_0 - x_1X_1 - x_2X_2 - \ldots - x_{n-1}X_{n-1})\, b \\
&\quad + \left(x^2 - x_0^2X_0 - x_1^2X_1 - x_2^2X_2 - \ldots - x_{n-1}^2X_{n-1}\right) c \\
&\quad + \ldots\ldots\ldots\ldots\ldots\ldots\ldots\ldots\ldots\ldots\ldots\ldots \\
&\quad + \left(x^{n-1} - x_0^{n-1}X_0 - x_1^{n-1}X_1 - x_2^{n-1}X_2 - \ldots - x_{n-1}^{n-1}X_{n-1}\right) h,
\end{aligned}
$$

and consequently

$$
u = X_0u_0 + X_1u_1 + X_2u_2 + \ldots + X_{n-1}u_{n-1}, \tag{3}
$$

as long as the quantities

$$
X_0,\ X_1,\ X_2,\ \ldots,\ X_{n-1}
$$

are subject to the equations of condition

$$
\left\{
\begin{aligned}
X_0 + \quad X_1 + \quad X_2 + \ldots + \quad X_{n-1} &= 1, \\
x_0X_0 + \quad x_1X_1 + \quad x_2X_2 + \ldots + x_{n-1}X_{n-1} &= x, \\
x_0^2X_0 + \quad x_1^2X_1 + \quad x_2^2X_2 + \ldots + x_{n-1}^2X_{n-1} &= x^2, \\
\ldots\ldots\ldots\ldots\ldots\ldots\ldots\ldots\ldots\ldots\ldots\ldots&\ldots, \\
x_0^{n-1}X_0 + x_1^{n-1}X_1 + x_2^{n-1}X_2 + \ldots + x_{n-1}^{n-1}X_{n-1} &= x^{n-1}.
\end{aligned}
\right. \tag{4}
$$

If we solve these new equations by the method described in Chapter III (§ I), we obtain the formulas

$$
\left\{
\begin{aligned}
X_0 &= \frac{(x-x_1)(x-x_2)\ldots(x-x_{n-1})}{(x_0-x_1)(x_0-x_2)\ldots(x_0-x_{n-1})}, \\
X_1 &= \frac{(x-x_0)(x-x_2)\ldots(x-x_{n-1})}{(x_1-x_0)(x_1-x_2)\ldots(x_1-x_{n-1})}, \\
&\ldots\ldots\ldots\ldots\ldots\ldots\ldots\ldots\ldots\ldots, \\
X_{n-1} &= \frac{(x-x_0)(x-x_1)\ldots(x-x_{n-2})}{(x_{n-1}-x_0)(x_{n-1}-x_1)\ldots(x_{n-1}-x_{n-2})},
\end{aligned}
\right. \tag{5}
$$

by virtue of which equation (3) is transformed into the formula of Lagrange.

[431] Moreover, Lagrange's formula is contained in another more general formula, to which we find ourselves drawn when we wish to determine not just an integer function, but a rational function of the variable x, given a certain number of particular values assumed to be known. To clarify the ideas, imagine that the rational function should be of the form

$$
u = \frac{a + bx + cx^2 + \ldots + hx^{n-1}}{\alpha + \beta x + \gamma x^2 + \ldots + \theta x^m}. \tag{6}
$$

In this case, the unknowns in the problem are the coefficients

$$a,\ b,\ c,\ \ldots,\ h \quad \text{and} \quad \alpha,\ \beta,\ \gamma,\ \ldots,\ \theta,$$

or better still, the ratios

$$\frac{a}{\alpha},\ \frac{b}{\alpha},\ \frac{c}{\alpha},\ \ldots,\ \frac{h}{\alpha} \quad \text{and} \quad \frac{\alpha}{\alpha},\ \frac{\beta}{\alpha},\ \frac{\gamma}{\alpha},\ \ldots,\ \frac{\theta}{\alpha}.$$

There are $n+m$ such ratios. It is easy to conclude from this that the function u is completely determined if we know $n+m$ particular values

$$u_0,\ u_1,\ u_2,\ \ldots,\ u_{n+m-1} \tag{7}$$

corresponding to the $n+m$ values

$$x_0,\ x_1,\ x_2,\ \ldots,\ x_{n+m-1} \tag{8}$$

of the variable x. We arrive at the same conclusions by showing that a second rational function of the form

$$\frac{a'+b'x+c'x^2+\ldots+h'x^{n-1}}{\alpha'+\beta'x+\gamma'x^2+\ldots+\theta'x^m} \tag{9}$$

cannot satisfy the same conditions as the first without being identically equal to it. Indeed, suppose that the fractions (6) and (9) become equal to each other for the particular values of x contained in series (8). Therefore, the equation

$$\left\{\begin{aligned} &(a+bx+\ldots+hx^{n-1})(\alpha'+\beta'x+\ldots+\theta'x^m)\\ -\ &(a'+b'x+\ldots+h'x^{n-1})(\alpha+\beta x+\ldots+\theta x^m)=0 \end{aligned}\right. \tag{10}$$

[432] remains true for $n+m$ values of the variable, while its degree remains less than $n+m$, so it must necessarily be an identity. From this it follows that we have identically

$$\left\{\begin{aligned} &\frac{a+bx+cx^2+\ldots+hx^{n-1}}{\alpha+\beta x+\gamma x^2+\ldots+\theta x^m}\\ &\quad=\frac{a'+b'x+c'x^2+\ldots+h'x^{n-1}}{\alpha'+\beta'x+\gamma'x^2+\ldots+\theta'x^m}. \end{aligned}\right. \tag{11}$$

Thus, we may solve the given question in only one way. We effectively resolve it by taking the general value of u to be the fraction

$$\frac{u_0u_1\ldots u_m\frac{(x-x_{m+1})(x-x_{m+2})\ldots(x-x_{m+n-1})}{(x_0-x_{m+1})\ldots(x_0-x_{m+n-1})\ldots(x_m-x_{m+1})\ldots(x_m-x_{m+n-1})}+\ldots}{u_0u_1\ldots u_{m-1}\frac{(x_0-x)(x_1-x)\ldots(x_{m-1}-x)}{(x_0-x_m)\ldots(x_0-x_{m+n-1})\ldots(x_{m-1}-x_m)\ldots(x_{m-1}-x_{m+n-1})}+\ldots},$$

in which the denominator must be replaced by 1 when we suppose that $m=0$ and the numerator by the product $u_0u_1\ldots u_m$ when we suppose that $n=1$. Given this, we find that for $m=0$,

$$(12)\qquad u = u_0\frac{(x-x_1)(x-x_2)\dots(x-x_{n-1})}{(x_0-x_1)(x_0-x_2)\dots(x_0-x_{n-1})}+\dots.$$

For $m = 1$,

$$(13)\qquad u = \frac{u_0u_1\frac{(x-x_2)(x-x_3)\dots(x-x_n)}{(x_0-x_2)(x_0-x_3)\dots(x_0-x_n)(x_1-x_2)(x_1-x_3)\dots(x_1-x_n)}+\dots}{u_0\frac{x_0-x}{(x_0-x_1)(x_0-x_2)\dots(x_0-x_n)}+u_1\frac{x_1-x}{(x_1-x_0)(x_1-x_2)\dots(x_1-x_n)}+\dots},$$

and so forth. For $n = 1$,

$$(14)\qquad u = \frac{u_0u_1\dots u_m}{u_0u_1\dots u_{m-1}\frac{(x_0-x)(x_1-x)\dots(x_{m-1}-x)}{(x_0-x_m)(x_1-x_m)\dots(x_{m-1}-x_m)}+\dots}.$$

In each of the preceding formulas, we may complete the numerator or the denominator of the fraction that represents the value of u without trouble by adding to the first term of this numerator or of this denominator all of those terms that we may derive with the assistance of one or several exchanges performed among the indices. For example, if we suppose that $m = 1$ and [433] $n = 2$, we find the completely expanded value of u to be

$$(15)\qquad u = \frac{u_0u_1\frac{x-x_2}{(x_0-x_2)(x_1-x_2)}+u_0u_2\frac{x-x_1}{(x_0-x_1)(x_2-x_1)}+u_1u_2\frac{x-x_0}{(x_1-x_0)(x_2-x_0)}}{u_0\frac{x_0-x}{(x_0-x_1)(x_0-x_2)}+u_1\frac{x_1-x}{(x_1-x_0)(x_1-x_2)}+u_2\frac{x_2-x}{(x_2-x_0)(x_2-x_1)}}.$$

It is worth noting that formula (12) is Lagrange's formula, and that to deduce formula (14) from it, it suffices to replace $n-1$ by m and then to take as the unknown the function $\frac{1}{u}$, assumed to be integer, in place of the function u.

Note VI – On figurate numbers.

[434] We call numbers *figurate* of the first order, second order, third order, etc. which serve as the coefficients of the successive powers of x in the expansions of the expressions

$$(1-x)^{-2}, \quad (1-x)^{-3}, \quad (1-x)^{-4}, \quad \ldots.$$

This definition gives an easy means to calculate them. Indeed, we have proved in Chapter VI (§ IV), that we have, for any real values of μ and for numerical values of x less than 1,

$$(1)\quad (1+x)^{\mu} = 1+\frac{\mu}{1}x+\frac{\mu(\mu-1)}{1\cdot 2}x^2+\ldots+\frac{\mu(\mu-1)\ldots(\mu-n+1)}{1\cdot 2\cdot 3\ldots n}x^n+\ldots.$$

If in the preceding equation we set $\mu=-(m+1)$, where m denotes any integer number, we find

$$(2)\quad \begin{cases} (1+x)^{-m-1} = 1-\dfrac{m+1}{1}x+\dfrac{(m+1)(m+2)}{1\cdot 2}x^2-\ldots \\ \qquad\qquad \pm\dfrac{(m+1)(m+2)\ldots(m+n)}{1\cdot 2\cdot 3\ldots n}x^n\pm\ldots. \end{cases}$$

Moreover, because we evidently have[1]

$$(3)\quad \begin{cases} \dfrac{(m+1)(m+2)\ldots(m+n)}{1\cdot 2\cdot 3\ldots n} = \dfrac{(m+1)(m+2)\ldots m(m+1)\ldots(m+n)}{(1\cdot 2\cdot 3\ldots m)(1\cdot 2\cdot 3\ldots n)} \\ \qquad\qquad = \dfrac{(n+1)(n+2)\ldots(n+m)}{1\cdot 2\cdot 3\ldots m}, \end{cases}$$

[435] it follows that equation (2) can be written as follows:

[1] This is evidently the familiar identity for binomial coefficients $\binom{m+n}{n}=\binom{m+n}{m}$.

$$(4)\begin{cases} (1+x)^{-m-1} = \dfrac{1\cdot 2\cdot 3\ldots m}{1\cdot 2\cdot 3\ldots m} - \dfrac{2\cdot 3\cdot 4\ldots (m+1)}{1\cdot 2\cdot 3\ldots m}x \\ \qquad + \dfrac{3\cdot 4\cdot 5\ldots (m+2)}{1\cdot 2\cdot 3\ldots m}x^2 - \ldots \mp \dfrac{n\,(n+1)\ldots (n+m-1)}{1\cdot 2\cdot 3\ldots m}x^{n-1} \\ \qquad \pm \dfrac{(n+1)\,(n+2)\ldots (n+m)}{1\cdot 2\cdot 3\ldots m}x^n \mp \ldots. \end{cases}$$

The numerical coefficients of the successive powers of x in the right-hand side of this last formula, namely

$$(5)\quad \frac{1\cdot 2\cdot 3\ldots m}{1\cdot 2\cdot 3\ldots m},\quad \frac{2\cdot 3\cdot 4\ldots (m+1)}{1\cdot 2\cdot 3\ldots m},\quad \ldots,\quad \frac{n\,(n+1)\ldots (n+m-1)}{1\cdot 2\cdot 3\ldots m},\quad \ldots,$$

are precisely the figurate numbers of order m. The sequence of these same numbers or series (5) extends to infinity. Its nth term, that is to say the fraction

$$\frac{n\,(n+1)\ldots (n+m-1)}{1\cdot 2\cdot 3\ldots m},$$

is at the same time the numerical coefficient of x^{n-1} in the expansion of $(1+x)^{-m-1}$ and the coefficient of x^m in the expansion of $(1+x)^{n+m-1}$. Moreover, if in series (5) we successively make

$$m=1,\quad m=2,\quad m=3,\quad \ldots,$$

we obtain: 1° the sequence of *natural* numbers, or figurate numbers of the first order,

$$1,\quad 2,\quad 3,\quad \ldots,\quad n,\quad \ldots;$$

2° the sequence of numbers which we call *triangular*, or figurate of the second order, namely

$$1,\quad 3,\quad 6,\quad 10,\quad \ldots,\quad \frac{n\,(n+1)}{1\cdot 2},\quad \ldots;$$

and 3° the sequence of numbers we call *pyramidal*, or figurate of the third order, namely

$$1,\quad 4,\quad 10,\quad 20,\quad \ldots,\quad \frac{n\,(n+1)\,(n+2)}{1\cdot 2\cdot 3},\quad \ldots,$$

....

[436] If we write these different sequences one below the other, preceding them by a first sequence composed of terms all equal to 1, and additionally placing them so that the first term of each is under the second term of the sequence immediately above it, we obtain the following table:

$$(6)\qquad \begin{cases} 1,\ 1,\ 1,\ 1,\ 1,\ \dots, \\ \quad 1,\ 2,\ 3,\ 4,\ \dots, \\ \qquad 1,\ 3,\ 6,\ \dots, \\ \qquad\quad 1,\ 4,\ \dots, \\ \qquad\qquad 1,\ \dots, \\ \qquad\qquad\quad \dots. \end{cases}$$

The numbers contained in the $(n+1)$st vertical column of this table are the coefficients of the nth power of a binomial. Pascal, in his *Traité du triangle arithmétique,*[2] first gave the law of formation for these very numbers. Newton later showed how the formula established from this law can be extended to fractional or negative powers.

Several remarkable properties of figurate numbers are deduced immediately from formula (4) of Chapter IV (§ III). Imagine, for example, that after replacing n by $n-1$ in this formula, we suppose that

$$x = m+1 \quad \text{and} \quad y = m'+1,$$

where m and m' are any two integer numbers. Then we find that

$$(7)\qquad \begin{cases} \dfrac{(m+m'+2)(m+m'+3)\dots(m+m'+n)}{1\cdot 2\cdot 3\dots(n-1)} \\ \quad = \dfrac{(m+1)(m+2)\dots(m+n-1)}{1\cdot 2\cdot 3\dots(n-1)} \\ \qquad + \dfrac{(m+1)(m+2)\dots(m+n-2)}{1\cdot 2\cdot 3\dots(n-2)}\dfrac{m'+1}{1} \\ \qquad + \dots\dots\dots\dots\dots\dots\dots\dots\dots\dots \\ \qquad + \dfrac{m+1}{1}\dfrac{(m'+1)(m'+2)\dots(m'+n-2)}{1\cdot 2\cdot 3\dots(n-2)} \\ \qquad + \dfrac{(m'+1)(m'+2)\dots(m'+n-1)}{1\cdot 2\cdot 3\dots(n-1)}. \end{cases}$$

[437] Then, by making $m' = 0$, we find that[3]

$$(8)\qquad \begin{cases} \dfrac{(m+2)(m+3)\dots(m+n)}{1\cdot 2\cdot 3\dots(n-1)} \\ \quad = \dfrac{(m+1)(m+2)\dots(m+n-1)}{1\cdot 2\cdot 3\dots(n-1)} \\ \qquad + \dfrac{(m+1)(m+2)\dots(m+n-2)}{1\cdot 2\cdot 3\dots(n-2)} \\ \qquad + \dots + \dfrac{m+1}{1} + 1. \end{cases}$$

[2] See [Pascal 1665].

[3] Formula (8) is sometimes called the "hockey stick pattern" in Pascal's Triangle.

Likewise, if after replacing n by $n-1$ in formula (4) (Chap. IV, § III), we instead make

$$x=m+1 \quad \text{and} \quad y=-\left(m'+1\right),$$

then we conclude that

$$(9) \qquad \begin{cases} \dfrac{(m-m')\,(m-m'+1)\ldots(m-m'+n-2)}{1\cdot 2\cdot 3\ldots(n-1)} \\ \quad = \dfrac{(m+1)\,(m+2)\ldots(m+n-1)}{1\cdot 2\cdot 3\ldots(n-1)} \\ \quad -\dfrac{(m+1)\,(m+2)\ldots(m+n-2)}{1\cdot 2\cdot 3\ldots(n-2)}\dfrac{m'+1}{1} \\ \quad +\ldots\ldots\ldots\ldots\ldots\ldots\ldots\ldots\ldots\ldots\ldots\ldots \\ \quad \mp\dfrac{m+1}{1}\dfrac{(m'+1)\,m'\ldots(m'-n+4)}{1\cdot 2\cdot 3\ldots(n-2)} \\ \quad \pm\dfrac{(m'+1)\,m'\ldots(m'-n+3)}{1\cdot 2\cdot 3\ldots(n-1)}. \end{cases}$$

When in the preceding equation we suppose that $m' \geq m$ and at the same time $n \geq m'+2$, we find that[4]

$$(10) \qquad \begin{cases} 0=\dfrac{(m+1)\ldots(m+n-1)}{1\cdot 2\cdot 3\ldots(n-1)}\dfrac{m'+1}{1}\dfrac{(m+1)\ldots(m+n-2)}{1\cdot 2\cdot 3\ldots(n-2)} \\ \quad +\dfrac{(m'+1)\,m'}{1\cdot 2}\dfrac{(m+1)\ldots(m+n-3)}{1\cdot 2\cdot 3\ldots(n-3)} \\ \quad -\ldots\ldots\ldots\ldots\ldots\ldots\ldots\ldots\ldots\ldots\ldots \\ \quad +\dfrac{m'+1}{1}\dfrac{(m+1)\ldots(m+n-m'-1)}{1\cdot 2\cdot 3\ldots(n-m'-1)} \\ \quad \pm\dfrac{(m+1)\ldots(m+n-m'-2)}{1\cdot 2\cdot 3\ldots(n-m'-2)}. \end{cases}$$

[438] Finally, because equations (8) and (10) can be written as follows

$$(11) \qquad \begin{cases} \dfrac{1\cdot 2\cdot 3\ldots m}{1\cdot 2\cdot 3\ldots m}+\dfrac{2\cdot 3\cdot 4\ldots(m+1)}{1\cdot 2\cdot 3\ldots m}+\ldots \\ \qquad +\dfrac{n\,(n+1)\ldots(n+m-1)}{1\cdot 2\cdot 3\ldots m}=\dfrac{n\,(n+1)\ldots(n+m)}{1\cdot 2\cdot 3\ldots(m+1)}, \end{cases}$$

and

---

[4] The numerator $m'+1$ in the first line of equation (10) was incorrectly written as $m+1$ in [Cauchy 1821, p. 534], but not noted its Errata. It was corrected in [Cauchy 1897, p. 437]. (tr.)

$$(12)\quad \begin{cases} 0=\dfrac{n\dots(n+m-1)}{1\cdot 2\cdot 3\dots m}-\dfrac{m'+1}{1}\dfrac{(n-1)\dots(n+m-2)}{1\cdot 2\cdot 3\dots m}+\dots \\ \qquad\mp\dfrac{m'+1}{1}\dfrac{(n-m')\dots(n+m-m'-1)}{1\cdot 2\cdot 3\dots m} \\ \qquad\pm\dfrac{(n-m'-1)\dots(n+m-m'-2)}{1\cdot 2\cdot 3\dots m}, \end{cases}$$

it is clear that they entail the two propositions which I am going to state:

Theorem I. — *If we form the sequence of figurate numbers of order m, and then if we add together the first n terms of this sequence, we obtain for the sum the nth figurate number of order* $m+1$.

Theorem II. — *If we denote by m and m′ two integer numbers subject to the condition*

$$m' \geq m,$$

and if in the expansion of $(1-x)^{m'+1}$ *we replace the successive powers of x by* $m'+2$ *consecutive terms taken from the sequence of figurate numbers of order m, we obtain a result equal to zero.*

Corollary I. — If we suppose that the different terms of the sequence

$$(13)\qquad a_0,\quad a_1,\quad a_2,\quad \dots,\quad a_n,\quad \dots$$

successively represent the natural numbers, the triangular numbers and the pyramidal numbers, we find in the first case that

$$(14)\qquad a_n-2a_{n-1}+a_{n-2}=0,$$

in the second case that

$$(15)\qquad a_n-3a_{n-1}+3a_{n-2}-a_{n-3}=0,$$

and in the third case

$$(16)\qquad a_n-4a_{n-1}+6a_{n-2}-4a_{n-3}+a_{n-4}=0.$$

[439] The first of the preceding equations becomes identical to formula (3) of Chapter XII (§ I).

Corollary II. — If we denote the figurate numbers of order m in general by

$$(13)\qquad a_0,\quad a_1,\quad a_2,\quad \dots,\quad a_n,\quad \dots,$$

then

$$(17)\qquad a_0,\quad a_1x,\quad a_2x^2,\quad \dots,\quad a_nx^n,\quad \dots$$

is a recurrent series for which the recurrence relation has for its terms the quantities

$$(18)\qquad 1,\quad -\frac{m+1}{1},\quad +\frac{(m+1)m}{1\cdot 2},\quad -\frac{(m+1)m(m-1)}{1\cdot 2\cdot 3},\quad +\ldots,$$

that is, the coefficients of successive powers of x in the expansion of $(1-x)^{m+1}$. Thus, for example, the series

$$1,\quad 3x,\quad 6x^2,\quad 10x^3,\quad \ldots,$$

in which the successive powers of x have for coefficients the triangular numbers, is recurrent and its recurrence relation is composed of the quantities

$$1,\quad -3,\quad +3\quad \text{and}\quad -1.$$

Among the principal properties of figurate numbers, we ought to point out those which equations (7) and (9) give when we put them in the following forms:

$$(19)\qquad \left\{\begin{aligned} &\frac{n(n+1)\ldots(n+m+m')}{1\cdot 2\cdot 3\ldots(m+m'+1)}\\ &\quad=\frac{n(n+1)\ldots(n+m-1)}{1\cdot 2\cdot 3\ldots m}\,\frac{1\cdot 2\cdot 3\ldots m'}{1\cdot 2\cdot 3\ldots m'}\\ &\qquad+\frac{(n-1)n\ldots(n+m-2)}{1\cdot 2\cdot 3\ldots m}\,\frac{2\cdot 3\cdot 4\ldots(m'+1)}{1\cdot 2\cdot 3\ldots m'}\\ &\qquad+\ldots\ldots\ldots\ldots\ldots\ldots\ldots\ldots\ldots\ldots\\ &\qquad+\frac{1\cdot 2\cdot 3\ldots m}{1\cdot 2\cdot 3\ldots m}\,\frac{n(n+1)\ldots(n+m'-1)}{1\cdot 2\cdot 3\ldots m'}\end{aligned}\right.$$

[440] and

$$(20)\qquad \left\{\begin{aligned} &\frac{n(n+1)\ldots(n+m-m'-2)}{1\cdot 2\cdot 3\ldots(m-m'-1)}\\ &\quad=\frac{n(n+1)\ldots(n+m-1)}{1\cdot 2\cdot 3\ldots m}-\frac{m'+1}{1}\,\frac{(n-1)n\ldots(n+m-2)}{1\cdot 2\cdot 3\ldots m}\\ &\qquad+\ldots\ldots\ldots\ldots\ldots\ldots\ldots\ldots\ldots\ldots\ldots\ldots\\ &\qquad\mp\frac{(m'+1)\ldots(m'-n+4)}{1\cdot 2\cdot 3\ldots(n-2)}\,\frac{2\cdot 3\cdot 4\ldots(m+1)}{1\cdot 2\cdot 3\ldots m}\\ &\qquad\pm\frac{(m'+1)\ldots(m'-n+3)}{1\cdot 2\cdot 3\ldots(n-1)}\,\frac{1\cdot 2\cdot 3\ldots m}{1\cdot 2\cdot 3\ldots m}.\end{aligned}\right.$$

We add that in the sequence of figurate numbers of order n, the $(n+1)$st term equals the sum of the squares of the coefficients which are contained in the nth power of a binomial. Indeed, if in formula (2) (Chap. IV, § III) we suppose that both $x=n$ and $y=n$, we find

$$(21)\quad \left\{ \begin{aligned} &\frac{2n\,(2n-1)\dots(n+1)}{1\cdot 2\cdot 3\dots(n-1)\,n} \\ &\quad = 1+\left(\frac{n}{1}\right)^2+\left[\frac{n\,(n-1)}{1\cdot 2}\right]^2+\dots+\left[\frac{n\,(n-1)}{1\cdot 2}\right]^2+\left(\frac{n}{1}\right)^2+1. \end{aligned} \right.$$

Note VII – On double series.

[441] Let

$$(1)\qquad \begin{cases} u_0, \ u_1, \ u_2, \ \ldots, \\ u'_0, \ u'_1, \ u'_2, \ \ldots, \\ u''_0, \ u''_1, \ u''_2, \ \ldots, \\ \ldots, \ldots, \ldots, \ldots \end{cases}$$

be any quantities arranged in horizontal and vertical lines in such a way that every series, horizontal or vertical, contains an infinity of terms. The system of all these quantities is what we can call a *double series,* and these quantities themselves are the different *terms* of the series, which has for its *general term*

$$u_n^{(m)},$$

where m and n denote any two integer numbers. Given this, imagine that we represent by

$$s_n^{(m)}$$

the sum of the terms of series (1) which are contained in the following table

$$(2)\qquad \begin{cases} u_0, & u_1, & u_2, & \ldots, u_{n-1}, \\ u'_0, & u'_1, & u'_2, & \ldots, u'_{n-1}, \\ u''_0, & u''_1, & u''_2, & \ldots, u''_{n-1}, \\ \ldots, & \ldots, & \ldots, & \ldots, \ldots, \\ u_0^{(m-1)}, & u_1^{(m-1)}, & u_2^{(m-1)}, & \ldots, u_{n-1}^{(m-1)}, \end{cases}$$

that is to say, of terms which carry at the same time a lower index smaller than n and an upper index smaller than m. If the sum of the remaining terms, taken in whatever order and whatever number as we might wish, becomes infinitely small for infinitely large values of m and n, it is clear that the sum $s_n^{(m)}$ – and all those sums which we can then derive by adding to $s_n^{(m)}$ some of the terms excluded from

Table (2) – converges, for increasing values [442] of m and n, towards a fixed limit s. In this case we say that series (1) is *convergent,* and that it has the limit s for its *sum*. In the contrary case, series (1) is *divergent* and it does not have a sum.

When, for infinitely large values of m and n, the terms excluded from Table (2), however many may be added together, never give anything but infinitely small sums, we can say as much *a fortiori* about the sums of the terms which belong to one or several horizontal or vertical rows of Table (1). It follows immediately from this remark that if the double series contained in Table (1) is convergent, each of the simple series contained in the rows, horizontal or vertical, of the same table is convergent as well. Under this hypothesis, denote by

$$s^{(m)}$$

the result which we obtain by adding the sum of the first m horizontal series of Table (1), that is to say, the m first terms of the simple series

$$(3) \qquad \begin{array}{lll} u_0+u_1+u_2+\ldots, & u'_0+u'_1+u'_2+\ldots, & u''_0+u''_1+u''_2+\ldots, \\ \ldots\ldots\ldots\ldots\ldots\ldots, & \ldots\ldots\ldots\ldots\ldots\ldots, & \ldots\ldots\ldots\ldots\ldots\ldots, \end{array}$$

and by

$$s_n$$

the result which we obtain by adding the sums of the first n vertical series, that is to say the first n terms of the simple series

$$(4) \qquad \begin{array}{lll} u_0+u'_0+u''_0+\ldots, & u_1+u'_1+u''_1+\ldots, & u_2+u'_2+u''_2+\ldots, \\ \ldots\ldots\ldots\ldots\ldots\ldots, & \ldots\ldots\ldots\ldots\ldots\ldots, & \ldots\ldots\ldots\ldots\ldots\ldots \end{array}$$

Then $s^{(m)}$ is evidently the limit of the expression $s_n^{(m)}$ for increasing values of n, and s_n is the limit of the same expression for increasing values of m. Consequently, it suffices to make m grow indefinitely in $s^{(m)}$ and n in s_n to make $s^{(m)}$ and s_n converge towards the limit s. We can thus state the following proposition:

Theorem I. – *Suppose that the double series contained in Table* (1) *is convergent, and denote by s the sum of this series. Then series* (3) *and* (4) *are likewise convergent and each of them have for their sum the quantity s.*

Now imagine that the numerical values of the quantities contained [443] in Table (1) are, respectively, denoted by

$$(5) \qquad \begin{cases} \rho_0, \ \rho_1, \ \rho_2, \ \ldots, \\ \rho'_0, \ \rho'_1, \ \rho'_2, \ \ldots, \\ \rho''_0, \ \rho''_1, \ \rho''_2, \ \ldots, \\ \ldots, \ \ldots, \ \ldots, \ \ldots. \end{cases}$$

The terms of Table (1) which are excluded from Table (2), however many of them we wish to add together, evidently give a sum less than or at most equal (ignoring the sign) to the sum of the corresponding terms of Table (5). Thus, if for infinitely large values of the numbers m and n, this last sum becomes infinitely small, and it is the same, *a fortiori*, as the first one. We can also express this by saying that if the double series contained in Table (5) is convergent, then series (1) is convergent as well. I add that we are completely assured of the convergence of the double series contained in Table (5) any time that the horizontal series of this table are convergent and their sums, namely

$$(6)\qquad \begin{array}{lll} \rho_0+\rho_1+\rho_2+\ldots, & \rho_0'+\rho_1'+\rho_2'+\ldots, & \rho_0''+\rho_1''+\rho_2''+\ldots, \\ \ldots\ldots\ldots\ldots, & \ldots\ldots\ldots\ldots, & \ldots\ldots\ldots\ldots, \end{array}$$

themselves form a simple convergent series. Indeed, under this hypothesis, let ε be a number as small as we may wish. We can choose m large enough that the addition of the sums

$$\begin{array}{ll} \rho_0^{(m)}+\rho_1^{(m)}+\rho_2^{(m)}+\ldots, & \rho_0^{(m+1)}+\rho_1^{(m+1)}+\rho_2^{(m+1)}+\ldots, \\ \ldots\ldots\ldots\ldots, & \ldots\ldots\ldots\ldots, \end{array}$$

and consequently that of the terms of Table (5) affected with an upper index greater than or equal to m, never produces a result greater than $\frac{1}{2}\varepsilon$. Moreover, if the number m is determined as we have just said, and because each of the horizontal series of Table (5) is convergent, we can also choose n large enough so that each of the sums

$$\begin{array}{llll} \rho_n & +\rho_{n+1} & +\rho_{n+2} & +\ldots, \\ \rho_n' & +\rho_{n+1}' & +\rho_{n+2}' & +\ldots, \\ \ldots\ldots & \ldots\ldots & \ldots\ldots & \ldots, \\ \rho_n^{(m-1)} & +\rho_{n+1}^{(m-1)} & +\rho_{n+2}^{(m-1)} & +\ldots \end{array}$$

is equal to or less than $\frac{1}{2m}\varepsilon$. In this case, the addition of the terms in [444] Table (5) which have an upper index less than m and a lower index at least equal to n never produces a result greater than $\frac{1}{2}\varepsilon$. When the two preceding conditions are satisfied, it is clear that in series (5), the addition of the terms with an upper index at least equal to m and a lower index at least equal to n cannot give anything but a sum at most equal to ε. Thus if we attribute infinitely large values to the numbers m and n, this sum becomes infinitely small, because we can make ε decrease below any assignable value. Thus the given hypothesis entails the convergence of series (5), and consequently that of series (1). By combining this principle with the first theorem, we deduce a new proposition which I am going to state.

Theorem II. — *Suppose that all the horizontal series of Table* (1) *are convergent and their sums, namely*

$$
(3) \qquad \begin{array}{lll} u_0+u_1+u_2+\ldots, & u'_0+u'_1+u'_2+\ldots, & u''_0+u''_1+u''_2+\ldots, \\ \ldots\ldots\ldots\ldots\ldots\ldots, & \ldots\ldots\ldots\ldots\ldots\ldots, & \ldots\ldots\ldots\ldots\ldots\ldots, \end{array}
$$

also form a convergent series, and that this double property of horizontal series remains true in the case where we replace each term of Table (1) *by its numerical value. Then we can affirm:* 1° *that all the vertical series are convergent;* 2° *that their sums, namely*

$$
(4) \qquad \begin{array}{lll} u_0+u'_0+u''_0+\ldots, & u_1+u'_1+u''_1+\ldots, & u_2+u'_2+u''_2+\ldots, \\ \ldots\ldots\ldots\ldots\ldots\ldots, & \ldots\ldots\ldots\ldots\ldots\ldots, & \ldots\ldots\ldots\ldots\ldots\ldots, \end{array}
$$

also form a convergent series; and 3° *finally that the sum of series* (4) *is precisely equal to that of series* (3).

Corollary I. — The preceding theorem remains true as well when we suppose that some of the series, horizontal or vertical, are composed of a finite number of terms. Indeed, each series of this kind can be considered as a convergent series indefinitely extended, but in which all the terms for which the rank surpasses a given given number vanish.

Corollary II. — Let

$$
(7) \qquad \begin{cases} u_0,\ u_1,\ u_2,\ u_3,\ \ldots, \\ v_0,\ v_1,\ v_2,\ v_3,\ \ldots \end{cases}
$$

be two convergent series which have, respectively, for their sums the two quantities s and s', and for which each remains convergent when we reduce [445] its various terms to their numerical values. If we form the table

$$
(8) \qquad \left\{ \begin{array}{lllll} u_0v_0, & u_1v_0, & u_2v_0, & u_3v_0, & \ldots, \\ & u_0v_1, & u_1v_1, & u_2v_1, & \ldots, \\ & & u_0v_2, & u_1v_2, & \ldots, \\ & & & u_0v_3, & \ldots, \\ & & & & \ldots, \end{array} \right.
$$

we recognize without trouble that the horizontal series of this table enjoy the properties stated in theorem II, and that their sums are, respectively,

$$
(9) \qquad v_0s, \quad v_1s, \quad v_2s, \quad v_3s, \quad \ldots.
$$

Consequently, by virtue of theorem II and its first corollary, the vertical series, namely

$$
(10) \qquad \begin{cases} u_0v_0, \quad u_0v_1+u_1v_0, \quad u_0v_2+u_1v_1+u_2v_0, \quad \ldots, \\ u_0v_n+u_1v_{n-1}+\ldots+u_{n-1}v_1+u_nv_0, \quad \ldots, \end{cases}
$$

form a new convergent series and the sum of this new series is equal to that of series (9), that is to say, evidently equal to the product ss'. Thus we find that we have returned to theorem VI of Chapter VI (§ III) by the consideration of double series.

Corollary III. — Let x be the sine of an arc contained between the limits $-\frac{\pi}{2}$ and $+\frac{\pi}{2}$, and let z be its tangent. We find that

$$z = \frac{x}{\sqrt{1-x^2}} = x\left(1-x^2\right)^{-\frac{1}{2}}.$$

Given this, and by virtue of formula (39) (Chap. IX, § II), for numerical values of z less than 1, we have

$$\arctan z = z - \frac{z^3}{3} + \frac{z^5}{5} - \frac{z^7}{7} + \dots.$$

We conclude that, for numerical values of x less than $\frac{1}{\sqrt{2}}$,

$$\begin{aligned}\arcsin x &= \arctan x\left(1-x^2\right)^{-\frac{1}{2}}\\ &= x\left(1-x^2\right)^{-\frac{1}{2}} - \frac{x^3}{3}\left(1-x^2\right)^{-\frac{3}{2}} + \frac{x^5}{5}\left(1-x^2\right)^{-\frac{5}{2}}\\ &\quad - \frac{x^7}{7}\left(1-x^2\right)^{-\frac{7}{2}} + \dots,\end{aligned}$$

[446] or what amounts to the same thing,

$$\begin{array}{rlllll}\arcsin x = x + & \frac{3}{2}\frac{x^3}{3} & + \frac{3\cdot 5}{2\cdot 4}\frac{x^5}{5} & + \frac{3\cdot 5\cdot 7}{2\cdot 4\cdot 6}\frac{x^7}{7} & + \dots \\ - & \frac{x^3}{3} & - \frac{5}{2}\frac{x^5}{5} & - \frac{5\cdot 7}{2\cdot 4}\frac{x^7}{7} & - \dots \\ & + & \frac{x^5}{5} & + \frac{7}{2}\frac{x^7}{7} & + \dots \\ & & - & \frac{x^7}{7} & - \dots \\ & & & & \dots .\end{array}$$

Because the horizontal series contained in the right-hand side of the preceding equation evidently satisfy the conditions stated in theorem II as long as the variable x maintains a numerical value less than $\frac{1}{\sqrt{2}}$, it follows that this equation can be written as follows:

$$\begin{aligned}\arcsin(x) = x &+ \left(\frac{3}{2}-1\right)\frac{x^3}{3} + \left(\frac{5\cdot 3}{2\cdot 4} - \frac{5}{2} + 1\right)\frac{x^5}{5}\\ &+ \left(\frac{7\cdot 5\cdot 3}{2\cdot 4\cdot 6} - \frac{7\cdot 5}{2\cdot 4} + \frac{7}{2} - 1\right)\frac{x^7}{7} + \dots\\ &\left(x = -\frac{1}{\sqrt{2}}, \quad x = +\frac{1}{\sqrt{2}}\right).\end{aligned}$$

Moreover, if in formula (5) of Chapter IV (§ III) we attribute to y the negative value -2 and to x one of the positive values 3, 5, 7, ..., we get successively

$$(11)\qquad \begin{cases} \dfrac{3}{2}-1=\dfrac{1}{2}, \\ \dfrac{3\cdot5}{2\cdot4}-\dfrac{5}{2}+1=\dfrac{1\cdot3}{2\cdot4}, \\ \dfrac{7\cdot5\cdot3}{2\cdot4\cdot6}-\dfrac{7\cdot5}{2\cdot4}+\dfrac{7}{2}-1=\dfrac{1\cdot3\cdot5}{2\cdot4\cdot6}, \\ \dots\dots\dots\dots\dots\dots, \end{cases}$$

and consequently we find finally

$$(12)\qquad \begin{cases} \arcsin x = x+\dfrac{1}{2}\dfrac{x^3}{3}+\dfrac{1\cdot3}{2\cdot4}\dfrac{x^5}{5}+\dfrac{1\cdot3\cdot5}{2\cdot4\cdot6}\dfrac{x^7}{7}+\dots \\ \left(x=-\frac{1}{\sqrt{2}},\quad x=+\frac{1}{\sqrt{2}}\right). \end{cases}$$

[447] It is easy to prove with the aid of infinitesimal Calculus that this last equation remains true not only between the limits $x=-\frac{1}{\sqrt{2}}$ and $x=+\frac{1}{\sqrt{2}}$, but also between the limits $x=-1$ and $x=+1$.

Corollary IV. — By virtue of formula (20) (Chap. VI, § IV), we have, for all values of x contained between the limits -1 and $+1$,

$$\frac{(1+x)^{\mu}-1}{\mu}=x-\frac{x^2}{2}(1-\mu)+\frac{x^3}{3}(1-\mu)\left(1-\frac{1}{2}\mu\right)+\dots,$$

or what amounts to the same thing,

$$\begin{aligned} \frac{(1+x)^{\mu}-1}{\mu} = {} & \frac{x}{1} \\ & -\frac{x^2}{2}+\mu\frac{x^2}{2} \\ & +\frac{x^3}{3}-\mu\left(1+\frac{1}{2}\right)\frac{x^3}{3}+\mu^2\left(\frac{1}{1\cdot2}\right)\frac{x^3}{3} \\ & -\frac{x^4}{4}+\mu\left(1+\frac{1}{2}+\frac{1}{3}\right)\frac{x^4}{4}-\mu^2\left(\frac{1}{1\cdot3}+\frac{1}{1\cdot2}+\frac{1}{2\cdot3}\right)\frac{x^4}{4} \\ & \qquad ++\mu^3\left(\frac{1}{1\cdot2\cdot3}\right)\frac{x^4}{4} \\ & +\dots\dots\dots\dots\dots\dots \end{aligned}$$

Because the horizontal series which make up the right-hand side of the preceding equation satisfy the conditions stated in theorem II, as long as the variable x maintains a numerical value less than 1, it follows that this equation can be written as follows:

$$
(13)\qquad \left\{
\begin{aligned}
\frac{(1+x)^{\mu}-1}{\mu} &= \frac{x}{1}-\frac{x^2}{2}+\frac{x^3}{3}-\frac{x^4}{4}+\dots\\
&+\mu\left[\frac{x^2}{2}-\left(1+\frac{1}{2}\right)\frac{x^3}{3}+\left(1+\frac{1}{2}+\frac{1}{3}\right)\frac{x^4}{4}-\dots\right]\\
&+\mu^2\left[\frac{1}{1\cdot 2}\frac{x^3}{3}-\left(\frac{1}{1\cdot 2}+\frac{1}{1\cdot 3}+\frac{1}{2\cdot 3}\right)\frac{x^4}{4}+\dots\right]\\
&+\dots\dots\dots\dots\dots\dots\dots\dots\dots\dots
\end{aligned}
\right.
$$
$$(x=-1,\quad x=+1).$$

However, we already have found (Chap. VI, § IV, problem I, corollary II)

$$
(14)\qquad \frac{(1+x)^{\mu}-1}{\mu} = \ln(1+x)+\frac{\mu}{2}\left[\ln(1+x)\right]^2+\dots,
$$

where ln is the characteristic of the Napierian logarithms. Formulas (13) and [448]
(14) ought to agree with each other (see theorem VI of Chapter VI, § IV), and we conclude that for all values of x contained between the limits -1 and $+1$,

$$
(15)\qquad \left\{
\begin{aligned}
\ln(1+x) &= x-\tfrac{x^2}{2}+\tfrac{x^3}{3}-\tfrac{x^4}{4}+\dots,\\
\tfrac{1}{2}\left[\ln(1+x)\right]^2 &= \tfrac{x^2}{2}-\left(1+\tfrac{1}{2}\right)\tfrac{x^3}{3}+\left(1+\tfrac{1}{2}+\tfrac{1}{3}\right)\tfrac{x^4}{4}-\dots\\
&\qquad \pm\left(1+\tfrac{1}{2}+\tfrac{1}{3}+\dots+\tfrac{1}{n-1}\right)\tfrac{x^n}{n}\mp\dots,\\
\tfrac{1}{2\cdot 3}\left[\ln(1+x)\right]^3 &= \tfrac{1}{1\cdot 2}\tfrac{x^3}{3}-\left(\tfrac{1}{1\cdot 2}+\tfrac{1}{1\cdot 3}+\tfrac{1}{2\cdot 3}\right)\tfrac{x^4}{4}+\dots,\\
&\dots\dots\dots\dots\dots\dots\dots\dots\dots\dots
\end{aligned}
\right.
$$

In what precedes, we have not considered anything but double series, convergent or divergent, for which the different terms are real quantities. However, what has been said with regard to these series can just as well be applied to the case where their terms become imaginary, provided that we write throughout *imaginary expression* in place of *quantity* and *modulus* in place of *numerical value.* With these modifications, theorems I and II still remain true. We prove this without trouble by applying the following principle:

The modulus of the sum of several imaginary expressions is always less than the sum of their moduli.[1]

To establish this principle, it suffices to observe that if we make

$$
\begin{aligned}
&\rho\left(\cos\theta+\sqrt{-1}\sin\theta\right)+\rho'\left(\cos\theta'+\sqrt{-1}\sin\theta'\right)+\dots\\
&\qquad = R\left(\cos T+\sqrt{-1}\sin T\right),
\end{aligned}
$$

where ρ, ρ', ... and R denote positive quantities, then we conclude that

[1] This is the triangle inequality for complex numbers.

$$\begin{aligned}R^2 &= (\rho\cos\theta+\rho'\cos\theta'+\ldots)^2+(\rho\cos\theta+\rho'\cos\theta'+\ldots)^2\\&=\rho^2+\rho'^2+\ldots+2\rho\rho'\cos(\theta-\theta')+\ldots\\&<\rho^2+\rho'^2+\ldots+2\rho\rho'+\ldots=(\rho+\rho'+\ldots)^2,\end{aligned}$$

and consequently

$$R<\rho+\rho'+\ldots.$$

Note VIII – On formulas that are used to convert the sines or cosines of multiples of an arc into polynomials, the different terms of which have the ascending powers of the sines or the cosines of the same arc as factors.

[449] The formulas in question here are those which we constructed in solving the two first problems stated in section V of Chapter VII, and which we gave the numbers (3), (4), (5), (6), (9), (10), (11) and (12). They give rise to the following remarks.

First, if in the calculations we used to establish formulas (3), (4), (5) and (6), we substitute equations (24) of Chapter IX (§ II) into equations (12) of Chapter VII (§ II), we recognize immediately that the same formulas remain true in the case where we replace the integer number m by any quantity μ while we suppose that the numerical value of z is less than $\frac{\pi}{4}$. Thus, under this hypothesis we have

$$(1)\qquad \begin{cases} \cos\mu z = 1 - \dfrac{\mu\cdot\mu}{1\cdot 2}\sin^2 z + \dfrac{(\mu+2)\,\mu\cdot\mu\,(\mu-2)}{1\cdot 2\cdot 3\cdot 4}\sin^4 z \\ \qquad - \dfrac{(\mu+4)\,(\mu+2)\,\mu\cdot\mu\,(\mu-2)\,(\mu-4)}{1\cdot 2\cdot 3\cdot 4\cdot 5\cdot 6}\sin^6 z + \ldots, \end{cases}$$

$$(2)\qquad \begin{cases} \sin\mu z = \cos z\left[\mu\sin z - \dfrac{(\mu+2)\,\mu\,(\mu-2)}{1\cdot 2\cdot 3}\sin^3 z\right. \\ \qquad \left. + \dfrac{(\mu+4)\,(\mu+2)\,\mu\,(\mu-2)\,(\mu-4)}{1\cdot 2\cdot 3\cdot 4\cdot 5}\sin^5 z - \ldots\right] \end{cases}$$

and

$$(3)\qquad \begin{cases} \cos\mu z = \cos z\left[1 - \dfrac{(\mu+1)\,(\mu-1)}{1\cdot 2}\sin^2 z\right. \\ \qquad \left. + \dfrac{(\mu+3)\,(\mu+1)\,(\mu-1)\,(\mu-3)}{1\cdot 2\cdot 3\cdot 4}\sin^4 z - \ldots\right], \end{cases}$$

$$(4)\qquad \begin{cases} \sin\mu z = \mu\sin z - \dfrac{(\mu+1)\,\mu\,(\mu-1)}{1\cdot 2\cdot 3}\sin^3 z \\ \qquad + \dfrac{(\mu+3)\,(\mu+1)\,\mu\,(\mu-1)\,(\mu-3)}{1\cdot 2\cdot 3\cdot 4\cdot 5}\sin^5 z - \ldots. \end{cases}$$

[450] Moreover, by virtue of the principles established in Chapter IX (§ II) and in the preceding note, we can expand not only $\cos\mu z$ and $\sin\mu z$, but also the right-

hand sides of formulas (1), (2), (3) and (4), according to ascending powers of μ, and because the coefficients of these powers ought thus to be the same in the left- and the right-hand sides of each formula, we obtain, by comparing two by two the coefficients in question, one series of equations among which we distinguish those which I am going to write:

$$\frac{1}{2}z^2 = \frac{\sin^2 z}{2} + \frac{2}{3}\frac{\sin^4 z}{4} + \frac{2\cdot 4}{3\cdot 5}\frac{\sin^6 z}{6} + \ldots \quad \text{and} \tag{5}$$

$$z = \sin z + \frac{1}{2}\frac{\sin^3 z}{3} + \frac{1\cdot 3}{2\cdot 4}\frac{\sin^5 z}{5} + \ldots. \tag{6}$$

We still suppose here that the variable z is contained between the limits $-\frac{\pi}{4}$ and $+\frac{\pi}{4}$. However, we easily prove, with the aid of infinitesimal Calculus, that, without changing equations (1), (2), (3), (4), (5), (6), ..., we can make the numerical value of z as large as $\frac{\pi}{2}$. We add that by taking $\sin z = x$, we make equation (6) coincide with formula (12) of Note VII, and equation (5) with the following:

$$(\arcsin x)^2 = x^2 + \frac{2}{3}\frac{x^4}{2} + \frac{2\cdot 4}{3\cdot 5}\frac{x^6}{3} + \frac{2\cdot 4\cdot 6}{3\cdot 5\cdot 7}\frac{x^8}{4} + \ldots.$$

This last formula is found in the *Mélanges d'Analyse,* published in 1815 by Mr. de Stainville,[1] lecturer at the École Royale Polytechnique.

Imagine for now that in the formulas in Chapter VII (§ V) already cited, we attribute an imaginary value to the variable z. We conclude without trouble from the principles developed in Chapter IX (§ III) that they are still exact. Suppose, for example, that

$$z = \sqrt{-1}\ln x,$$

where ln is the characteristic of the Napierian logarithms. Because under this hypothesis we have

$$\cos z = \frac{1}{2}\left(e^{\ln x} + e^{-\ln x}\right) = \frac{1}{2}\left(x + \frac{1}{x}\right) \quad \text{and}$$

$$\sin z = \frac{\sqrt{-1}}{2}\left(e^{\ln x} - e^{-\ln x}\right) = \frac{\sqrt{-1}}{2}\left(x - \frac{1}{x}\right),$$

[451] and in general

$$\cos nz = \frac{1}{2}\left(x^n + \frac{1}{x^n}\right) \quad \text{and} \quad \sin nz = \frac{\sqrt{-1}}{2}\left(x^n - \frac{1}{x^n}\right),$$

where n denotes any integer number. From equations (3), (4), (5) and (6) (Chapter VII, § V), we get: 1° for even values of m,

[1] See [Stainville 1815, pp. 406–408].

$$(7)\quad \begin{cases} x^m + \dfrac{1}{x^m} = 2\left[1 + \dfrac{m\cdot m}{2\cdot 4}\left(x - \dfrac{1}{x}\right)^2 + \dfrac{(m+2)\,m\cdot m\,(m-2)}{2\cdot 4\cdot 6\cdot 8}\left(x - \dfrac{1}{x}\right)^4\right. \\ \qquad\left. + \dfrac{(m+4)\,(m+2)\,m\,(m-2)\,(m-4)}{2\cdot 4\cdot 6\cdot 8\cdot 10\cdot 12}\left(x - \dfrac{1}{x}\right)^6 + \dots\right], \end{cases}$$

$$(8)\quad \begin{cases} x^m - \dfrac{1}{x^m} = \left(x + \dfrac{1}{x}\right)\left[\dfrac{m}{2}\left(x - \dfrac{1}{x}\right) + \dfrac{(m+2)\,m\,(m-2)}{2\cdot 4\cdot 6}\left(x - \dfrac{1}{x}\right)^3\right. \\ \qquad\left. + \dfrac{(m+4)\,(m+2)\,m\,(m-2)\,(m-4)}{2\cdot 4\cdot 6\cdot 8\cdot 10}\left(x - \dfrac{1}{x}\right)^5 + \dots\right]; \end{cases}$$

and 2° for odd values of m,

$$(9)\quad \begin{cases} x^m + \dfrac{1}{x^m} = \left(x + \dfrac{1}{x}\right)\left[1 + \dfrac{(m+1)\,(m-1)}{2\cdot 4}\left(x - \dfrac{1}{x}\right)^2\right. \\ \qquad\left. + \dfrac{(m+3)\,(m+1)\,(m-1)\,(m-3)}{2\cdot 4\cdot 6\cdot 8}\left(x - \dfrac{1}{x}\right)^4 + \dots\right], \end{cases}$$

$$(10)\quad \begin{cases} x^m - \dfrac{1}{x^m} = 2\left[\dfrac{m}{2}\left(x - \dfrac{1}{x}\right) + \dfrac{(m+1)\,m\,(m-1)}{2\cdot 4\cdot 6}\left(x - \dfrac{1}{x}\right)^3\right. \\ \qquad\left. + \dfrac{(m+3)\,(m+1)\,m\,(m-1)\,(m-3)}{2\cdot 4\cdot 6\cdot 8\cdot 10}\left(x - \dfrac{1}{x}\right)^5 + \dots\right]. \end{cases}$$

Formulas (9), (10), (11) and (12) of section V (Chap. VII) give analogous results.

Now we return to formula (3) of the same section. By virtue of this formula, for even values of m, $\cos mz$ is an integer function of $\sin z$ of degree m, and because this function, as well as $\cos mz$, ought to vanish for all values of z contained in the sequence

$$-\frac{(m-1)\,\pi}{2m},\ \dots,\ -\frac{3\pi}{2m},\ -\frac{\pi}{2m},\ +\frac{\pi}{2m},\ +\frac{3\pi}{2m},\ \dots,\ +\frac{(m-1)\,\pi}{2m},$$

[452] it is clear that it is divisible by each of the binomial factors

$$\begin{gathered} \sin z + \sin\frac{(m-1)\,\pi}{2m},\ \dots,\ \sin z + \sin\frac{3\pi}{2m},\ \sin z + \sin\frac{\pi}{2m}, \\ \sin z - \sin\frac{(m-1)\,\pi}{2m},\ \dots,\ \sin z - \sin\frac{3\pi}{2m},\ \sin z - \sin\frac{\pi}{2m}. \end{gathered}$$

Consequently it is equal to a the product of all these binomial factors by the numerical coefficient of $\sin^m z$, namely

$$(-1)^{\frac{m}{2}} \frac{(m+m-2)\ldots(m+2)\,m\,(m-2)\ldots(m-m+2)}{1\cdot 2\cdot 3\ldots(m-1)\cdot m}$$
$$= (-1)^{\frac{m}{2}}\, 2^{m-1}.$$

Thus we have, for even values of m,

$$(11)\quad \begin{cases} \cos mz = 2^{m-1}\left(\sin^2\dfrac{\pi}{2m} - \sin^2 z\right)\left(\sin^2\dfrac{3\pi}{2m} - \sin^2 z\right)\ldots \\ \quad\ldots\left(\sin^2\dfrac{(m-1)\,\pi}{2m} - \sin^2 z\right). \end{cases}$$

By similar reasoning, from formulas (4), (5) and (6) (Chap. VIII, § V) we get: 1° for even values of m,

$$(12)\quad \begin{cases} \sin mz = 2^{m-1}\sin z\cos z\left(\sin^2\dfrac{2\pi}{2m} - \sin^2 z\right)\left(\sin^2\dfrac{4\pi}{2m} - \sin^2 z\right)\ldots \\ \quad\ldots\left(\sin^2\dfrac{(m-2)\,\pi}{2m} - \sin^2 z\right); \end{cases}$$

and 2° for odd values of m,

$$(13)\quad \begin{cases} \cos mz = 2^{m-1}\cos z\left(\sin^2\dfrac{\pi}{2m} - \sin^2 z\right)\left(\sin^2\dfrac{3\pi}{2m} - \sin^2 z\right)\ldots \\ \quad\ldots\left(\sin^2\dfrac{(m-2)\,\pi}{2m} - \sin^2 z\right) \end{cases}$$

and

$$(14)\quad \begin{cases} \sin mz = 2^{m-1}\sin z\left(\sin^2\dfrac{2\pi}{2m} - \sin^2 z\right)\left(\sin^2\dfrac{4\pi}{2m} - \sin^2 z\right)\ldots \\ \quad\ldots\left(\sin^2\dfrac{(m-1)\,\pi}{2m} - \sin^2 z\right). \end{cases}$$

In the preceding four equations, if we reduce the constant part of each binomial factor to 1 by writing, for example,

$$1 - \frac{\sin^2 z}{\sin^2\frac{\pi}{2m}} \quad \text{in place of} \quad \sin^2\frac{\pi}{2m} - \sin^2 z,$$

the numerical factors of the right-hand sides evidently become equal to those of the terms in formulas (3), (4), (5) and (6) of Chapter VII (§ V), which are independent of $\sin z$ or contain its first power, that is to say equal to 1 or to the number m. Consequently, we find: [453] 1° for even values of m,

$$(15)\quad \begin{cases} \cos mz = \left(1-\dfrac{\sin^2 z}{\sin^2 \frac{\pi}{2m}}\right)\left(1-\dfrac{\sin^2 z}{\sin^2 \frac{3\pi}{2m}}\right)\cdots \\ \qquad \cdots\left(1-\dfrac{\sin^2 z}{\sin^2 \frac{(m-1)\pi}{2m}}\right), \end{cases}$$

$$(16)\quad \begin{cases} \sin mz = m\sin z\cos z\left(1-\dfrac{\sin^2 z}{\sin^2 \frac{2\pi}{2m}}\right)\left(1-\dfrac{\sin^2 z}{\sin^2 \frac{4\pi}{2m}}\right)\cdots \\ \qquad \cdots\left(1-\dfrac{\sin^2 z}{\sin^2 \frac{(m-2)\pi}{2m}}\right); \end{cases}$$

and 2° for odd values of m,

$$(17)\quad \begin{cases} \cos mz = \cos z\left(1-\dfrac{\sin^2 z}{\sin^2 \frac{\pi}{2m}}\right)\left(1-\dfrac{\sin^2 z}{\sin^2 \frac{3\pi}{2m}}\right)\cdots \\ \qquad \cdots\left(1-\dfrac{\sin^2 z}{\sin^2 \frac{(m-2)\pi}{2m}}\right), \end{cases}$$

$$(18)\quad \begin{cases} \sin mz = m\sin z\left(1-\dfrac{\sin^2 z}{\sin^2 \frac{2\pi}{2m}}\right)\left(1-\dfrac{\sin^2 z}{\sin^2 \frac{4\pi}{2m}}\right)\cdots \\ \qquad \cdots\left(1-\dfrac{\sin^2 z}{\sin^2 \frac{(m-1)\pi}{2m}}\right). \end{cases}$$

Moreover, if we observe that in general we have

$$\sin^2 b - \sin^2 a = \frac{\cos 2a - \cos 2b}{2},$$

we recognize without trouble that equations (11), (12), (13) and (14) can be replaced by these which follow

$$(19)\quad \begin{cases} \cos mz = 2^{\frac{m}{2}-1}\left(\cos 2z - \cos\dfrac{\pi}{m}\right)\left(\cos 2z - \cos\dfrac{3\pi}{m}\right)\cdots \\ \qquad \cdots\left(\cos 2z - \cos\dfrac{(m-1)\pi}{m}\right), \\ \sin mz = 2^{\frac{m}{2}-1}\sin 2z\left(\cos 2z - \cos\dfrac{2\pi}{m}\right)\left(\cos 2z - \cos\dfrac{4\pi}{m}\right)\cdots \\ \qquad \cdots\left(\cos 2z - \cos\dfrac{(m-2)\pi}{m}\right), \end{cases}$$

$$(20)\quad \begin{cases} \cos mz = 2^{\frac{m-1}{2}} \cos z \left(\cos 2z - \cos\dfrac{\pi}{m}\right)\left(\cos 2z - \cos\dfrac{3\pi}{m}\right)\cdots \\ \qquad \cdots\left(\cos 2z - \cos\dfrac{(m-2)\pi}{m}\right), \\ \sin mz = 2^{\frac{m-1}{2}} \sin z \left(\cos 2z - \cos\dfrac{2\pi}{m}\right)\left(\cos 2z - \cos\dfrac{4\pi}{m}\right)\cdots \\ \qquad \cdots\left(\cos 2z - \cos\dfrac{(m-1)\pi}{m}\right), \end{cases}$$

where the first two equations apply in the case where m is an even number and the last two in the case where m is an odd number.

The 12 preceding equations remain true whether the values attributed to the variable z are real or imaginary. Thus, we can replace this variable by $\frac{\pi}{2} - z$, by $\sqrt{-1}\ln x$, ... In the first case, we obtain several new equations corresponding to [454] formulas (9), (10), (11) and (12) of Chapter VII (§ V). In the second case, equations (19) and (20) give, respectively, for even values of m,

$$(21)\quad \begin{cases} x^m + \dfrac{1}{x^m} = \left(x^2 - 2\cos\dfrac{\pi}{m} + \dfrac{1}{x^2}\right)\left(x^2 - 2\cos\dfrac{3\pi}{m} + \dfrac{1}{x^2}\right)\cdots \\ \qquad \cdots\left(x^2 - 2\cos\dfrac{(m-1)\pi}{m} + \dfrac{1}{x^2}\right), \\ x^m - \dfrac{1}{x^m} = \left(x^2 - \dfrac{1}{x^2}\right)\left(x^2 - 2\cos\dfrac{2\pi}{m} + \dfrac{1}{x^2}\right)\cdots \\ \qquad \cdots\left(x^2 - 2\cos\dfrac{(m-2)\pi}{m} + \dfrac{1}{x^2}\right), \end{cases}$$

and for odd values of m,[2]

$$(22)\quad \begin{cases} x^m + \dfrac{1}{x^m} = \left(x + \dfrac{1}{x}\right)\left(x^2 - 2\cos\dfrac{\pi}{m} + \dfrac{1}{x^2}\right)\cdots \\ \qquad \cdots\left(x^2 - 2\cos\dfrac{(m-2)\pi}{m} + \dfrac{1}{x^2}\right), \\ x^m - \dfrac{1}{x^m} = \left(x - \dfrac{1}{x}\right)\left(x^2 - 2\cos\dfrac{2\pi}{m} + \dfrac{1}{x^2}\right)\cdots \\ \qquad \cdots\left(x^2 - 2\cos\dfrac{(m-1)\pi}{m} + \dfrac{1}{x^2}\right), \end{cases}$$

which agrees with the results obtained in Chapter X (§ II).

It remains for us to point out several rather remarkable consequences of equations (11) and (15), (12) and (16), (13) and (17), and (14) and (18). When we expand their right-hand sides according to ascending powers of $\sin z$, the numerical coefficients

[2] In [Cauchy 1897, p. 454] the cosine term of the last factor of equation (22) has a numerator of $(m-2)\pi$. The numerator is correctly written as $(m-1)\pi$ in [Cauchy 1821, p. 555] (tr.)

of these powers evidently must be the same as in formulas (3), (4), (5) and (6) of Chapter VII (§ V). From this single observation, we deduce immediately several new equations which are satisfied by the sines of the arcs

$$\frac{\pi}{2m},\quad \frac{2\pi}{2m},\quad \frac{3\pi}{2m},\quad \frac{4\pi}{2m},\quad \ldots.$$

We find, for example, for even values of m,

$$(23)\qquad \begin{cases} 1 = 2^{m-1}\sin^2\dfrac{\pi}{2m}\sin^2\dfrac{3\pi}{2m}\ldots\sin^2\dfrac{(m-1)\pi}{2m}, \\ m = 2^{m-1}\sin^2\dfrac{2\pi}{2m}\sin^2\dfrac{4\pi}{2m}\ldots\sin^2\dfrac{(m-2)\pi}{2m} \end{cases}$$

and

$$(24)\qquad \begin{cases} \frac{m\cdot m}{1\cdot 2} = \dfrac{1}{\sin^2\frac{\pi}{2m}} + \dfrac{1}{\sin^2\frac{3\pi}{2m}} + \ldots + \dfrac{1}{\sin^2\frac{(m-1)\pi}{2m}}, \\ \frac{(m+2)(m-2)}{1\cdot 2\cdot 3} = \dfrac{1}{\sin^2\frac{2\pi}{2m}} + \dfrac{1}{\sin^2\frac{4\pi}{2m}} + \ldots + \dfrac{1}{\sin^2\frac{(m-2)\pi}{2m}}, \end{cases}$$

[455] and for odd values of m,

$$(25)\qquad \begin{cases} 1 = 2^{m-1}\sin^2\dfrac{\pi}{2m}\sin^2\dfrac{3\pi}{2m}\ldots\sin^2\dfrac{(m-2)\pi}{2m}, \\ m = 2^{m-1}\sin^2\dfrac{2\pi}{2m}\sin^2\dfrac{4\pi}{2m}\ldots\sin^2\dfrac{(m-1)\pi}{2m} \end{cases}$$

and

$$(26)\qquad \begin{cases} \frac{(m+1)(m-1)}{1\cdot 2} = \dfrac{1}{\sin^2\frac{\pi}{2m}} + \dfrac{1}{\sin^2\frac{3\pi}{2m}} + \ldots + \dfrac{1}{\sin^2\frac{(m-2)\pi}{2m}}, \\ \frac{(m+1)(m-1)}{1\cdot 2\cdot 3} = \dfrac{1}{\sin^2\frac{2\pi}{2m}} + \dfrac{1}{\sin^2\frac{4\pi}{2m}} + \ldots + \dfrac{1}{\sin^2\frac{(m-1)\pi}{2m}}. \end{cases}$$

I add that if we multiply both sides of each of equations (24) and (26) by $\left(\frac{\pi}{m}\right)^2$, we conclude by making m grow indefinitely that[3]

$$(27)\qquad \frac{\pi^2}{8} = 1 + \frac{1}{9} + \frac{1}{25} + \frac{1}{49} + \ldots \quad \text{and}$$

$$(28)\qquad \frac{\pi^2}{6} = 1 + \frac{1}{4} + \frac{1}{9} + \frac{1}{16} + \frac{1}{25} + \ldots.$$

Indeed, to clarify the ideas, consider the second of equations (24). Multiplying both sides by $\left(\frac{\pi}{m}\right)^2$, we find

---

[3] These series give another solution to one of the most important problems of the 18th century, the so-called "Basel problem." Euler gave the first solution in [Euler 1740].

$$(29)\quad \begin{cases} \dfrac{\pi^2}{6}\left(1-\dfrac{4}{m^2}\right) = \dfrac{\left(\frac{\pi}{m}\right)^2}{\sin^2\frac{\pi}{m}} + \dfrac{1}{4}\dfrac{\left(\frac{2\pi}{m}\right)^2}{\sin^2\frac{2\pi}{m}} + \dfrac{1}{9}\dfrac{\left(\frac{3\pi}{m}\right)^2}{\sin^2\frac{3\pi}{m}} + \dots \\ \qquad + \dfrac{1}{\left(\frac{m}{2}-1\right)^2}\dfrac{\left[\frac{(m-2)\pi}{2m}\right]^2}{\sin^2\frac{(m-2)\pi}{2m}}. \end{cases}$$

First, let n be an integer number less than $\frac{m}{2}$. As usual, indicate by the notation $M(a,b)$ an average value between the quantities a and b. Finally, we observe that the ratio $\frac{x}{\sin x}$ is still (see page 45)[4] contained between the limits 1 and $\frac{1}{\cos x}$ and that consequently we have, for numerical values of x less than $\frac{\pi}{2}$,

$$\frac{x}{\sin x} = \frac{\frac{1}{2}x}{\sin\frac{1}{2}x\cos\frac{1}{2}x}\frac{1}{\cos\frac{1}{2}x} < \frac{1}{\cos^2\frac{1}{2}x} < \frac{1}{\cos^2\frac{\pi}{4}} = 2.$$

[456] The right-hand side of equation (29) is evidently the sum of the two polynomials

$$\frac{\left(\frac{\pi}{m}\right)^2}{\sin^2\frac{\pi}{m}} + \frac{1}{4}\frac{\left(\frac{2\pi}{m}\right)^2}{\sin^2\frac{2\pi}{m}} + \dots + \frac{1}{n^2}\frac{\left(\frac{n\pi}{m}\right)^2}{\sin^2\frac{n\pi}{m}}$$

and

$$\frac{1}{(n+1)^2}\frac{\left[\frac{(n+1)\pi}{m}\right]^2}{\sin^2\frac{(n+1)\pi}{m}} + \frac{1}{(n+2)^2}\frac{\left[\frac{(n+2)\pi}{m}\right]^2}{\sin^2\frac{(n+2)\pi}{m}} + \dots + \frac{1}{\left(\frac{m}{2}-1\right)^2}\frac{\left[\frac{(m-2)\pi}{2m}\right]^2}{\sin^2\frac{(m-2)\pi}{2m}},$$

where the first of these, by virtue of equation (11) of the Preliminaries, can be presented in the form

$$\left(1+\frac{1}{4}+\frac{1}{9}+\dots+\frac{1}{n^2}\right)M\left(1,\frac{1}{\cos^2\frac{n\pi}{m}}\right),$$

while the second, composed of $\frac{m}{2}-n-1$ terms all less than $\frac{4}{n^2}$, remains contained between the limits 0 and $\frac{2m}{n^2}$. Given this, equation (29) becomes

$$\frac{\pi^2}{6}\left(1-\frac{4}{m^2}\right) = \left(1+\frac{1}{4}+\frac{1}{9}+\dots+\frac{1}{n^2}\right)M\left(1,\frac{1}{\cos^2\frac{n\pi}{m}}\right) + \frac{2m}{n^2}M(0,1),$$

and then we immediately conclude

[4] [Cauchy 1821, p. 63, Cauchy 1897, p. 66].

$$(30)\qquad \begin{cases} 1+\frac{1}{4}+\frac{1}{9}+\ldots+\frac{1}{n^2}=\frac{\pi^2}{6}\left(1-\frac{4}{m^2}\right)M\left(1,\cos^2\frac{n\pi}{m}\right) \\ \qquad\qquad\qquad\qquad -\frac{2m}{n^2}M(0,1). \end{cases}$$

This last formula remains true whatever the integer numbers m and n may be, provided we have $\frac{1}{2}m > n$. In addition, it is easy to see that if we always take for $\frac{1}{2}m$ the smallest of the integers greater than n^a (where a denotes a number contained between 1 and 2), the ratios $\frac{n}{m}$ and $\frac{m}{n^2}$ converge together, for increasing values of n, towards the limit zero, and the right-hand side of formula (30) converges towards the limit $\frac{\pi^2}{6}$. The left-hand side ought to have the [457] same limit as the right-hand side, and it follows that: 1° the series

$$1,\quad \frac{1}{4},\quad \frac{1}{9},\quad \frac{1}{16},\quad \ldots,\quad \frac{1}{n^2},\quad \ldots$$

is convergent, as we already know (see the corollary to theorem III, Chap. VI, § II); and 2° this series has for its sum $\frac{\pi^2}{6}$.

Because equation (28) is thus proved, we can divide its two sides by 4 to get

$$\frac{\pi^2}{24}=\frac{1}{4}+\frac{1}{16}+\frac{1}{36}+\ldots.$$

As a consequence, we have

$$\frac{\pi^2}{6}-\frac{\pi^2}{24}=1+\frac{1}{9}+\frac{1}{25}+\ldots.$$

This new formula agrees with equation (27), which we could deduce directly from the first of equations (24) or (26).

Before ending this note, we remark that to establish the eight formulas (3), (4), (5), (6), (9), (10), (11) and (12) of Chapter VII (§ V), it suffices to prove the four last ones, and that we can do that very succinctly by expanding equations (10) of Chapter IX (§ I), namely

$$(31)\qquad 1+z\cos\theta+z^2\cos2\theta+z^3\cos3\theta+\ldots=\frac{1-z\cos\theta}{1-2z\cos\theta+z^2}$$
$$(z=-1,\quad z=+1),$$

and

$$(32)\qquad z\sin\theta+z^2\sin2\theta+z^3\sin3\theta+\ldots=\frac{z\sin\theta}{1-2z\cos\theta+z^2}$$
$$(z=-1,\quad z=+1).$$

Consider, for example, equation (32). For numerical values of z less than 1, we get[5]

$$\begin{aligned}
&z\sin\theta+z^2\sin 2\theta+z^3\sin 3\theta+\ldots+z^{2n}\sin 2n\theta+z^{2n+1}\sin(2n+1)\theta+\ldots\\
&\quad=\frac{1}{1+z^2}\frac{z\sin\theta}{1-\frac{2z\cos\theta}{1+z^2}}\\
&\quad=\sin\theta\left[z\left(1+z^2\right)^{-1}+2z^2\cos\theta\left(1+z^2\right)^{-2}+4z^3\cos\theta\left(1+z^2\right)^{-3}+\ldots\right]\\
&\quad=\sin\theta\,\{z-z^3+z^5-\ldots\pm z^{2n+1}\mp\ldots\\
&\qquad+\cos\theta\left(2z^2-4z^4+\ldots\mp 2nz^{2n}\pm\ldots\right)\\
&\qquad+\cos^2\theta\left[\frac{2\cdot 4}{1\cdot 2}z^3-\frac{4\cdot 6}{1\cdot 2}z^5+\ldots\mp\frac{2n(2n+2)}{1\cdot 2}z^{2n+1}\pm\ldots\right]\\
&\qquad+\cos^3\theta\left[\frac{2\cdot 4\cdot 6}{1\cdot 2\cdot 3}z^4-\ldots\pm\frac{(2n-2)2n(2n+2)}{1\cdot 2\cdot 3}z^{2n}\mp\ldots\right]\\
&\qquad+\ldots\ldots\ldots\ldots\ldots\ldots\ldots\ldots\ldots\ldots\ldots\ldots\}.
\end{aligned}$$

[458] Consequently, by equating the coefficients of like powers of z, we find

$$\sin 2n\theta=(-1)^{n+1}\sin\theta\left[2n\cos\theta-\tfrac{(2n-2)2n(2n+2)}{1\cdot 2\cdot 3}\cos^3\theta+\ldots\right], \tag{33}$$

and

$$\sin(2n+1)\theta=(-1)^n\sin\theta\left[1-\frac{2n(2n+2)}{1\cdot 2}\cos^2\theta+\ldots\right]. \tag{34}$$

In these last formulas, if we replace θ by z and $2n$ or $2n+1$ by m, we obtain precisely equations (10) and (11) of Chapter VII (§ V). Equations (9) and (12) of the same section are deduced from formula (31) by a similar calculation.

[5] The second – sign on the fourth line of this calculation was + in [Cauchy 1897, p. 457]. It was correctly given in [Cauchy 1821, p. 560]. (tr.)

Note IX – On products composed of an infinite number of factors.

[459] Denote by

$$(1) \qquad u_0, \quad u_1, \quad u_2, \quad \ldots, \quad u_n, \quad \ldots$$

an infinite sequence of terms, positive or negative, each of which is greater than -1. If the quantities

$$(2) \qquad \ln(1+u_0), \ \ln(1+u_1), \ \ln(1+u_2), \ \ldots, \ \ln(1+u_n), \ \ldots$$

(where ln is the characteristic of the Napierian logarithms), form a convergent series for which the sum is equal to s, then the product

$$(3) \qquad (1+u_0)(1+u_1)(1+u_2)\ldots(1+u_{n-1})$$

evidently converges for increasing values of the integer n towards a finite limit different from zero, equal to e^s. On the other hand, if series (2) is divergent, product (3) does not converge towards a finite limit different from zero. In the first case, we agree to indicate the limit of the product under consideration by writing the product of its first factors followed by ..., as we see here,

$$(4) \qquad (1+u_0)(1+u_1)(1+u_2)\ldots.$$

The same notation can still be used in the case where this limit vanishes.

For series (2) to be convergent, it is first necessary that as the number n grows indefinitely, each of the expressions

$$\ln(1+u_n), \quad \ln(1+u_{n+1}), \quad \ln(1+u_{n+2}), \quad \ldots,$$

and consequently that each of the quantities

$$u_n, \quad u_{n+1}, \quad u_{n+2}, \quad \ldots$$

[460] becomes infinitely small. If this condition is satisfied, then because in general we have

$$\ln(1+x) = x - \frac{x^2}{2} + \frac{x^3}{3} - \frac{x^4}{4} + \ldots \tag{5}$$
$$(x=-1, \quad x=+1),$$

we find that, for very large values of n,

$$\text{(6)} \quad \begin{cases} \ln(1+u_n) \quad = u_n - \frac{1}{2}u_n^2 + \frac{1}{3}u_n^3 - \ldots \qquad\qquad = u_n - \frac{1}{2}u_n^2(1 \pm \varepsilon_n), \\ \ln(1+u_{n+1}) = u_{n+1} - \frac{1}{2}u_{n+1}^2 + \frac{1}{3}u_{n+1}^3 - \ldots = u_{n+1} - \frac{1}{2}u_{n+1}^2(1 \pm \varepsilon_{n+1}), \\ \ldots\ldots\ldots\ldots\ldots\ldots\ldots\ldots\ldots\ldots\ldots\ldots\ldots\ldots\ldots, \end{cases}$$

where $\pm\varepsilon_n, \pm\varepsilon_{n+1}, \ldots$ again denote infinitely small quantities. Then by representing by m any integer number whatsoever and by $1 \pm \varepsilon$ an average among the factors $1 \pm \varepsilon_n$, $1 \pm \varepsilon_{n+1}$, ..., we conclude that

$$\text{(7)} \quad \begin{cases} \ln(1+u_n) + \ln(1+u_{n+1}) + \ldots + \ln(1+u_{n+m-1}) \\ \quad = u_n + u_{n+1} + \ldots + u_{n+m-1} - \frac{1}{2}\left(u_n^2 + u_{n+1}^2 + \ldots + u_{n+m-1}^2\right)(1 \pm \varepsilon). \end{cases}$$

Imagine now that in the preceding formula we make the number m increase beyond all limits. Series (2) is convergent or divergent depending on whether or not both sides of the formula converge towards a fixed limit. Given this, the inspection of just the right-hand side suffices to establish the proposition that I am going to state.

Theorem I. — *If series* (1) *and the following*

$$u_0^2, \quad u_1^2, \quad u_2^2, \quad \ldots, \quad u_n^2, \quad \ldots \tag{8}$$

are both convergent, then series (2) *is convergent as well, and consequently product* (3) *converges, for increasing values of n, towards a finite limit different from zero. However, if series* (1) *is convergent and series* (8) *is divergent, the right-hand side of formula* (7) *then has for its limit negative infinity, and so product* (3) *necessarily converges towards the limit zero.*

Corollary I. — If series (2) is convergent and has all its terms positive, or if it remains convergent when we reduce its various terms [461] to their numerical values, we are evidently assured of the convergence of series (8) and as a consequence product (3) has for its limit a finite quantity different from zero. This is what happens, for example, if the product in question reduces to one of the following:

$$(1+1)\,(1+\tfrac{1}{2^2})\,(1+\tfrac{1}{3^2})\ldots(1+\tfrac{1}{n^2}),$$
$$(1+1)\,(1-\tfrac{1}{2^2})\,(1+\tfrac{1}{3^2})\ldots(1\pm\tfrac{1}{n^2}),$$
$$(1+x^2)\,(1+\tfrac{x^2}{2^2})\,(1+\tfrac{x^2}{3^2})\ldots(1+\tfrac{x^2}{n^2}).$$

Corollary II. — Because the series

$$1, \quad -\frac{1}{\sqrt{2}}, \quad +\frac{1}{\sqrt{3}}, \quad -\frac{1}{\sqrt{4}}, \quad \ldots$$

is convergent, while the squares of its various terms, namely

$$1, \quad \frac{1}{2}, \quad \frac{1}{3}, \quad \frac{1}{4}, \quad \ldots,$$

form a divergent series, it follows from theorem I that the product

$$(1+1)\left(1-\frac{1}{\sqrt{2}}\right)\left(1+\frac{1}{\sqrt{3}}\right)\left(1-\frac{1}{\sqrt{4}}\right)\cdots$$

has zero for its limit.

Corollary III. — Theorem I evidently remains true even in the case where among the first terms of series (1), some of the terms remain less than -1. However, when we admit this new hypothesis, we ought to replace the logarithms of negative quantities in series (2) with the logarithms of their numerical values. Given this, it is clear for increasing values of n that the product

$$(1-x^2)\left(1-\frac{x^2}{2^2}\right)\left(1-\frac{x^2}{3^2}\right)\cdots\left(1-\frac{x^2}{n^2}\right)$$

converges, whatever the value of x, towards a finite limit different from zero.

Corollary IV. — Whenever series (1) is convergent, product (3) converges for increasing values of n towards a finite limit which might reduce to zero.

[462] When the limit of product (3) is finite without being zero, we can not always assign it an exact value. In the small number of products of this kind which correspond to a known limit, we ought to point out the following

$$(1-x^2)\left(1-\frac{x^2}{2^2}\right)\left(1-\frac{x^2}{3^2}\right)\cdots\left(1-\frac{x^2}{n^2}\right), \tag{9}$$

which we are now going to consider.

If we set $x = \frac{\pi}{n}$ and make n grow indefinitely, then product (9) converges towards a finite limit represented by the notation

$$\left(1-\frac{z^2}{\pi^2}\right)\left(1-\frac{z^2}{2^2\pi^2}\right)\left(1-\frac{z^2}{3^2\pi^2}\right)\cdots. \tag{10}$$

To determine this limit, it suffices to recall equation (16) or (18) of the preceding note. To clarify the ideas, consider equation (16). If we write throughout $\frac{z}{m}$ in place of z, we find, for even values of m,

$$\begin{cases} \sin z = m \sin\dfrac{z}{m}\cos\dfrac{z}{m}\left(1-\dfrac{\sin^2\frac{z}{m}}{\sin^2\frac{\pi}{m}}\right)\left(1-\dfrac{\sin^2\frac{z}{m}}{\sin^2\frac{2\pi}{m}}\right)\cdots \\ \qquad \cdots\left(1-\dfrac{\sin^2\frac{z}{m}}{\sin^2\frac{(m-2)\pi}{2m}}\right), \end{cases} \tag{11}$$

and consequently (by supposing that the numerical value of z is less than π and the number m is equal to or greater than 2),

$$(12)\quad \begin{cases} \ln \dfrac{\sin z}{m \sin \frac{z}{m} \cos \frac{z}{m}} = \ln\left(1 - \dfrac{\sin^2 \frac{z}{m}}{\sin^2 \frac{\pi}{m}}\right) + \ln\left(1 - \dfrac{\sin^2 \frac{z}{m}}{\sin^2 \frac{2\pi}{m}}\right) + \dots \\ \qquad + \ln\left(1 - \dfrac{\sin^2 \frac{z}{m}}{\sin^2 \frac{(m-2)\pi}{2m}}\right). \end{cases}$$

Moreover, let n be a number less than $\frac{1}{2}m$ and $1+\alpha$ be an average quantity among the ratios

$$\frac{\ln\left(1 - \frac{\sin^2 \frac{z}{m}}{\sin^2 \frac{\pi}{m}}\right)}{\ln\left(1 - \frac{z^2}{\pi^2}\right)}, \quad \frac{\ln\left(1 - \frac{\sin^2 \frac{z}{m}}{\sin^2 \frac{2\pi}{m}}\right)}{\ln\left(1 - \frac{z^2}{2^2\pi^2}\right)}, \quad \dots, \quad \frac{\ln\left(1 - \frac{\sin^2 \frac{z}{m}}{\sin^2 \frac{n\pi}{m}}\right)}{\ln\left(1 - \frac{z^2}{n^2\pi^2}\right)},$$

[463] and let $1+\beta$ be another average quantity among the expressions

$$-\frac{\ln\left(1 - \frac{\sin^2 \frac{z}{m}}{\sin^2 \frac{(n+1)\pi}{m}}\right)}{\left(\frac{\sin^2 \frac{z}{m}}{\sin^2 \frac{(n+1)\pi}{m}}\right)}, \quad -\frac{\ln\left(1 - \frac{\sin^2 \frac{z}{m}}{\sin^2 \frac{(n+2)\pi}{m}}\right)}{\left(\frac{\sin^2 \frac{z}{m}}{\sin^2 \frac{(n+2)\pi}{m}}\right)}, \quad \dots, \quad -\frac{\ln\left(1 - \frac{\sin^2 \frac{z}{m}}{\sin^2 \frac{(m-2)\pi}{2m}}\right)}{\left(\frac{\sin^2 \frac{z}{m}}{\sin^2 \frac{(m-2)\pi}{2m}}\right)}.$$

The right-hand side of equation (12) is evidently the sum of the two polynomials

$$\ln\left(1 - \frac{\sin^2 \frac{z}{m}}{\sin^2 \frac{\pi}{m}}\right) + \ln\left(1 - \frac{\sin^2 \frac{z}{m}}{\sin^2 \frac{2\pi}{m}}\right) + \dots + \ln\left(1 - \frac{\sin^2 \frac{z}{m}}{\sin^2 \frac{n\pi}{m}}\right) \quad \text{and}$$

$$\ln\left(1 - \frac{\sin^2 \frac{z}{m}}{\sin^2 \frac{(n+1)\pi}{m}}\right) + \ln\left(1 - \frac{\sin^2 \frac{z}{m}}{\sin^2 \frac{(n+2)\pi}{m}}\right) + \dots + \ln\left(1 - \frac{\sin^2 \frac{z}{m}}{\sin^2 \frac{(m-2)\pi}{2m}}\right),$$

where the first one can be presented in the form

$$\left[\ln\left(1 - \frac{z^2}{\pi^2}\right) + \ln\left(1 - \frac{z^2}{2^2\pi^2}\right) + \dots + \ln\left(1 - \frac{z^2}{n^2\pi^2}\right)\right](1+\alpha),$$

while the second takes the form of the product

$$-\frac{\sin^2 \frac{z}{m}}{\left(\frac{\pi}{m}\right)^2} \left\{ \frac{1}{(n+1)^2} \frac{\left(\frac{(n+1)\pi}{m}\right)^2}{\sin^2 \frac{(n+1)\pi}{m}} + \frac{1}{(n+2)^2} \frac{\left(\frac{(n+2)\pi}{m}\right)^2}{\sin^2 \frac{(n+2)\pi}{m}} + \dots \right.$$
$$\left. + \frac{1}{\left(\frac{m}{2} - 1\right)^2} \frac{\left(\frac{(n-2)\pi}{2m}\right)^2}{\sin^2 \frac{(m-2)\pi}{2m}} \right\} (1+\beta),$$

which (by virtue of the principles established in the preceding note) we can reduce to

$$-\frac{2m}{n^2}\frac{\sin^2\frac{z}{m}}{\left(\frac{\pi}{m}\right)^2}(1+\beta)M(0,1).$$

[464] Given this, equation (12) becomes

$$(13)\qquad \begin{cases} \ln\dfrac{\sin z}{m\sin\frac{z}{m}\cos\frac{z}{m}} = \left[\ln\left(1-\frac{z^2}{\pi^2}\right)+\ln\left(1-\frac{z^2}{2^2\pi^2}\right)+\ldots\right.\\ \qquad \left.+\ln\left(1-\frac{z^2}{n^2\pi^2}\right)\right](1+\alpha)-\dfrac{2m}{n^2}\dfrac{\sin^2\frac{z}{m}}{\left(\frac{\pi}{m}\right)^2}(1+\beta)M(0,1). \end{cases}$$

For brevity, let

$$(14)\qquad \frac{1}{1+\alpha}=1+\gamma \quad\text{and}\quad \frac{\sin^2\frac{z}{m}}{\left(\frac{\pi}{m}\right)^2}\frac{1+\beta}{1+\alpha}=\frac{z^2}{\pi^2}(1+\delta).$$

Then returning from logarithms to numbers, we conclude

$$(15)\qquad \begin{cases} \left(1-\dfrac{z^2}{\pi^2}\right)\left(1-\dfrac{z^2}{2^2\pi^2}\right)\cdots\left(1-\dfrac{z^2}{n^2\pi^2}\right)\\ \quad = \left(\dfrac{\sin z}{m\sin\frac{z}{m}\cos\frac{z}{m}}\right)^{1+\gamma} e^{\frac{2mz^2}{n^2\pi^2}(1+\delta)M(0,1)}. \end{cases}$$

Now let the value of n be chosen arbitrarily. Suppose that we take for $\frac{1}{2}m$ the integer number immediately greater than n^a (where a denotes a number, fractional or irrational, contained between 1 and 2). When the value of n is very large, the quantities $\frac{n}{m}$, $\frac{m}{n^2}$, α, β, γ and δ are infinitely small and the product

$$m\sin\frac{z}{m}\cos\frac{z}{m}=\frac{\sin\frac{z}{m}}{\frac{z}{m}}z\cos\frac{z}{m}$$

differs very little from z, and consequently the right-hand side of equation (15) approaches indefinitely the limit

$$\frac{\sin z}{z}.$$

The left-hand side must converge towards the same limit, so we necessarily have[1]

$$(16)\qquad \left(1-\frac{z^2}{\pi^2}\right)\left(1-\frac{z^2}{2^2\pi^2}\right)\left(1-\frac{z^2}{3^2\pi^2}\right)\cdots=\frac{\sin z}{z}.$$

[1] Euler used this indentity in his first solution of the Basel Problem in [Euler 1740], but his derivation was flawed. This derivation justifies Euler's proof.

[465] Thus we find that this last formula is proved in the case where the numerical value of z remains less than π. Then the quantities for which we have taken the logarithms are all positive. But the given proof holds as well for numerical values of z greater than π when we agree to replace the logarithm of each negative quantity by the logarithm of its numerical value. Consequently, equation (16) remains true whatever the real value attributed to the variable z. We do not even have to exclude the case where we suppose that

$$z = \pm k\pi,$$

where k denotes any integer number, because under this hypothesis, both sides of the equation vanish at the same time.

Equation (16), once established, immediately leads to several others. Thus, for example, for any real values of the variables x, y and z, we get

$$(17)\quad \begin{cases} \sin z = z\left(1-\frac{z^2}{\pi^2}\right)\left(1-\frac{z^2}{2^2\pi^2}\right)\left(1-\frac{z^2}{3^2\pi^2}\right)\cdots \\ \quad = z\left(1-\frac{z}{\pi}\right)\left(1+\frac{z}{\pi}\right)\left(1-\frac{z}{2\pi}\right)\left(1+\frac{z}{2\pi}\right)\left(1-\frac{z}{3\pi}\right)\left(1+\frac{z}{3\pi}\right)\cdots \end{cases}$$

and

$$(18)\qquad \frac{\sin x}{\sin y} = \frac{x}{y}\,\frac{\pi-x}{\pi-y}\,\frac{\pi+x}{\pi+y}\,\frac{2\pi-x}{2\pi-y}\,\frac{2\pi+x}{2\pi+y}\,\frac{3\pi-x}{3\pi-y}\,\frac{3\pi+x}{3\pi+y}\cdots.$$

If we make $z = \frac{\pi}{2}$ in equation (17), we find

$$1 = \frac{\pi}{2}\,\frac{1}{2}\,\frac{3}{2}\,\frac{3}{4}\,\frac{5}{4}\,\frac{5}{6}\,\frac{7}{6}\cdots,$$

and consequently we obtain the expansion of $\frac{\pi}{2}$ into factors, discovered by the geometer Wallis,[2] namely

$$(19)\qquad \frac{\pi}{2} = \frac{2}{1}\,\frac{2}{3}\,\frac{4}{3}\,\frac{4}{5}\,\frac{6}{5}\,\frac{6}{7}\,\frac{8}{7}\,\frac{8}{9}\cdots.$$

Likewise, by taking $z = \frac{\pi}{4}$, we find

$$(20)\qquad \frac{\pi}{4} = \frac{1}{\sqrt{2}}\,\frac{4}{3}\,\frac{4}{5}\,\frac{8}{7}\,\frac{8}{9}\,\frac{12}{11}\,\frac{12}{13}\,\frac{16}{15}\,\frac{16}{17}\cdots.$$

[466] In equation (18), if we set both

$$x = \frac{\pi}{2} - z \quad \text{and} \quad y = \frac{\pi}{2},$$

we conclude that

$$(21)\quad \begin{cases} \cos z = \left(1-\frac{2z}{\pi}\right)\left(1+\frac{2z}{\pi}\right)\left(1-\frac{2z}{3\pi}\right)\left(1+\frac{2z}{3\pi}\right)\left(1-\frac{2z}{5\pi}\right)\left(1+\frac{2z}{5\pi}\right)\cdots \\ \quad = \left(1-\frac{4z^2}{\pi^2}\right)\left(1-\frac{4z^2}{3^2\pi^2}\right)\left(1-\frac{4z^2}{5^2\pi^2}\right)\cdots. \end{cases}$$

---

[2] John Wallis, (1616–1703); see [Wallis 1656].

We can deduce the same formula from equation (15) or (17) (preceding note), by replacing z by $\frac{z}{m}$, then making the number m converge towards the limit ∞. Finally, if we suppose that the numerical value of z in equation (16) is less than π, we observe that we can get

$$(22)\quad \begin{cases} \ln\dfrac{\sin z}{z} = \ln\left(1-\dfrac{z^2}{\pi^2}\right)+\ln\left(1-\dfrac{z^2}{2^2\pi^2}\right)+\ln\left(1-\dfrac{z^2}{3^2\pi^2}\right)+\ldots \\ \qquad = -\dfrac{z^2}{\pi^2}\left(1+\dfrac{1}{2^2}+\dfrac{1}{3^2}+\ldots\right) \\ \qquad\quad -\dfrac{1}{2}\dfrac{z^4}{\pi^4}\left(1+\dfrac{1}{2^4}+\dfrac{1}{3^4}+\ldots\right) \\ \qquad\quad -\dfrac{1}{3}\dfrac{z^6}{\pi^6}\left(1+\dfrac{1}{2^6}+\dfrac{1}{3^6}+\ldots\right) \\ \qquad\quad -\ldots\ldots\ldots\ldots\ldots\ldots\ldots\ldots \end{cases}$$

Moreover, because under this hypothesis we have

$$\frac{\sin z}{z} < 1,$$

we get

$$(23)\quad \begin{cases} \ln\dfrac{\sin z}{z} = \ln\left(1-\dfrac{z^2}{1\cdot2\cdot3}+\dfrac{z^4}{1\cdot2\cdot3\cdot4\cdot5}-\dfrac{z^6}{1\cdot2\cdot3\cdot4\cdot5\cdot6\cdot7}+\ldots\right) \\ \qquad = -\dfrac{z^2}{1\cdot2\cdot3}\left(1-\dfrac{z^2}{4\cdot5}+\dfrac{z^4}{4\cdot5\cdot6\cdot7}-\ldots\right) \\ \qquad\quad -\dfrac{1}{2}\left(\dfrac{z^2}{1\cdot2\cdot3}\right)^2\left(1-\dfrac{z^2}{4\cdot5}+\ldots\right)^2 \\ \qquad\quad -\dfrac{1}{3}\left(\dfrac{z^2}{1\cdot2\cdot3}\right)^3(1-\ldots)^3 \\ \qquad\quad -\ldots\ldots\ldots\ldots\ldots\ldots\ldots \end{cases}$$

[467] Consequently (by virtue of the principles established in Chapter VI and in Note VII),

$$(24)\qquad \ln\frac{\sin z}{z} = -\frac{1}{6}\frac{2z^2}{1\cdot2}-\frac{1}{2}\frac{1}{30}\frac{2^3z^4}{1\cdot2\cdot3\cdot4}-\frac{1}{3}\frac{1}{42}\frac{2^5z^6}{1\cdot2\cdot3\cdot4\cdot5\cdot6}-\ldots.$$

The comparison of the coefficients of the same powers of z in formulas (22) and (24) gives the equations

$$
(25) \quad \begin{cases}
1+\frac{1}{2^2}+\frac{1}{3^2}+\frac{1}{4^2}+\ldots=\frac{1}{6}\frac{2\pi^2}{1\cdot 2} & =\frac{\pi^2}{6}, \\
1+\frac{1}{2^4}+\frac{1}{3^4}+\frac{1}{4^4}+\ldots=\frac{1}{30}\frac{2^3\pi^4}{1\cdot 2\cdot 3\cdot 4} & =\frac{\pi^4}{90}, \\
1+\frac{1}{2^6}+\frac{1}{3^6}+\frac{1}{4^6}+\ldots=\frac{1}{42}\frac{2^5\pi^6}{1\cdot 2\cdot 3\cdot 4\cdot 5\cdot 6} & =\frac{\pi^2}{945}, \\
\dots\dots\dots\dots\dots\dots\dots\dots\dots\dots\dots\dots\dots,
\end{cases}
$$

where the first of these agrees with formula (28) of Note VIII. The numerical factors $\frac{1}{6}, \frac{1}{30}, \frac{1}{42}, \ldots$ which enter into the right-hand sides of these equations are what we call the *Bernoulli numbers*.[3] We add that if we denote any even number by $2m$, we have in general

$$
(26) \quad \begin{cases}
1+\frac{1}{3^{2m}}+\frac{1}{5^{2m}}+\frac{1}{7^{2m}}+\ldots \\
\quad =1+\frac{1}{2^{2m}}+\frac{1}{3^{2m}}+\frac{1}{4^{2m}}+\ldots-\frac{1}{2^{2m}}\left(1+\frac{1}{2^{2m}}+\frac{1}{3^{2m}}+\ldots\right) \\
\quad =\left(1-\frac{1}{2^{2m}}\right)\left(1+\frac{1}{2^{2m}}+\frac{1}{3^{2m}}+\frac{1}{4^{2m}}+\ldots\right).
\end{cases}
$$

In the preceding analysis, we have only considered products for which all the factors are real quantities and series for which all the terms are real. However, we ought to remark: 1° that by virtue of the principles established in Chapter IX [see equation (37) of § II and equation (26) of § III], formula (5) remains true even in the case where the variable x becomes imaginary, provided that its modulus remains less than 1; 2° that the ratio

$$
\frac{\sin z}{z} = 1-\frac{z^2}{1\cdot 2\cdot 3}+\frac{z^4}{1\cdot 2\cdot 3\cdot 4\cdot 5} \ldots
$$

converges towards 1 whenever the real or imaginary value attributed [468] to the variable z indefinitely approaches zero; and 3° finally that equations (15), (16), (17) and (18) of Note VIII remain true for real values as well as for imaginary values of z. On the basis of these remarks, we soon come to recognize how we ought to modify the propositions and the formulas proved above in the case where the expressions

$$
u_0, \quad u_1, \quad u_2, \quad \ldots, \quad x, \quad y, \quad z
$$

are imaginary. Thus, for example, with the aid of formulas (6), we establish without difficulty the following proposition, analogous to corollary I of theorem I:

Theorem II. — *Suppose that series* (1) *is imaginary and remains convergent when we reduce its different terms to their respective moduli. Then product* (3) *nec-*

[3] Named for Jakob (I) Bernoulli (1654–1705). Bernoulli numbers first appeared in [Bernoulli 1713]. They were first called "Bernoulli numbers" in [Euler 1755].

essarily converges, for increasing values of n, towards a finite limit, real or imaginary.

Moreover, we easily prove that equations (17) and (21) remain true when we attribute to z any imaginary value $u+v\sqrt{-1}$, from which it follows that: 1° that we can express by products composed of an infinite number of factors the imaginary expressions

$$(27)\qquad \begin{cases} \dfrac{e^v+e^{-v}}{2}\sin u+\sqrt{-1}\dfrac{e^v-e^{-v}}{2}\cos u \quad \text{and} \\ \dfrac{e^v+e^{-v}}{2}\cos u-\sqrt{-1}\dfrac{e^v-e^{-v}}{2}\sin u, \end{cases}$$

and the squares of their moduli, namely

$$(28)\qquad \begin{cases} \left(\frac{e^v+e^{-v}}{2}\right)^2\sin^2 u+\left(\frac{e^v-e^{-v}}{2}\right)^2\cos^2 u=\frac{e^{2v}+e^{-2v}}{2}-\cos 2u \quad \text{and} \\ \left(\frac{e^v+e^{-v}}{2}\right)^2\cos^2 u+\left(\frac{e^v-e^{-v}}{2}\right)^2\sin^2 u=\frac{e^{2v}+e^{-2v}}{2}+\cos 2u; \end{cases}$$

and 2° that the expressions

$$(29)\qquad \begin{cases} \arctan\left(\dfrac{e^v-e^{-v}}{e^v+e^{-v}}\cot u\right) \quad \text{and} \\ \arctan\left(\dfrac{e^v-e^{-v}}{e^v+e^{-v}}\tan u\right) \end{cases}$$

[469] are, respectively, equal to the two sums

$$(30)\qquad \begin{cases} \arctan\dfrac{v}{u}-\arctan\dfrac{v}{\pi-u}+\arctan\dfrac{v}{\pi+u} \\ \qquad -\arctan\dfrac{v}{2\pi-u}+\arctan\dfrac{v}{2\pi+u}-\dots \quad \text{and} \\ \arctan\dfrac{2v}{\pi-2u}-\arctan\dfrac{2v}{\pi+2u}+\arctan\dfrac{2v}{3\pi-2u} \\ \qquad -\arctan\dfrac{2v}{3\pi+2u}+\arctan\dfrac{2v}{5\pi+2u}+\dots, \end{cases}$$

augmented or dimished by a multiple of the circumference 2π. In addition, because expressions (29) and sums (30) are continuous functions of v which always vanish with this variable, we can be sure that the multiple of which we have just spoken reduces to zero.

If we suppose in particular that $u=0$, we find

$$(31)\qquad \begin{cases} \dfrac{e^v-e^{-v}}{2}=v\left(1+\dfrac{v^2}{\pi^2}\right)\left(1+\dfrac{v^2}{2^2\pi^2}\right)\left(1+\dfrac{v^2}{3^2\pi^2}\right)\cdots \quad \text{and} \\ \dfrac{e^v+e^{-v}}{2}=\left(1+\dfrac{2^2v^2}{\pi^2}\right)\left(1+\dfrac{2^2v^2}{3^2\pi^2}\right)\left(1+\dfrac{2^2v^2}{5^2\pi^2}\right)\cdots. \end{cases}$$

By taking $u = \frac{\pi}{4}$, we also find that

$$(32)\qquad \begin{cases} \arctan \dfrac{e^v - e^{-v}}{e^v + e^{-v}} = & \arctan \dfrac{4v}{\pi} - \arctan \dfrac{4v}{3\pi} \\ & + \arctan \dfrac{4v}{5\pi} - \arctan \dfrac{4v}{7\pi} + \dots, \end{cases}$$

and by taking $u = v$,

$$(33)\qquad \begin{cases} \dfrac{e^{2v} + e^{-2v}}{2} - \cos 2v = 2v^2 \left(1 + \dfrac{2^2 v^4}{\pi^4}\right)\left(1 + \dfrac{2^2 v^4}{2^4 \pi^4}\right)\left(1 + \dfrac{2^2 v^4}{3^4 \pi^4}\right)\cdots \\ \quad \text{and} \\ \dfrac{e^{2v} + e^{-2v}}{2} + \cos 2v = \left(1 + \dfrac{2^2 v^4}{\pi^4}\right)\left(1 + \dfrac{2^2 v^4}{3^4 \pi^4}\right)\left(1 + \dfrac{2^2 v^4}{5^4 \pi^4}\right)\cdots. \end{cases}$$

Finally, if we suppose that the numerical value of $\frac{4v}{\pi}$ in formula (32) is less than 1, then the two sides of this formula can be expanded according to the ascending powers of v, and the comparison of coefficients of the same powers in the expansions in question [470] gives the equations

$$(34)\qquad \begin{cases} 1 - \dfrac{1}{3} + \dfrac{1}{5} - \dfrac{1}{7} + \dots = \dfrac{\pi}{4}, \\ 1 - \dfrac{1}{3^3} + \dfrac{1}{5^3} - \dfrac{1}{7^3} + \dots = \dfrac{\pi^3}{32}, \\ 1 - \dfrac{1}{3^5} + \dfrac{1}{5^5} - \dfrac{1}{7^5} + \dots = \dfrac{5\pi^5}{1536}, \\ \dots\dots\dots\dots\dots\dots, \end{cases}$$

where the first of these equations coincides with equation (40) of Chapter IX (§ II).

Now imagine that we divide expressions (29) and sums (30) by v and then we make the variable v converge towards the limit zero. By passing to the limits, we find that

$$(35)\qquad \begin{cases} \cot u = \dfrac{1}{u} - \dfrac{1}{\pi - u} + \dfrac{1}{\pi + u} - \dfrac{1}{2\pi - u} + \dfrac{1}{2\pi + u} - \dots \\ = \dfrac{1}{u} - 2u\left(\dfrac{1}{\pi^2 - u^2} + \dfrac{1}{2^2\pi^2 - u^2} + \dfrac{1}{3^2\pi^2 - u^2} + \dots\right) \end{cases}$$

and

$$(36)\qquad \begin{cases} \tan u = \dfrac{1}{\frac{\pi}{2} - u} - \dfrac{1}{\frac{\pi}{2} + u} + \dfrac{1}{\frac{3\pi}{2} - u} - \dfrac{1}{\frac{3\pi}{2} + u} + \dfrac{1}{\frac{5\pi}{2} - u} - \dots \\ = 2u\left[\dfrac{1}{\left(\frac{\pi}{2}\right)^2 - u^2} + \dfrac{1}{\left(\frac{3\pi}{2}\right)^2 - u^2} + \dfrac{1}{\left(\frac{5\pi}{2}\right)^2 - u^2} + \dots\right]. \end{cases}$$

Moreover, because we generally have, for numerical values of u less than those of a,

$$\frac{1}{a^2-u^2}=\frac{1}{a^2}\left(1-\frac{u^2}{a^2}\right)^{-1}=\frac{1}{a^2}+\frac{u^2}{a^4}+\frac{u^4}{a^6}+\dots,$$

we get from formulas (35) and (36), by supposing that the numerical value of u is less than $\frac{\pi}{2}$, that

$$(37)\qquad \begin{cases} \cot u = \frac{1}{u} - \frac{2u}{\pi^2}\left(1+\frac{1}{2^2}+\frac{1}{3^2}+\frac{1}{4^2}+\dots\right) \\ \qquad - \frac{2u^3}{\pi^4}\left(1+\frac{1}{2^4}+\frac{1}{3^4}+\frac{1}{4^4}+\dots\right) \\ \qquad - \frac{2u^5}{\pi^6}\left(1+\frac{1}{2^6}+\frac{1}{3^6}+\frac{1}{4^6}+\dots\right) \\ \qquad - \dots\dots\dots\dots\dots\dots \end{cases}$$

[471] and

$$(38)\qquad \begin{cases} \tan u = \frac{2^3 u}{\pi^2}\left(1+\frac{1}{3^2}+\frac{1}{5^2}+\frac{1}{7^2}+\dots\right) \\ \qquad + \frac{2^5 u^3}{\pi^4}\left(1+\frac{1}{3^4}+\frac{1}{5^4}+\frac{1}{7^4}+\dots\right) \\ \qquad + \frac{2^7 u^5}{\pi^6}\left(1+\frac{1}{3^6}+\frac{1}{5^6}+\frac{1}{7^6}+\dots\right) \\ \qquad + \dots\dots\dots\dots\dots\dots \end{cases}$$

Consequently, by virtue of equations (25) and (26), we have

$$(39)\qquad \cot u = \frac{1}{u} - \frac{1}{6}\frac{2^2 u}{1\cdot 2} - \frac{1}{30}\frac{2^4 u^3}{1\cdot 2\cdot 3\cdot 4} - \frac{1}{42}\frac{2^6 u^5}{1\cdot 2\cdot 3\cdot 4\cdot 5\cdot 6} - \dots$$

and

$$(40)\qquad \begin{cases} \tan u = \frac{1}{6}\left(2^2-1\right)\frac{2^2 u}{1\cdot 2} + \frac{1}{30}\left(2^4-1\right)\frac{2^4 u^3}{1\cdot 2\cdot 3\cdot 4} \\ \qquad\qquad + \frac{1}{42}\left(2^6-1\right)\frac{2^6 u^5}{1\cdot 2\cdot 3\cdot 4\cdot 5\cdot 6} + \dots. \end{cases}$$

If we replace u by $\frac{1}{2}u$ and then add these last two equations together, we obtain the expansion of

$$\cot\frac{1}{2}u+\tan\frac{1}{2}u=\frac{\cos\frac{1}{2}u}{\sin\frac{1}{2}u}+\frac{\sin\frac{1}{2}u}{\cos\frac{1}{2}u}=\frac{1}{\sin\frac{1}{2}u\cos\frac{1}{2}u}=2\csc u$$

into a series, and we conclude that

$$
(41) \quad \begin{cases} \csc u = \dfrac{1}{u} + \dfrac{1}{6}(2-1)\dfrac{2u}{1\cdot 2} + \dfrac{1}{30}(2^3-1)\dfrac{2u^3}{1\cdot 2\cdot 3\cdot 4} \\ \qquad\qquad + \dfrac{1}{42}\left(2^5-1\right)\dfrac{2u^5}{1\cdot 2\cdot 3\cdot 4\cdot 5\cdot 6} + \dots. \end{cases}
$$

There is no need to dwell any further on the consequences of formula (17). On this subject, we can consult the excellent Work of Euler, entitled *Introductio in Analysin infinitorum*.[4]

[4] See [Euler 1748, § 158–187].

Page Concordance of the 1821 and 1897 Editions

Notes

References

[D'Alembert 1754] D'Alembert, Jean le Rond, "Différentiel," in *Encyclopédie ou dictionnaire raisonné des sciences, des arts et des métiers*, ed. Denis Diderot, Jean le Rond d'Alembert, vol. 4, pp. 985–989, Briasson et al., Paris, 1754.

[Belhoste 1991] Belhoste, Bruno, *Augustin-Louis Cauchy: A biography*, tr. Frank Ragland, Springer-Verlag, New York, 1991.

[Berkeley 1734] Berkeley, George, *The analyst, or a discourse addressed to an infidel mathematician*, Tonson, London, 1734.

[Bernoulli 1713] Bernoulli, Jakob, *Ars conjectandi*, Thurnisi, Basel, 1713.

[Bolzano 1817] Bolzano, Bernard, *Rein analytischer Beweis der Lehrsatzes dass zwischen je zwey Werthen, die ein entgegengesetztes Resultat gewahren wenigstens einer reele Wurzel der Gleichung liege*, Prague, 1817.

[Bottazzini 1986] Bottazzini, Umberto, *The higher calculus: A history of real and complex analysis from Euler to Weierstrass*, tr. W. Van Egmond, Springer-Verlag, New York, 1986.

[Bottazzini 1990] Bottazzini, Umberto, "Geometrical Rigour and 'modern' analysis." An introduction to Cauchy's *Cours d'analyse*, pp. xi–clxvii, in Bottazzini's 1990 facsimile edition of [Cauchy 1821].

[Buée 1806] M. Buée, "Mémoire sur les quantités imaginaries," *Philosophical Transactions of the Royal Society of London*, **96** (1806), pp. vi–88.

[Burden and Faires 2001] Burden, Richard L., and Douglas J. Faires, *Numerical analysis*, Brooks Cole, 2001.

[Cauchy 1813] Cauchy, Augustin-Louis, "Recherches sur les polyèdres, Sur les polygones et les polyèdres," *Journal de l'École Polytechnique*, **IX** (1813), pp. 68–98. Reprinted in *Oeuvres complètes*, 13 (2.I), pp. 7–38, Gauthier-Villars, Paris, 1905.

[Cauchy 1815] Cauchy, Augustin-Louis, "Démonstration générale du théorème de Fermat sur les nombres polygones," *Mémoires de l'Institut*, **14** (1813–15), pp. 172–220. Reprinted in *Oeuvres complètes*, 18 (2.VI), pp. 320–353, Gauthier-Villars, Paris, 1887.

[Cauchy 1817] Cauchy, Augustin-Louis, "Seconde note sur les racines imaginaires des equations," *Bulletin de la Société philomathique*, 1817, pp. 161–164. Reprinted in *Oeuvres complètes*, 14 (2.II), pp. 217–222, Gauthier-Villars, Paris, 1958.

[Cauchy 1821] Cauchy, Augustin-Louis, *Cours d'analyse de l'école royale polytechnique*, de Bure, Paris, 1821.

[Cauchy 1823] Cauchy, Augustin-Louis, *Résumée des leçons données a l'école royale polytechnique sur le calcul infinitésimal*, de Bure, Paris, 1823. Reprinted in *Oeuvres complètes*, 16 (2.IV), pp. 5–261, Gauthier-Villars, Paris, 1899.

[Cauchy 1828] Cauchy, Augustin-Louis, *Lehrbuch der algebraischen analysis*, tr. C. L. B. Huzler, Bornträger, Königsberg, 1828.

[Cauchy 1864] Cauchy, Augustin-Louis, *Algebraicheskie analiz*, tr. F. Ewald, B. Grigoriev and A. Ilin, Leipzig, 1864.

[Cauchy 1885] Cauchy, Augustin-Louis, *Algebraische analysis*, tr. C. Itzigohn, Springer-Verlag, 1885.

[Cauchy 1897] Cauchy, Augustin-Louis, *Cours d'analyse de l'école royale polytechnique*, in *Oeuvres complètes*, 15 (2.III), Gauthier-Villars, Paris, 1897.

[Cauchy 1994] Cauchy, Augustin-Louis, *Curso de análisis*, tr. C. Alvares, J. Dhombres, UNAM, Mexico, 1994.

[Chapelle 1765] Chapelle, Jean-Baptiste de la, "Limite," in *Encyclopédie ou dictionnaire raisonné des sciences, des arts et des métiers*, ed. Denis Diderot, Jean le Rond d'Alembert, vol. 9, , p. 542, Briasson et al., Paris, 1765.

[Cramer 1750] Cramer, Gabriel, *Introduction à l'analyses des lignes courbes algbriques*, Bousquet, Lausanne, 1750.

[D'Alembert] see under "A."

[DSB Cauchy] Freudenthal, Hans, "Cauchy," *Dictionary of scientific biography*, vol. 3, pp. 131–148, Scribners, New York, 1971.

[Dunham 1990] Dunham, William, *Journey through genius*, Wiley, New York, 1990.

[Euler 1740] Euler, Leonhard, "De summis serierum reciprocarum," *Commentarii academiae scientorum Petropolitanae*, **7** (1734/35) 1740, pp. 123–134. Reprinted in *Opera omnia*, Ser. I, vol. 14, pp. 73–86, Teubner, Leipzig, 1924.

[Euler 1748] Euler, Leonhard, *Introductio in analysin infinitorum*, Bosquet, Lausanne, 1748. Reprinted in *Opera omnia*, Ser. I, vols. 8–9, Teubner, Leipzig, 1922 and 1945. English translation by John Blanton, Springer-Verlag, New York, 1988 and 1990.

[Euler 1751] Euler, Leonhard, "De la controverse entre Mrs. Leibnitz & Bernoulli sur les logarithmes des nombres négatifs et imaginaires," *Mémoires de l'acad. Berlin*, **5** (1749) 1751, pp. 139–179. Reprinted in *Opera omnia*, Ser. I, vol. 17, pp. 195–232, Teubner, Leipzig, 1915.

[Euler 1755] Euler, Leonhard, *Institutiones calculi differentialis*, Acad. Petrop., St. Petersburg, 1755. Reprinted in *Opera omnia*, Ser. I, vol. 10, Teubner, Leipzig, 1913. English translation of Part I by John Blanton, Springer-Verlag, NY, 2000.

[Euler 1758] Euler, Leonhard, "De numeris qui sunt aggregata duorum quadratorum" (E228), *Novi Commentarii Petropolitanae*, **4** (1758), pp. 3–40. Reprinted in *Opera omnia*, Ser. I, vol. 2, pp. 295–327, Teubner, Leipzig, 1915.

[Euler 1769] Euler, Leonhard *Institutionun calculi integralis*, Imperial Academy of Sciences, St. Petersburg, 1768–1770. Reprinted in *Opera omnia*, Ser. I, vols. 11–13, Teubner, Leipzig, 1913–14.

[Euler 1774] Euler, Leonhard, "Summatio progressionem $\sin\varphi^\lambda + \sin 2\varphi^\lambda + \sin 3\varphi^\lambda + \cdots + \sin n\varphi^\lambda, \cos\varphi^\lambda + \cos 2\varphi^\lambda + \cos 3\varphi^\lambda + \cdots + \cos n\varphi^\lambda$" (E447), *Novi Commentarii Petropolitanae*, **18** (1774), pp. 8–11, 24–36. Reprinted in *Opera omnia*, Ser. I, vol. 15, pp. 168–184, Teubner, Leipzig, 1927.

[Ferraro 2008] Ferraro, Giovanni, *The rise and development of the theory of series up to the early 1820s*, Springer-Verlag, New York, 2008.

[Fourier 1818] Fourier, Jean Baptiste Joseph, "Question d'analyse algébrique," *Bulletin de la Société philomathique*, 1818, pp. 61–67.

[Freudenthal 1971a] Freudenthal, Hans, see [DSB Cauchy].

[Freudenthal 1971b] Freudenthal, Hans, "Did Cauchy plagiarize bolzano?" *Archive for History of Exact Sciences*, **7** (1971), pp. 375–392.

[Galuzzi 2001] Galuzzi, Massimo, "Galois' note on the approximative solution of numerical equations (1830)," *Archive for the History of Exact Sciences*, **56, 1** (November 2001), pp. 29–37.

[Gelbaum 2003] Gelbaum, Bernard L., Olmsted, John M. H., *Counterexamples in Analysis*, Dover, Mineola, NY, 2003. Reprint of Holden-Day, San Francisco, 1964.

[Gilain 1989] Gilain, Christian, "Cauchy et le cours d'analyse de l'ecole polytechnique," *Bulletin de la Société des Amis de la Bibliothèque de l'Ecole polytechnique*, **5**, 1989.

[Grabiner 2005] Grabiner, Judith V., *The origins of Cauchy's rigorous calculus*, Dover, New York, 2005. Originally published MIT Press, Cambridge, 1981.

[Grattan-Guinness 1970a] Grattan-Guinness, Ivor, *The development of the foundations of mathematical analysis from Euler to Riemann*, MIT Press, Cambridge, 1970.

[Grattan-Guinness 1970b] Grattan-Guinness, Ivor, "Balzano, Cauchy and the 'new analysis' of the early nineteenth century," *Archive for History of Exact Sciences*, **6** (1970), pp. 372–400.

[Grattan-Guinness 1990] Grattan-Guinness, Ivor, *Convolutions in French mathematics, 1800–1840: From the calculus and mechanics to mathematical analysis and mathematical physics*, 3 vols., Birkhäuser, Basel, 1990.

[Grattan-Guinness 2005] Grattan-Guinness, Ivor, Cooke, Roger, eds., *Landmark writings in western mathematics 1640–1940*, Elsevier, Amsterdam, 2005.

[Jahnke 2003] Jahnke, Hans Neils, ed., *A history of analysis*, History of mathematics, vol. 24, American Mathematical Society, Providence, RI, 2003.

[Kline 1990] Kline, Morris, *Mathematical thought from ancient to modern times*, new edition, Oxford University Press, 1990.

[L'Huilier 1787] L'Huilier, Simon Antoine Jean, *Exposition élémentaire des principes des calculs supérieurs*, Decker, Berlin, 1787.

[Lagrange 1769] Lagrange, Joseph-Louis, "Sur la résolution des équations numériques," *Mémoires de l'acad. Berlin*, **23** (1767) 1769, pp. 311–352. Reprinted in *Oeuvres de Lagrange*, vol. 2, pp. 539–580, Gauthier-Villars, Paris, 1868.

[Lagrange 1797] Lagrange, Joseph-Louis, *Théorie des fonctions analytiques, contenant les principes du calcul différentiel, dégagés de toute considération d'infiniment petits et d'évanouissans de limites ou de fluxions, et réduits l'analyse algébrique des quantités finies*, Imprimérie de la République, Paris, An V (1797).

[Lambert 1768] Lambert, Johann Heinrich, "Mémoire sur quelques propriétés remarquables des quantités transcendentes circulaires et logarithmiques," *Mémoires de l'acad. Berlin* **17** (1761) 1768, pp. 265–322.

[Legendre 1808] Legendre, Adrien-Marie, *Théorie des nombres*, Courcier, Paris, 1808.

[Legendre 1816] Legendre, Adrien-Marie, *Supplément a l'essai sur la théorie des nombres, seconde edition*, Courcier, Paris, February 1816.

[Lützen 2003] Lützen, Jesper, "The foundation of analysis in the 19th century," in [Jahnke 2003], pp. 155–196.

[Maclaurin 1742] Maclaurin, Colin, *A treatise of fluxions. In two books*, Ruddimans, Edinburgh, 1742.

[Nicholson 1820] Nicholson, Peter, *Essay on involution and evolution: Particularly applied to the operation of extracting the roots of equations and numbers, according to a process entirely arithmetical*, Davis and Dickson, 1820.

[Oeuvres] Cauchy, Augustin-Louis, *Oeuvres complètes*, 27 vols., Gauthier-Villars, Paris, 1882–1974.

[Pascal 1665] Pascal, Blaise, *Traité du triangle arithmétique avec quelques autres petits traités sur la même matière*, G. Desprez, Paris, 1665.

[Rolle 1703] Rolle, Michel, *Remarques de M. Rolle ... touchant le problesme general des tangentes*, Bourdot, Paris, 1703.

[Sandifer 2007] Sandifer, C. Edward, *How Euler did it*, Mathematical Assoc. of America, Washington, DC, 2007.

[Servois 1814] Servois, Francois Joseph, "Essai sur un nouveau mode d'exposition des principes du calcul differentiel," *Annales de mathématiques*, **5** (1814–1815), p. 93–140.

[Smith 1958] Smith, David Eugene, *History of modern mathematics*, Dover, New York, 1958 (reprint of 1925 edition).

[Stainville 1815] Stainville, M. J. de, *Mélanges d'analyse algébrique et de géométrie*, Ve. Courcier, Paris, 1815.

[Van Brummelen 2009] Van Brummelen, Glen, *The mathematics of the heavens and the earth: The early history of trigonometry*, Princeton University Press, 2009.

[Wallis 1656] Wallis, John, *Arithmetica infinitorum*, Oxford, 1656.

[Windred 1933] Windred, G., "The history of mathematical time: I," *Isis* **19** (1933), pp. 121–153.

Index

MIX
Papier aus verantwortungsvollen Quellen
Paper from responsible sources
FSC® C105338

If you have any concerns about our products,
you can contact us on
ProductSafety@springernature.com

In case Publisher is established outside the EU,
the EU authorized representative is:
Springer Nature Customer Service Center GmbH
Europaplatz 3, 69115 Heidelberg, Germany

Printed by Libri Plureos GmbH
in Hamburg, Germany